Lecture Notes in Computer Science

**Lecture Notes in Artificial Intelligence**     **14178**

Founding Editor

Jörg Siekmann

Series Editors

Randy Goebel, *University of Alberta, Edmonton, Canada*
Wolfgang Wahlster, *DFKI, Berlin, Germany*
Zhi-Hua Zhou, *Nanjing University, Nanjing, China*

The series Lecture Notes in Artificial Intelligence (LNAI) was established in 1988 as a topical subseries of LNCS devoted to artificial intelligence.

The series publishes state-of-the-art research results at a high level. As with the LNCS mother series, the mission of the series is to serve the international R & D community by providing an invaluable service, mainly focused on the publication of conference and workshop proceedings and postproceedings.

Xiaochun Yang · Heru Suhartanto ·
Guoren Wang · Bin Wang · Jing Jiang · Bing Li ·
Huaijie Zhu · Ningning Cui
Editors

# Advanced Data Mining and Applications

19th International Conference, ADMA 2023
Shenyang, China, August 21–23, 2023
Proceedings, Part III

 Springer

*Editors*
Xiaochun Yang
Northeastern University
Shenyang, China

Guoren Wang
Beijing Institute of Technology
Beijing, China

Jing Jiang
University of Technology Sydney
Sydney, NSW, Australia

Huaijie Zhu
Sun Yat-sen University
Guangzhou, China

Heru Suhartanto
The University of Indonesia
Depok, Indonesia

Bin Wang
Northeastern University
Shenyang, China

Bing Li
Agency for Science, Technology
and Research (A*STAR)
Singapore, Singapore

Ningning Cui
Anhui University
Hefei, China

ISSN 0302-9743 ISSN 1611-3349 (electronic)
Lecture Notes in Artificial Intelligence
ISBN 978-3-031-46670-0 ISBN 978-3-031-46671-7 (eBook)
https://doi.org/10.1007/978-3-031-46671-7

LNCS Sublibrary: SL7 – Artificial Intelligence

This Springer imprint is published by the registered company Springer Nature Switzerland AG
The registered company address is: Gewerbestrasse 11, 6330 Cham, Switzerland

Paper in this product is recyclable.

# Preface

The 19th International Conference on Advanced Data Mining and Applications (ADMA 2023) was held in Shenyang, China, during August 21–23, 2023. Researchers and practitioners from around the world came together at this leading international forum to share innovative ideas, original research findings, case study results, and experienced insights into advanced data mining and its applications. With the ever-growing importance of appropriate methods in these data-rich times, ADMA has become a flagship conference in this field. ADMA 2023 received a total of 503 submissions from 22 countries across five continents. After a rigorous double-blind review process involving 318 reviewers, 216 regular papers were accepted to be published in the proceedings, 123 were selected to be delivered as oral presentations at the conference, 85 were selected as poster presentations, and 8 were selected as industry papers. This corresponds to a full oral paper acceptance rate of 24.4%. The Program Committee (PC), composed of international experts in relevant fields, did a thorough and professional job of reviewing the papers submitted to ADMA 2023, and each paper was reviewed by an average of 2.97 PC members. With the growing importance of data in this digital age, papers accepted at ADMA 2023 covered a wide range of research topics in the field of data mining, including pattern mining, graph mining, classification, clustering and recommendation, multi-objective, optimization, augmentation, and database, data mining theory, image, multimedia and time series data mining, text mining, web and IoT applications, finance and healthcare. It is worth mentioning that ADMA 2023 was organized as a physical-only event, allowing for in-person gatherings and networking. We thank the PC members for completing the review process and providing valuable comments within tight schedules. The high-quality program would not have been possible without the expertise and dedication of our PC members. Moreover, we would like to take this valuable opportunity to thank all authors who submitted technical papers and contributed to the tradition of excellence at ADMA. We firmly believe that many colleagues will find the papers in these proceedings exciting and beneficial for advancing their research. We would like to thank Microsoft for providing the CMT system, which is free to use for conference organization, Springer for their long-term support, the host institution, Northeastern University, for their hospitality and support, Niu Translation and Shuangzhi Bo for their sponsorship. We are grateful for the guidance of the steering committee members, Osmar R. Zaiane, Chengqi Zhang, Michael Sheng, Guodong Long, Xue Li, Jianxin Li, and Weitong Chen. With their leadership and support, the conference ran smoothly. We also would like to acknowledge the support of the other members of the organizing committee. All of them helped to make ADMA 2023 a success. We appreciate the local arrangements, registration and finance management from the local arrangement chairs, registration management chairs and finance chairs Kui Di, Baoyan Song, Junchang Xin, Donghong Han, Guoqiang Ma, Yuanguo Bi, and Baiyou Qiao, the time and effort of the proceedings chairs, Bing Li, Huaijie Zhu, and Ningning Cui, the effort in advertising the conference by the publicity chairs and social network and social media coordination chairs, Xin Wang, Yongxin

Tong, Lina Wang, and Sen Wang, and the effort of managing the Tutorial sessions by the tutorial chairs, Zheng Zhang and Shuihua Wang, We would like to give very special thanks to the web chair, industry chairs, and PhD school chairs Faming Li, Chi Man Pun, Sen Wang, Linlin Ding, M. Emre Celebi, and Zheng Zhang, for creating a successful and memorable event. We also thank sponsorship chair Hua Shao for his sponsorship. Finally, we would like to thank all the other co-chairs who have contributed to the conference.

August 2023

Xiaochun Yang
Bin Wang
Jing Jiang

# Organization

## Chair of the Steering Committee

Xue Li         University of Queensland, Australia

## Steering Committee

Osmar R. Zaiane      University of Alberta, Canada
Chengqi Zhang       Sydney University of Technology, Australia
Michael Sheng       Macquarie University, Australia
Guodong Long       Sydney University of Technology, Australia
Xue Li         University of Queensland, Australia
Jianxin Li        Deakin University, Australia
Weitong Chen       Adelaide University, Australia

## Honor Chairs

Xingwei Wang       Northeastern University, China
Xuemin Lin        Shanghai Jiao Tong University, China
Ge Yu         Northeastern University, China

## General Chairs

Xiaochun Yang       Northeastern University, China
Heru Suhartanto      The University of Indonesia, Indonesia
Guoren Wang       Beijing Institute of Technology, China

## Program Chairs

Bin Wang        Northeastern University, China
Jing Jiang        University of Technology Sydney, Australia

## Local Arrangement Chairs

| | |
|---|---|
| Kui Di | Northeastern University, China |
| Baoyan Song | Liaoning University, China |
| Junchang Xin | Northeastern University, China |

## Registration Management Chairs

| | |
|---|---|
| Donghong Han | Northeastern University, China |
| Guoqiang Ma | Northeastern University, China |
| Yuanguo Bi | Northeastern University, China |

## Finance Chair

| | |
|---|---|
| Baiyou Qiao | Northeastern University, China |

## Sponsorship Chair

| | |
|---|---|
| Hua Shao | Shenyang Huaruibo Information Technology Co., Ltd., China |

## Publicity Chairs

| | |
|---|---|
| Xin Wang | Tianjin University, China |
| Yongxin Tong | Beihang University, China |
| Lina Wang | Wuhan University, China |

## Social Network and Social Media Coordination Chair

| | |
|---|---|
| Sen Wang | University of Queensland, Australia |

## Proceeding Chairs

| | |
|---|---|
| Bing Li | Agency for Science, Technology and Research (A*STAR), Singapore |
| Huaijie Zhu | Sun Yat-sen University, China |
| Ningning Cui | Anhui University, China |

## Tutorial Chairs

| | |
|---|---|
| Zheng Zhang | Harbin Institute of Technology, Shenzhen, China |
| Shuihua Wang | University of Leicester, UK |

## Web Chair

| | |
|---|---|
| Faming Li | Northeastern University, China |

## Industry Chairs

| | |
|---|---|
| Chi Man Pun | University of Macau, China |
| Sen Wang | University of Queensland, Australia |
| Linlin Ding | Liaoning University, China |

## PhD School Chairs

| | |
|---|---|
| M. Emre Celebi | University of Central Arkansas, USA |
| Zheng Zhang | Harbin Institute of Technology, Shenzhen, China |

## Program Committee

## Meta Reviewers

| | |
|---|---|
| Bohan Li | Nanjing University of Aeronautics and Astronautics, China |
| Can Wang | Griffith University, Australia |
| Chaokun Wang | Tsinghua University, China |
| Cheqing Jin | East China Normal University, China |
| Guodong Long | University of Technology Sydney, Australia |
| Hongzhi Wang | Harbin Institute of Technology, China |
| Huaijie Zhu | Sun Yat-sen University, China |
| Jianxin Li | Deakin University, Australia |
| Jun Gao | Peking University, China |
| Lianhua Chi | La Trobe University, Australia |
| Lin Yue | University of Newcastle, Australia |
| Tao Shen | University of Technology Sydney, Australia |

| Wei Emma Zhang | University of Adelaide, Australia |
| Weitong Chen | Adelaide University, Australia |
| Xiang Lian | Kent State University, USA |
| Xiaoling Wang | East China Normal University, China |
| Xueping Peng | University of Technology Sydney, Australia |
| Xuyun Zhang | Macquarie University, Australia |
| Yanjun Zhang | Deakin University, Australia |
| Zheng Zhang | Harbin Institute of Technology, Shenzhen, China |

## Reviewers

| Abdulwahab Aljubairy | Macquarie University, Australia |
| Adita Kulkarni | SUNY Brockport, USA |
| Ahoud Alhazmi | Macquarie University, Australia |
| Akshay Peshave | GE Research, USA |
| Alex Delis | Univ. of Athens, Greece |
| Alexander Zhou | Hong Kong University of Science and Technology, China |
| Baoling Ning | Heilongjiang University, China |
| Bin Zhao | Nanjing Normal University, China |
| Bing Li | Institute of High Performance Computing, A*STAR, Singapore |
| Bo Tang | Southern University of Science and Technology, China |
| Carson Leung | University of Manitoba, Canada |
| Changdong Wang | Sun Yat-sen University, China |
| Chao Zhang | Tsinghua University, China |
| Chaokun Wang | Tsinghua University, China |
| Chaoran Huang | University of New South Wales, Australia |
| Chen Wang | Chongqing University, China |
| Chengcheng Yang | East China Normal University, China |
| Chenhao Zhang | University of Queensland, Australia |
| Cheqing Jin | East China Normal University, China |
| Chuan Ma | Zhejiang Lab, China |
| Chuan Xiao | Osaka University and Nagoya University, Japan |
| Chuanyu Zong | Shenyang Aerospace University, China |
| Congbo Ma | University of Adelaide, Australia |
| Dan He | University of Queensland, Australia |
| David Broneske | German Centre for Higher Education Research and Science Studies, Germany |

| | |
|---|---|
| Dechang Pi | Nanjing University of Aeronautics and Astronautics, China |
| Derong Shen | Northeastern University, China |
| Dima Alhadidi | University of New Brunswick, Canada |
| Dimitris Kotzinos | ETIS, France |
| Dong Huang | South China Agricultural University, China |
| Dong Li | Liaoning University, China |
| Dong Wen | University of New South Wales, Australia |
| Dongxiang Zhang | Zhejiang University, China |
| Dongyuan Tian | Jilin University, China |
| Dunlu Peng | University of Shanghai for Science and Technology, China |
| Eiji Uchino | Yamaguchi University, Japan |
| Ellouze Mourad | University of Sfax, Tunisia |
| Elsa Negre | LAMSADE, Paris-Dauphine University, France |
| Faming Li | Northeastern University, China |
| Farid Nouioua | Université Mohamed El Bachir El Ibrahimi de Bordj Bou Arréridj, Algeria |
| Genoveva Vargas-Solar | CNRS, France |
| Gong Cheng | Nanjing University, China |
| Guanfeng Liu | Macquarie University, Australia |
| Guangquan Lu | Guangxi Normal University, China |
| Guangyan Huang | Deakin University, Australia |
| Guannan Dong | University of Macau, China |
| Guillaume Guerard | ESILV, France |
| Guodong Long | University of Technology Sydney, Australia |
| Haïfa Nakouri | ISG Tunis, Tunisia |
| Hailong Liu | Northwestern Polytechnical University, China |
| Haojie Zhuang | University of Adelaide, Australia |
| Haoran Yang | University of Technology Sydney, Australia |
| Haoyang Luo | Harbin Institute of Technology (Shenzhen), China |
| Hongzhi Wang | Harbin Institute of Technology, China |
| Huaijie Zhu | Sun Yat-sen University, China |
| Hui Yin | Deakin University, Australia |
| Indika Priyantha Kumara Dewage | Tilburg University, The Netherlands |
| Ioannis Konstantinou | University of Thessaly, Greece |
| Jagat Challa | BITS Pilani, India |
| Jerry Chun-Wei Lin | Western Norway University of Applied Sciences, Norway |
| Jiabao Han | NUDT, China |
| Jiajie Xu | Soochow University, China |
| Jiali Mao | East China Normal University, China |

| | |
|---|---|
| Jianbin Qin | Shenzhen University, China |
| Jianhua Lu | Southeast University, China |
| Jianqiu Xu | Nanjing University of Aeronautics and Astronautics, China |
| Jianxin Li | Deakin University, Australia |
| Jianxing Yu | Sun Yat-sen University, China |
| Jiaxin Jiang | National University of Singapore, Singapore |
| Jiazun Chen | Peking University, China |
| Jie Shao | University of Electronic Science and Technology of China, China |
| Jie Wang | Indiana University, USA |
| Jilian Zhang | Jinan University, China, China |
| Jingang Yu | Shenyang Institute of Computing Technology, Chinese Academy of Sciences |
| Jing Du | University of New South Wales, Australia |
| Jules-Raymond Tapamo | University of KwaZulu-Natal, South Africa |
| Jun Gao | Peking University, China |
| Junchang Xin | Northeastern University, China |
| Junhu Wang | Griffith University, Australia |
| Junshuai Song | Peking University, China |
| Kai Wang | Shanghai Jiao Tong University, China |
| Ke Deng | RMIT University, Australia |
| Kun Han | University of Queensland, Australia |
| Kun Yue | Yunnan University, China |
| Ladjel Bellatreche | ISAE-ENSMA, France |
| Lei Duan | Sichuan University, China |
| Lei Guo | Shandong Normal University, China |
| Lei Li | Hong Kong University of Science and Technology (Guangzhou), China |
| Li Li | Southwest University, China |
| Lin Guo | Changchun University of Science and Technology, China |
| Lin Mu | Anhui University, China |
| Linlin Ding | Liaoning University, China |
| Lizhen Cui | Shandong University, China |
| Long Yuan | Nanjing University of Science and Technology, China |
| Lu Chen | Swinburne University of Technology, Australia |
| Lu Jiang | Northeast Normal University, China |
| Lukui Shi | Hebei University of Technology, China |
| Maneet Singh | IIT Ropar, India |
| Manqing Dong | Macquarie University, Australia |

| Mariusz Bajger | Flinders University, Australia |
| Markus Endres | University of Applied Sciences Munich, Germany |
| Mehmet Ali Kaygusuz | Middle East Technical University, Turkey |
| Meng-Fen Chiang | University of Auckland, New Zealand |
| Ming Zhong | Wuhan University, China |
| Minghe Yu | Northeastern University, China |
| Mingzhe Zhang | University of Queensland, Australia |
| Mirco Nanni | CNR-ISTI Pisa, Italy |
| Misuk Kim | Sejong University, South Korea |
| Mo Li | Liaoning University, China |
| Mohammad Alipour Vaezi | Virginia Tech, USA |
| Mourad Nouioua | Mohamed El Bachir El Ibrahimi University, Bordj Bou Arreridj, Algeria |
| Munazza Zaib | Macquarie University, Australia |
| Nabil Neggaz | Université des Sciences et de la Technologie d'Oran Mohamed Boudiaf, Algeria |
| Nicolas Travers | Léonard de Vinci Pôle Universitaire, Research Center, France |
| Ningning Cui | Anhui University, China |
| Paul Grant | Charles Sturt University, Australia |
| Peiquan Jin | University of Science and Technology of China, China |
| Peng Cheng | East China Normal University, China |
| Peng Peng | Hunan University, China |
| Peng Wang | Fudan University, China |
| Pengpeng Zhao | Soochow University, China |
| Philippe Fournier-Viger | Shenzhen University, China |
| Ping Lu | Beihang University, China |
| Pinghui Wang | Xi'an Jiaotong University, China |
| Qiang Yin | Shanghai Jiao Tong University, China |
| Qing Liao | Harbin Institute of Technology (Shenzhen), China |
| Qing Liu | Data61, CSIRO, Australia |
| Qing Xie | Wuhan University of Technology, China |
| Quan Chen | Guangdong University of Technology, China |
| Quan Z. Sheng | Macquarie University, Australia |
| Quoc Viet Hung Nguyen | Griffith University, Australia |
| Rania Boukhriss | University of Sfax, Tunisia |
| Riccardo Cantini | University of Calabria, Italy |
| Rogério Luís Costa | Polytechnic of Leiria, Portugal |
| Rong Zhu | Alibaba Group, China |
| Ronghua Li | Beijing Institute of Technology, China |
| Rui Zhou | Swinburne University of Technology, Australia |

| | |
|---|---|
| Rui Zhu | Shenyang Aerospace University, China |
| Sadeq Darrab | Otto von Guericke University Magdeburg, Germany |
| Saiful Islam | Griffith University, Australia |
| Sayan Unankard | Maejo University, Thailand |
| Senzhang Wang | Central South University, China |
| Shan Xue | University of Wollongong, Australia |
| Shaofei Shen | University of Queensland, Australia |
| Shi Feng | Northeastern University, China |
| Shiting Wen | Zhejiang University, China |
| Shiyu Yang | Guangzhou University, China |
| Shouhong Wan | University of Science and Technology of China, China |
| Shuhao Zhang | Singapore University of Technology and Design, Singapore |
| Shuiqiao Yang | UNSW, Australia |
| Shuyuan Li | Beihang University, China |
| Silvestro Roberto Poccia | University of Turin, Italy |
| Sonia Djebali | Léonard de Vinci Pôle Universitaire, Research Center, France |
| Suman Banerjee | IIT Jammu, India |
| Tao Qiu | Shenyang Aerospace University, China |
| Tao Zhao | National University of Defense Technology, China |
| Tarique Anwar | University of York, UK |
| Thanh Tam Nguyen | Griffith University, Australia |
| Theodoros Chondrogiannis | University of Konstanz, Germany |
| Tianrui Li | Southwest Jiaotong University, China |
| Tianyi Chen | Peking University, China |
| Tieke He | Nanjing University, China |
| Tiexin Wang | Nanjing University of Aeronautics and Astronautics, China |
| Tiezheng Nie | Northeastern University, China |
| Uno Fang | Deakin University, Australia |
| Wei Chen | University of Auckland, New Zealand |
| Wei Deng | Southwestern University of Finance and Economics, China |
| Wei Hu | Nanjing University, China |
| Wei Li | Harbin Engineering University, China |
| Wei Liu | University of Macau, Sun Yat-sen University, China |
| Wei Shen | Nankai University, China |
| Wei Song | Wuhan University, China |

| | |
|---|---|
| Weijia Zhang | University of Newcastle, Australia |
| Weiwei Ni | Southeast University, China |
| Weixiong Rao | Tongji University, China |
| Wen Zhang | Wuhan University, China |
| Wentao Li | Hong Kong University of Science and Technology (Guangzhou), China |
| Wenyun Li | Harbin Institute of Technology (Shenzhen), China |
| Xi Guo | University of Science and Technology Beijing, China |
| Xiang Lian | Kent State University, USA |
| Xiangguo Sun | Chinese University of Hong Kong, China |
| Xiangmin Zhou | RMIT University, Australia |
| Xiangyu Song | Swinburne University of Technology, Australia |
| Xianmin Liu | Harbin Institute of Technology, China |
| Xianzhi Wang | University of Technology Sydney, Australia |
| Xiao Pan | Shijiazhuang Tiedao University, China |
| Xiaocong Chen | University of New South Wales, Australia |
| Xiaofeng Gao | Shanghai Jiaotong University, China |
| Xiaoguo Li | Singapore Management University, Singapore |
| Xiaohui (Daniel) Tao | University of Southern Queensland, Australia |
| Xiaoling Wang | East China Normal University, China |
| Xiaowang Zhang | Tianjin University, China |
| Xiaoyang Wang | University of New South Wales, Australia |
| Xiaojun Xie | Nanjing Agricultural University, China |
| Xin Cao | University of New South Wales, Australia |
| Xin Wang | Southwest Petroleum University, China |
| Xinqiang Xie | Neusoft, China |
| Xiuhua Li | Chongqing University, China |
| Xiujuan Xu | Dalian University of Technology, China |
| Xu Yuan | Harbin Institute of Technology, Shenzhen, China |
| Xu Zhou | Hunan University, China |
| Xupeng Miao | Carnegie Mellon University, USA |
| Xuyun Zhang | Macquarie University, Australia |
| Yajun Yang | Tianjin University, China |
| Yanda Wang | Nanjing University of Aeronautics and Astronautics, China |
| Yanfeng Zhang | Northeastern University, China |
| Yang Cao | Hokkaido University, China |
| Yang-Sae Moon | Kangwon National University, South Korea |
| Yanhui Gu | Nanjing Normal University, China |
| Yanjun Shu | Harbin Institute of Technology, China |
| Yanlong Wen | Nankai University, China |

| | |
|---|---|
| Yanmei Hu | Chengdu University of Technology, China |
| Yao Liu | University of New South Wales, Australia |
| Yawen Zhao | University of Queensland, Australia |
| Ye Zhu | Deakin University, Australia |
| Yexuan Shi | Beihang University, China |
| Yicong Li | University of Technology Sydney, Australia |
| Yijia Zhang | Jilin University, China |
| Ying Zhang | Nankai University, China |
| Yingjian Li | Harbin Institute of Technology, Shenzhen, China |
| Yingxia Shao | BUPT, China |
| Yishu Liu | Harbin Institute of Technology, Shenzhen, China |
| Yishu Wang | Northeastern University, China |
| Yixiang Fang | Chinese University of Hong Kong, Shenzhen, China |
| Yixuan Qiu | The University of Queensland, Australia |
| Yong Zhang | Tsinghua University, China |
| Yongchao Liu | Ant Group, China |
| Yongpan Sheng | Southwest University, China |
| Yongqing Zhang | Chengdu University of Information Technology, China |
| Youwen Zhu | Nanjing University of Aeronautics and Astronautics, China |
| Yu Gu | Northeastern University, China |
| Yu Liu | Huazhong University of Science and Technology, China |
| Yu Yang | Hong Kong Polytechnic University, China |
| Yuanbo Xu | Jilin University, China |
| Yucheng Zhou | University of Technology Sydney, Australia |
| Yue Tan | University of Technology Sydney, Australia |
| Yunjun Gao | Zhejiang University, China |
| Yunzhang Huo | Hong Kong Polytechnic University, China |
| Yurong Cheng | Beijing Institute of Technology, China |
| Yutong Han | Dalian Minzu University, China |
| Yutong Qu | University of Adelaide, Australia |
| Yuwei Peng | Wuhan University, China |
| Yuxiang Zeng | Hong Kong University of Science and Technology, China |
| Zesheng Ye | University of New South Wales, Sydney, Australia |
| Zhang Anzhen | Shenyang Aerospace University, China |
| Zhaojing Luo | National University of Singapore, Singapore |
| Zhaonian Zou | Harbin Institute of Technology, China |
| Zheng Liu | Nanjing University of Posts and Telecommunications, China |

| | |
|---|---|
| Zhengyi Yang | University of New South Wales, Australia |
| Zhenying He | Fudan University, China |
| Zhihui Wang | Fudan University, China |
| Zhiwei Zhang | Beijing Institute of Technology, China |
| Zhixin Li | Guangxi Normal University, China |
| Zhongnan Zhang | Xiamen University, China |
| Ziyang Liu | Tsinghua University, China |

# Contents – Part III

## Text Classification

## Graph

# Pharmaceutical Data Analysis

# Drug-Target Interaction Prediction Based on Drug Subgraph Fingerprint Extraction Strategy and Subgraph Attention Mechanism

Lizhi Wang[1,2,3], Xiaolong Zhang[1,2,3]([✉]), Xiaoli Lin[1,2,3], and Jing Hu[1,2,3]

[1] College of Computer Science and Technology, Wuhan University of Science and Technology, Wuhan, Hubei, China
{xiaolong.zhang,linxiaoli,hujing}@wust.edu.cn
[2] Hubei Key Laboratory of Intelligent Information Processing and Realtime Industrial System, Wuhan, Hubei, China
[3] Institute of Big Data Science and Engineering, Wuhan University of Science and Technology, Wuhan, Hubei, China

**Abstract.** Drug discovery is a major focus of modern research, and predicting drug-target interactions is one of the strategies to facilitate this research process. Traditional laboratory methods have long time cycles and are relatively costly, so the use of high-precision virtual screening is essential. Previous virtual screens have only considered the sequence structure of the drug or the graph structure of the drug, without considering the role of the drug subgraph structure. Therefore, this paper adopts a method for extracting drug subgraphs and designs a strategy for extracting drug subgraph fingerprints. Then, the fingerprint vector data of the drugs are trained by the graph neural network to produce drug information vectors containing the spatial structure of the drug subgraph. Finally, the subgraph attention mechanism is used to update the information on protein targets, facilitating the model to extract more information about the target site. The experimental results show that the model can predict drug-target interactions well, outperforming the state-of-the-art model in all metrics. And it is also effective in both classification prediction and affinity regression prediction.

**Keywords:** Drug-target interactions · Drug subgraph structure · Fingerprint extraction · Attention mechanism

## 1 Introduction

Drugs to treat disease are not only those that are on the market but also those that have not yet been discovered. Studying the potential relationship between drugs and proteins can speed up the drug discovery process [1]. Affinity is an important indicator of whether a drug can react with proteins, and the higher the affinity, the easier it is for a drug to react with proteins [2]. Affinity can be determined by chemical assays, but this is time-consuming and inefficient [3]. Therefore, computational methods for virtual screening have become an effective way to alleviate this problem. This drug-protein response is

X. Yang et al. (Eds.): ADMA 2023, LNAI 14178, pp. 3–17, 2023.
https://doi.org/10.1007/978-3-031-46671-7_1

called a drug-target interaction (DTI). It can be a classification problem or a regression task [4]. There are many virtual methods for predicting drug-target interactions. DOCK [5] and AutoDock [6] is a prediction method based on molecular docking. This method is usually used for regression tasks. However, molecular docking methods have their limitations and usually require high-resolution 3D structures of proteins. Machine learning techniques have been widely used to predict drug-target interactions [7]. Effective methods include Random Forest (RF) [8], Support Vector Machine (SVM) [9], k-Nearest Neighbour (k-NN) [10], etc.

For drug and target sequences, Deep Learning (DL) can extract the feature information in them very well, and it is a powerful method for conducting DTI research today [11]. A common type of deep learning network is the convolutional neural network (CNN). One-dimensional convolutional methods can be used to encode protein or drug sequences before further processing in multiple perceptron layers [12, 13]. Convolutional Neural Networks (CNN), as a type of deep learning network, can extract sequence information that contains spatial structure. The sequences of drugs and targets contain spatial constructs, and one-dimensional convolution can process the sequence and spatial information in them [12, 13]. For example, in DeepDTA [4], the sequence information of both the drug and target is extracted by CNN. The work of WideDTA [14] builds on DeepDTA and uses CNN to extract the information of the common sequence of target and drug. Recurrent neural network (RNN) is also a deep learning network that is very good at processing sequence information similar to SMILES [15]. For example, Bidirectional Long Short-Term Memory (BiLSTM) is a type of RNN that extracts sequence information using bidirectional hidden units. Chen et al. [16] used BiLSTM to extract features from proteins, and RNNs can also be combined with CNNs to form hybrid networks [17, 18]. However, drugs are not simply one-dimensional sequences; they have spatial structures. Relying on RNNs and CNNs alone cannot fully extract the spatial structure information of drugs, and also ignores the potential relationship of drugs to targets.

If the atoms of a drug molecule are treated as vertices and the connecting bonds as edges, the drug can be viewed as a graph of arbitrary size and shape. The use of graph neural networks (GNNs) to train graphical data structures for drugs has now become mainstream [19, 20]. For example, GraphDTA [21] also constructs graphs describing drugs and uses GNNs to extract drug target features. For the graphical structure information of drugs, GNNs can effectively extract the spatial information of drugs [22]. However, most GNNs currently use the graph structure of the drug but ignore the sub-graph structure of the drug.

In summary, many current prediction methods use the drug's graph structure but ignore the role of the functional groups in the drug's subgraph in the reaction with the target. Moreover, many methods ignore the potential relationship of drugs to targets.

Therefore, this paper proposes a fingerprint extraction strategy using the graph and subgraph structure of the drug. The drug fingerprints are trained and fused by graph neural networks to obtain a comprehensive information vector of the drug. Then, the targets are allowed to act on the graph and subgraph structure of the drug by the attention mechanism to obtain the target information vector containing the drug information. Finally, the drug information vector and the target information vector are fed into the fully connected layer for prediction. The main innovative work in this paper is as follows:

1. To extract more drug information, a method is proposed to extract the fingerprint of the drug subgraph structure, and the extracted drug fingerprint contains the spatial information of the drug.
2. Using subgraph attention, the weights of each part of the target information vector are redistributed. So that the protein target information vector contains the drug information.

## 2 Method

### 2.1 Neural Network Prediction Model

This subsection briefly describes the procedure of the experiment. First, the graph fingerprint and the subgraph fingerprint are extracted from the drug SMILES, and after being trained by the GNN network respectively, the two obtained vectors are joined together in a certain proportion and called the drug synthesis feature vector. Then, the protein target feature vector is obtained by redistributing the weights of each part of the protein target vector according to the drug feature vector through the subgraph attention mechanism. Finally, the drug synthesis feature vector and the protein target feature vector are spliced in a 1:1 ratio and then fed into the multi-layer fully connected layer for prediction to obtain the output results. The specific flowchart is shown in Fig. 1.

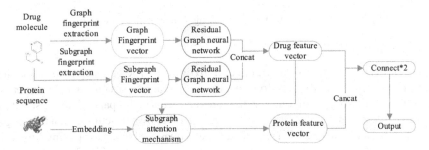

**Fig. 1.** Overall framework flow chart

### 2.2 Data Pre-processing

Three benchmark datasets are used in this paper. These three datasets are the DUD-E dataset, the human dataset, and the bindingDB dataset. These benchmark datasets have been extensively evaluated in previous comparative performance studies.

The data set is stored as a ternary set. The first part is the SMILES of the drug, such as "C1CC(C1)O". The use of SMILES strings to represent drug sequences is more common internationally. The second part is the protein sequence, which represents the long chain of amino acids in English letters [23]. The third part is the actual measurements of the drug-target interactions. The measurements in the DUD-E dataset and the human dataset are either 0 or 1, representing whether the two can react or not. The measurement in the bindingDB dataset is a floating point number representing the affinity between the two. The data set is divided into two parts: the training set and the test set. The data set is first perturbed and then divided into a training set and a test set in a 4:1 ratio.

## 2.3   Drug Feature Extraction

**Fingerprinting of Drug Graph Structure.** Using the RDKit toolkit, hydrogen is first added to the drug (SMILES does not originally contain hydrogen atom information) and then atoms and bonds are obtained as shown in Fig. 2 a. Then, based on the information from the individual atoms and bonds, the drug fingerprint with a radius of 2 is formed [1]. The fingerprint of a drug's graphical structure is centered on an atom and information about its neighboring atoms of depth two is collected and combined to form the fingerprint of that atom. Meanwhile, the adjacency matrix of each drug is extracted from the drug data in the paper and used for the subsequent training of the graph neural network.

a                                      b

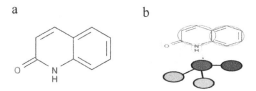

**Fig. 2.** Graph structure of drug and Drug subgraph structure extraction

**Fingerprint Extraction of Drug Subgraph Structure.** The first step is to extract the subgraph structure of the drug. A drug consists of several ring structures and atom pairs. In total, there are two types of drug substructures extracted in this paper; if a ring structure in a drug has less than three atoms in common with other ring structures, then this ring is one of the drug substructures; if a pair of atom pairs does not belong to any ring structure, then this pair of atoms is one of the drug substructures [24]. See Fig. 2 b. Let E be the set of bonds of the drug, G be the set of simple ring structures of the drug, and C be the set of substructures of the drug. E and G can be obtained using functions from the RDkit toolkit. First, traverse the elements in E. Suppose traverse to $e_i$ and add to the set of substructures C if it does not belong to any ring. Then traverse each ring in G. Suppose the traversal reaches $v_i$, $v_j$ (different rings) and determine whether the two rings have more than three atoms in common. If there are more than three, merge the two rings into one large ring temp and replace $v_i$, $v_j$ with temp. Finally, ensure the uniqueness of the elements in G and add each substructure in G to the set of substructures C. See Algorithm 1.

Algorithm 1: Drug subgraph structure extraction algorithm

Inputs: SMILES sequences for all drugs in the dataset.
Output: The set of drug substructures for this dataset C.

Begin:
  (1) Use RDKit to get the set of drug bonds E
  (2) Use RDKit to get the set of drug simple rings G
  (3) For each bond $e_i$ in E, do
        **If $e_i$ does not belong to any ring then**
           add $e_i$ to the set of substructures C
     ·End
  (4) For $v_i$ , $v_j$ (not the same ring) in G, do
       inter:   $v_i \cap v_j$
       **If the length of inter $>=$ 3 then**
          temp = merge $v_i$, $v_j$ to one unique ring
          $v_i$=temp
          $v_j$=temp
   End
  (5) **Ensure that the elements in G are unique (remove duplicates).**
    Add the elements in G to the set of substructures C
  (6) Return the set of drug substructures C
End

---

Algorithm 2: Drug subgraph structure fingerprint extraction algorithm

Inputs: Subgraph structure of drug $C_o$
Output: Drug subgraph structure fingerprint library F

Begin:
  (1) Use RDKit to get the set of bonds of the drug subgraph structure $E_g$
  (2) For each functional group $C_{oi}$ in $C_o$, do
        The set of functional groups connected to Coi is obtained from the set of bonds $E_g$
        NeighborI = the set of functional groups connected to $C_{oi}$
        **$F_{oi}$ = {$C_{oi}$ : NeighborI}**
        **If F does not have $F_{oi}$ then**
           Add $F_{oi}$ to the fingerprint library F
     End
  (3) Return drug subgraph structure fingerprint library F
End

In the second step, the subgraph structure of the drug is turned into a fingerprint. After the drug has gone through the steps of Algorithm 1, the subgraph structure vocabulary C is obtained. The subgraph structure of the drug can be represented using the elements in C. Let $C_o$ be the subgraph structure of the drug. The set $E_g$ of bonds in $C_o$ is obtained by RDKit. Iterate through the functional groups in $C_o$. Assume that the traversal goes to $C_{oi}$. The functional groups connected to $C_{oi}$ are obtained by $E_g$ and deposited into

NeighborI. Next, {$C_{oi}$: NeighborI} is deposited into $F_{oi}$. {$C_{oi}$: NeighborI} denotes a tuple that combines information from both. If $F_{oi}$ does not exist in the fingerprint library F, $F_{oi}$ is deposited into F as a new fingerprint. Finally, the drug subgraph structure fingerprint library F is obtained. The detailed algorithm is shown in Algorithm 2. 42798 fingerprints are extracted from the DUD-E dataset, 4767 fingerprints are extracted from the human dataset, and 2310 fingerprints are extracted from the bindingDB dataset.

**Feature Fusion Based on Residual Graph Neural Network.** The subgraph of a drug has functional groups as nodes, the bonds connecting functional groups as edges, and the functional groups contain information such as the rings of the drug. The subgraph adjacency matrix describes which functional groups are connected in the drug. The atoms in the structure of a compound graph are the nodes and the chemical bonds are the edges. The information about the atomic node includes characteristics such as its own type, whether it is an aromatic atom or not. The graph adjacency matrix describes which atoms are connected. To pass information to neighboring nodes, message passing is widely used. This is shown below:

$$c_i^{(l+1)} = G\left(c_i^{(l)}, \sum m(c_j^{(l)})\right) \tag{1}$$

where $c_i^{(l)}$ is the fingerprint feature of the drug graph or drug subgraph. The fingerprint feature contains information about the atom node itself and the atoms surrounding it. G is a fusion function. m is a function that aggregates the information of the atoms surrounding atom i. The exact form of the formula is given below:

$$\sum m = f\left(W_g A c_i^{(l)}\right) \tag{2}$$

where $W_g$ is the weight matrix, A is the two-dimensional adjacency matrix of the drug and $f$ ensures the non-linearity of the input matrix.

Furthermore, after the GNN iteration is completed, the initial fingerprint feature is added again as a residual in the final drug graph feature or drug subgraph feature. This is shown as follows:

$$c_i^{(l+1)} = c_i^{(l)} + c_i^{(1)} \tag{3}$$

where $c_i^{(1)}$ is the fingerprint feature of the drug graph or subgraph of the first layer as the initial iteration vector. Finally, the summation is used to aggregate the updated fingerprint feature to form the final composite feature:

$$c = \sum_{i \in compound} c_i^{(l)} \tag{4}$$

finally, the drug atom information processed as described above is aggregated to obtain the drug feature vector $c$.

Figure 3 shows the GNN model. The GNN-trained drug graph fingerprint vector and the drug subgraph fingerprint vector are stitched together in a 60:40 ratio to form a drug information vector. This ratio is obtained experimentally.

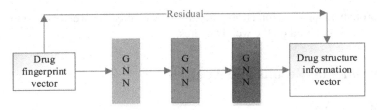

**Fig. 3.** Residual GNN model

## 2.4 Protein Target Feature Extraction

**Semantic Segmentation of Amino Acid Sequences.** This experiment uses the n-gram approach to segment protein sequences. This method takes each n letter of a protein sequence as a group and eventually aggregates them into a word vector with semantics [25]. For example, assume n is 3, "MSPLN" can be segmented into "MSP", "SPL" and "PLN". This technique has previously been used for feature extraction in Natural Language Processing (NLP) and is also very effective for protein features.

**Subgraph Attention Mechanism.** After the drug has been trained by the GNN model, a comprehensive drug information vector is obtained. The comprehensive drug information vector contains information about the drug subgraph structure. The target information is updated by a subgraph attention mechanism (See Fig. 4).

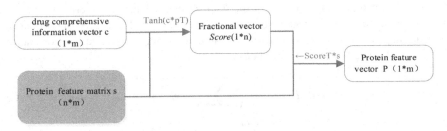

**Fig. 4.** Subgraph attention mechanism

The initial amino acid sequence of the protein is segmented using an n-gram followed by an embedding action to obtain a two-dimensional information matrix. Next, an attention mechanism is used to determine which part of the protein has a greater influence on the interaction.

$$score(c_e, P) = tanh\left(h_c * h_P^T\right) \qquad (5)$$

where *score* is the weight of each element of the protein information matrix in the interaction. $h_c$ and $h_p$ are the features that have been processed non-linearly. The Tanh function ensures that the interval of the score is between $-1$ and $1$, which makes it easy to calculate. After obtaining the weight scores, the protein feature vector can be expressed by the following equation.

$$p_e = \sum_{i \in protein} W_s score(c_e, P)^T h_p \qquad (6)$$

The protein feature vector is obtained by reassigning the weights of the protein information matrix based on the weight scores.

## 3  Experimental Results and Analysis

### 3.1  Evaluation Metrics

Since two of the three datasets in this experiment (DUD-E, human) are classification tasks and one dataset (bindingDB) is a regression task, the loss functions used for the datasets are different. The bindingDB dataset uses minimized mean squared error (MSE). MSE is one of the most commonly used loss functions for regression tasks. The formula is given below:

$$L_{MSE} = \frac{1}{n} \sum_{i=1}^{n} (o_i - y_i)^2 + \frac{\lambda}{2} \|\theta\|^2 \tag{7}$$

The other two datasets use minimized cross-entropy loss function with the following formula.

$$L_{CE} = -\sum_{i=1}^{n} (y_i log(\sigma(o_i)) + (1 - y_i)log(1 - \sigma(o_i))) + \frac{\lambda}{2} \|\theta\|^2 \tag{8}$$

The regression task makes a prediction of the drug-target interaction and then compares it to the actual measurements in the data set and calculates the error in it. This error can be reflected by the root mean square error (RMSE). The degree of correlation between the predicted and measured values can be reflected by the Pearson correlation coefficient (R). The formula is given below:

$$RMSE(y, p) = \sqrt{\frac{1}{n} \sum_{i=1}^{n} (y_i - p_i)^2} \tag{9}$$

$$R(y, p) = \frac{\sum_{i=1}^{n} (y_i - \bar{y})(p_i - \bar{p})}{\sqrt{\sum_{i=1}^{n} (y_i - \bar{y})^2} \sqrt{\sum_{i-1}^{n} (p_i - \bar{p})^2}} \tag{10}$$

The metrics for the classification task are mainly the area under the receiver operating characteristic curve (AUC), precision, and recall. In addition, the enrichment metric (RE) from [1] is applied to the DUD-E dataset.

### 3.2  Dataset

To test the generalization of the model, this paper contains two classification prediction datasets and one affinity regression prediction dataset, which are described in detail in the following introduction.

The DUD-E dataset originally had 22645 positive and 1407145 negative samples. The work in this paper used the dataset from reference [17]. A total of 91,239 data samples are collected, of which 45,609 are negative samples and 45,629 are positive samples.

The human data set is one of the data sets that has been applied to drug-target interactions in recent years. It was obtained by screening in [8]. The dataset used in the experiments of this paper is derived from [13]. This dataset collect 6212 samples, of which 2843 samples are negative and 3369 samples are positive.

The experiment uses not only the categorical prediction data, but also the bindingDB dataset used for affinity prediction. This shows a strong generalisation of the experimental model. The regression experiment are compared using the dataset from [1]. It contains 5000 records.

### 3.3  Experimental Settings

This experiment uses many hyperparameters to speed up the convergence of the model, and the experimental data is more accurate. The embedding dimension of the graph structure of the drug is 60 and the embedding dimension of the substructure of the drug is 40. The initial embedding dimension of the protein is 100. Next, the number of layers of the GNN is 3. The learning rate is 0.001, which makes the pace of the gradient feedback spanning smaller in order to find the local optimum solution more accurately. Finally, a discard rate of 0.2 is used to mitigate the problem of experimental overfitting. All hyperparameters are determined by several experiments.

### 3.4  Experiment

**Model Ablation Experiment.** To verify the effectiveness of the model approach and to better understand which part of the model plays a more critical role, ablation experiments are performed on three data sets.

As can be seen in Fig. 5 a. In the DUD-E dataset, the AUC of the model using GNN alone is 0.974. The AUC of the model with GNN plus attention mechanism is 0.985, The AUC of the model with GNN plus fingerprint is 0.990, and the AUC of the model with GNN plus fingerprint and attention mechanism is 0.999.

As can be seen in Fig. 5 b. In the human dataset, the AUC of the model using GNN alone is 0.964. The AUC of the model with GNN plus attention mechanism is 0.970, the

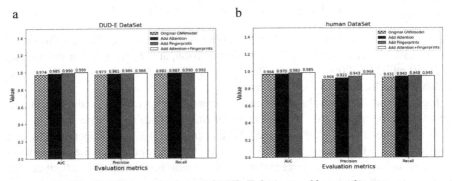

**Fig. 5.** Ablation experiment of DUD-E data set and human data set

AUC of the model with GNN plus fingerprint is 0.980, and the AUC of the model with GNN plus fingerprint and attention mechanism is 0.985.

As can be seen in Fig. 6, on the bindingDB dataset, the RMSE of the model using GNN alone is 0.74 and R is 0.896. The RMSE of the model with GNN plus the attention mechanism is 0.74 and R is 0.896. The RMSE of the model with GNN plus the fingerprint information is 0.708 and R is 0.904. The RMSE of the model with GNN plus the attention mechanism and fingerprint information is 0.691 and R is 0.908.

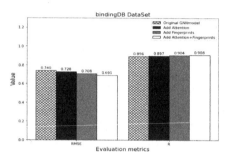

**Fig. 6.** Ablation experiment of bindingDB data set

The experimental results show that both the subgraph attention mechanism and the fingerprinting approach play an important role. For the overall model, the fingerprinting approach is slightly more important, indicating that the drug subgraph structure contains a large amount of drug information.

**Drug Synthesis Vector Ratio Comparison Experiment.** In the methods section, this paper obtains a drug graph fingerprint vector and drug subgraph fingerprint vector, which is finally stitched together at a certain scale to form a comprehensive drug feature vector. In this subsection, comparative experiments of the proportions are shown and correlation analysis is performed.

As shown in Fig. 7 a, the 60:40 ratio is optimal for the DUD-E dataset. The 60:40 ratio is 0.3% higher in AUC, 0.6% higher in Precision, and 0.4% higher in Recall than the 50:50 ratio. The 60:40 ratio is 0.2% higher in AUC, 0.3% higher in Precision, and 0.2% higher in Recall than the 70:30 ratio.

As shown in Fig. 7 b, the 60:40 ratio is optimal for the human dataset. The 60:40 ratio is 0.3% higher in AUC, 0.7% higher in Precision, and 1.2% higher in Recall than the 50:50 ratio. The 60:40 ratio is 0.7% higher in AUC, 3.3% higher in Precision, and 0.2% higher in Recall than the 70:30 ratio.

Figure 8 shows the results of the ratio comparison experiment for the bindingDB dataset. It can be seen that the values of all evaluation metrics are optimal at the ratio of 60:40. Compared to the 50:50 ratio, the RMSE decreases by 3% and R increases by 1%; compared to the 70:30 ratio, the RMSE decreases by 0.7% and R increases by 0.1%.

Taken together, the results suggest that a 60:40 ratio is optimal. This ratio indicates that the subgraph structure of the drug is also important in drug-target interaction.

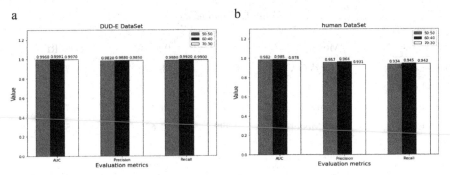

**Fig. 7.** Comparison experiment of DUD-E data set and human data set

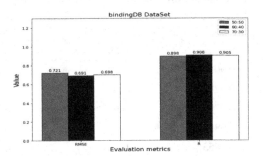

**Fig. 8.** Comparison experiment of bindingDB data set

**Classification Prediction Experiment.** In this paper, the classification prediction on the DUD-E dataset and human dataset both achieve good results. The specific metrics are compared as follows:

Experiments have been prepared on the DUD-E dataset to compare several different methods. These include the docking method Vina [26], as well as deep learning methods such as 3DCNN [27], PocktGCN [28], DrugVQA [17], GanDTI [1] and the work of Chen W [16]. As shown in Table 1. The AUC of the model in this experiment is about 99.9%, which is higher than all other models. The RE value of the model, as shown in Table 1, can reach 91.63 at 0.5% threshold and 86.53 at 1.0% threshold, which is also quite an improvement over deep learning models such as GanDTI.

**Table 1.** Performance of the different methods evaluated on the DUD-E dataset

| Type | Model | AUC | 0.5% RE | 1.0% RE |
|------|-------|-----|---------|---------|
| Docking | Vina | 0.716 | 9.139 | 7.321 |
| Deep-Learning | 3D CNN | 0.868 | 42.559 | 26.655 |
| Deep-Learning | PocketGCN | 0.886 | 44.406 | 29.748 |

(continued)

**Table 1.** (*continued*)

| Type | Model | AUC | 0.5% RE | 1.0% RE |
|---|---|---|---|---|
| Deep-Learning | DrugVQA | 0.972 | 88.71 | 58.71 |
| Deep-Learning | Chen W | 0.989 | – | – |
| Deep-Learning | GanDTI | 0.997 | 71.13 | 68.78 |
| Deep-Learning | OurModel | **0.999** | **91.63** | **86.53** |

**Table 2.** Performance of the different methods evaluated on the human dataset

| Method | AUC | Precision | Recall |
|---|---|---|---|
| KNN | 0.86 | 0.80 | 0.82 |
| RF | 0.90 | 0.82 | 0.84 |
| Tsubaki | 0.978 | 0.929 | 0.929 |
| DeepVQA | 0.979 | 0.954 | **0.961** |
| GanDTI | 0.983 | 0.960 | 0.933 |
| OurModel | **0.9853** | **0.964** | 0.945 |

To further validate the effectiveness of the present method and model for the binary classification task, the experiment also prepares the human dataset to test the present method. This dataset is much smaller than DUD-E. The experimental results are compared with previously available models, among which are KNN (K-nearest neighbor) [10], RF (random forest) [8], DeepVQA [17], and GanDTI [1].GanDTI is the best model among them. As shown in Table 2. The AUC is 98.53%, which is 0.23% higher than the GanDTI model, the accuracy rate is about 96.4%, which is 0.4% higher than the GanDTI model, and the recall rate is 94.5%, which is 1.2% higher than the GanDTI model.

**Affinity Regression Prediction Experiment.** This section uses the bindingDB dataset to validate the validity of the model on affinity regression prediction. Figure 9 shows that most of the prediction points are close to the y = x regression line, which is relatively good. The specific data and comparisons are as follows:

The experimental results are compared with other advanced models such as Deep-Affinity [29], DeepDTA [4], MONN [12], and GanDTI [1]. As shown in Table 3. It can be seen that the RMSE and R for this task are much better than most of the models mentioned above. The RMSE of the model is 0.691, which is 3% lower than the best of these models, GanDTI, and the R of the model is 0.908, which is 0.8% higher than GanDTI. The results provide strong evidence that the method in this paper is well suited to the task of predicting binding affinity.

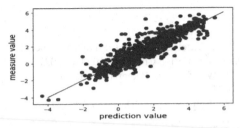

**Fig. 9.** Visualization image of regression curve of bindingDB dataset

**Table 3.** Performance of the different methods evaluated on the bindingDB dataset

| Method | RMSE | R |
|---|---|---|
| DeepAffinity | 0.74 | 0.84 |
| DeepDTA | 0.782 | 0.848 |
| MONN | 0.764 | 0.858 |
| GanDTI | 0.721 | 0.900 |
| OurModel | **0.691** | **0.908** |

## 4  Conclusion

In the present work, not only the graph structure of the drug but also the sub-graph structure of the drug is used. In this paper, the fingerprint information of the drug is extracted and a further graph neural network is used to extract more drug features. In addition, the sequence information of protein targets are first extracted using the semantic segmentation method. Then, the weights of the protein target information matrix are redistributed by the subgraph attention mechanism to obtain the target information vector. Finally, the feature vectors of the drug and target are fed into the fully connected layer for predictive analysis. In terms of datasets, the experiments used two classification prediction datasets and one affinity prediction dataset, demonstrating the high generalisability of the model. The final prediction results perform well for both classification and affinity regression tasks.

**Acknowledgements.** The authors thank the members of Machine Learning and Artificial Intelligence Laboratory, School of Computer Science and Technology, Wuhan University of Science and Technology, for their helpful discussion within seminars. This work was supported by National Natural Science Foundation of China (No. 61972299, 61502356).

# References

1. Wang, S., Shan, P., Zhao, Y., et al.: GanDTI: a multi-task neural network for drug-target interaction prediction. Comput. Biol. Chem. **92**, 107476 (2021)
2. Feinberg, E.N., Sur, D., Wu, Z., et al.: PotentialNet for molecular property prediction. ACS Cent. Sci. **4**(11), 1520–1530 (2018)
3. Chen, H., Engkvist, O., Wang, Y., et al.: The rise of deep learning in drug discovery. Drug Discov. Today **23**(6), 1241–1250 (2018)
4. Öztürk, H., Özgür, A., Ozkirimli, E.: DeepDTA: deep drug–target binding affinity prediction. Bioinformatics **34**(17), i821–i829 (2018)
5. Lang, P.T., Brozell, S.R., Mukherjee, S., et al.: DOCK 6: combining techniques to model RNA–small molecule complexes. Rna1 **5**(6), 1219–1230 (2009)
6. Morris, G.M., Huey, R., Lindstrom, W., et al.: AutoDock4 and AutoDockTools4: automated docking with selective receptor flexibility. J. Comput. Chem. **30**(16), 2785–2791 (2009)
7. Zhang, X., Lin, X., Zhao, J., et al.: Efficiently predicting hot spots in PPIs by combining random forest and synthetic minority over-sampling technique. IEEE/ACM Trans. Comput. Biol. Bioinf. **16**(3), 774–781 (2018)
8. Liu, H., Sun, J., Guan, J., et al.: Improving compound–protein interaction prediction by building up highly credible negative samples. Bioinformatics **31**(12), i221–i229 (2015)
9. Yamanishi, Y., Araki, M., Gutteridge, A., et al.: Prediction of drug–target interaction networks from the integration of chemical and genomic spaces. Bioinformatics **24**(13), i232–i240 (2008)
10. Fokoue, A., Sadoghi, M., Hassanzadeh, O., Zhang, P.: Predicting drug-drug interactions through large-scale similarity-based link prediction. In: Sack, H., Blomqvist, E., d'Aquin, M., Ghidini, C., Ponzetto, S., Lange, C. (eds.) ESWC 2016. LNCS, vol. 9678, pp. 774–789. Springer, Cham (2016). https://doi.org/10.1007/978-3-319-34129-3_47
11. Ye, Q., Zhang, X., Lin, X.: Drug-target interaction prediction via graph auto-encoder and multi-subspace deep neural networks. IEEE/ACM Trans. Comput. Biol. Bioinform. (2022)
12. Li, S., Wan, F., Shu, H., et al.: MONN: a multi-objective neural network for predicting compound-protein interactions and affinities. Cell Syst. **10**(4), 308–322 (2020)
13. Tsubaki, M., Tomii, K., Sese, J.: Compound–protein interaction prediction with end-to-end learning of neural networks for graphs and sequences. Bioinformatics **35**(2), 309–318 (2019)
14. Öztürk, H., Ozkirimli, E., Özgür, A.: WideDTA: prediction of drug-target binding affinity. arXiv preprint arXiv:1902.04166 (2019)
15. Wan, F., Zhu, Y., Hu, H., et al.: DeepCPI: a deep learning-based framework for large-scale in silico drug screening. Genomics Proteomics Bioinform. **17**(5), 478–495 (2019)
16. Chen, W., Chen, G., Zhao, L., et al.: Predicting drug–target interactions with deep-embedding learning of graphs and sequences. J. Phys. Chem. A **125**(25), 5633–5642 (2021)
17. Zheng, S., Li, Y., Chen, S., et al.: Predicting drug–protein interaction using quasi-visual question answering system. Nat. Mach. Intell. **2**(2), 134–140 (2020)
18. Gao, K.Y., Fokoue, A., Luo, H., et al.: Interpretable drug target prediction using deep neural representation. IJCAI **2018**, 3371–3377 (2018)
19. Karlov, D.S., Sosnin, S., Fedorov, M.V., et al.: GraphDelta: MPNN scoring function for the affinity prediction of protein–ligand complexes. ACS Omega **5**(10), 5150–5159 (2020)
20. Lim, J., Ryu, S., Park, K., et al.: Predicting drug–target interaction using a novel graph neural network with 3D structure-embedded graph representation. J. Chem. Inf. Model. **59**(9), 3981–3988 (2019)
21. Nguyen, T., Le, H., Quinn, T.P., et al.: GraphDTA: predicting drug–target binding affinity with graph neural networks. Bioinformatics **37**(8), 1140–1147 (2021)

22. Jiang, M., Wang, S., Zhang, S., et al.: Sequence-based drug-target affinity prediction using weighted graph neural networks. BMC Genomics 23(1), 1–17 (2022)

23. Lin, X., Zhang, X., Xu, X.: Efficient classification of hot spots and hub protein interfaces by recursive feature elimination and gradient boosting. IEEE/ACM Trans. Comput. Biol. Bioinf. 17(5), 1525–1534 (2019)

24. Jin, Y., Lu, J., Shi, R., et al.: EmbedDTI: enhancing the molecular representations via sequence embedding and graph convolutional network for the prediction of drug-target interaction. Biomolecules 11(12), 1783 (2021)

25. Dong, Q.W., Wang, X., Lin, L.: Application of latent semantic analysis to protein remote homology detection. Bioinformatics 22(3), 285–290 (2006)

26. Trott, O., Olson, A.J.: AutoDock Vina: improving the speed and accuracy of docking with a new scoring function, efficient optimization, and multithreading. J. Comput. Chem. 31(2), 455–461 (2010)

27. Ragoza, M., Hochuli, J., Idrobo, E., et al.: Protein–ligand scoring with convolutional neural networks. J. Chem. Inf. Model. 57(4), 942–957 (2017)

28. Torng, W., Altman, R.B.: Graph convolutional neural networks for predicting drug-target interactions. J. Chem. Inf. Model. 59(10), 4131–4149 (2019)

29. Karimi, M., Wu, D., Wang, Z., et al.: DeepAffinity: interpretable deep learning of compound–protein affinity through unified recurrent and convolutional neural networks. Bioinformatics 35(18), 3329–3338 (2019)

# Soft Prompt Transfer for Zero-Shot and Few-Shot Learning in EHR Understanding

Yang Wang[1,2(✉)], Xueping Peng[1(✉)], Tao Shen[1], Allison Clarke[2],
Clement Schlegel[2], Paul Martin[2], and Guodong Long[1]

[1] Australian AI Institute, Faculty of Engineering and IT, University of Technology,
Sydney, Australia
{xueping.peng,tao.shen,guodong.long}@uts.edu.au

[2] Health Economics and Research Division,Australian Government Department of
Health and Aged Care,Canberra, Australia
Yang.Wang-17@student.uts.edu.au,
{alvin.wang,allison.clarke,clement.schlegel,paul.martin}@health.gov.au

**Abstract.** Electronic Health Records (EHRs) are a rich source of information that can be leveraged for various medical applications, such as disease inference, treatment recommendation, and outcome analysis. However, the complexity and heterogeneity of EHR data, along with the limited availability of well-labeled samples, present significant challenges to the development of efficient and adaptable models for EHR tasks (such as rare or novel disease prediction or inference). In this paper, we propose Soft prompt transfer for Electronic Health Records (SptEHR), a novel pipeline designed to address these challenges. Specifically, SptEHR consists of three main stages: (1) self-supervised pre-training on raw EHR data for an EHR-centric transformer-based foundation model, (2) supervised multi-task continual learning from existing well-labeled tasks to further refine the foundation model and learn transferable task-specific soft prompts, and (3) further improve zero-shot and few-shot ability via prompt transfer. Specifically, the transformer-based foundation model learned from stage one captures domain-specific knowledge. Then the multi-task continual training in stage two improves model adaptability and performance on EHR tasks. Finally, stage three leverages soft prompt transfer which is based on the similarity between the new and the existing tasks, to effectively address new tasks without requiring additional/extensive training. The effectiveness of the SptEHR has been validated on the benchmark dataset - MIMIC-III.

**Keywords:** Prompt learning · Transfer learning · Multi-task continual training · Transformer · Electronic Health Record (EHR)

## 1  Introduction

Transformer, combined with transfer learning, self-supervised learning, multi-task learning, and prompt learning, has demonstrated great success in health-care applications. However, transformer-based models typically require large-scale labeled training data, which are often inaccessible or very expensive in real

X. Yang et al. (Eds.): ADMA 2023, LNAI 14178, pp. 18–32, 2023.
https://doi.org/10.1007/978-3-031-46671-7_2

clinical settings, such as in forecasting the progression of conditions for rare or novel diseases (i.e. Covid-19).

Given this, the primary motivation behind our work is to leverage soft prompt transfer techniques for improving the understanding and utilization of Electronic Health Records (EHRs) in various downstream tasks, especially in few-shot or low-resource settings. However, EHR data presents several challenges, including:

– Data complexity: EHR data is inherently complex, containing diverse types of information such as demographics, diagnoses, medications, laboratory test results, and clinical notes. This complexity necessitates an effective method to capture and integrate the information from different data sources for improved downstream task performance.
– Data sparsity and irregularity: EHR data is often sparse in some specific tasks (such as rare disease inference) and irregular due to the nature of healthcare delivery, with patients receiving care at different time intervals and varying amounts of available data. This sparsity and irregularity require a robust model that can handle missing or incomplete data.
– Noisy and biased labels: The quality of EHR data labels can be noisy and biased, arising from factors like data entry errors, inconsistencies in coding practices, and the influence of healthcare providers' subjective judgments. Developing a model that can overcome these challenges and generalize well to unseen tasks is essential.

In this work, we present a technically sound approach, Soft prompt transfer for EHR (SptEHR), designed to tackle the challenges inherent in EHR tasks. Specifically, SptEHR consists of three main stages: (1) self-supervised pre-training on raw EHR data for an EHR-centric transformer-based foundation model, (2) supervised multi-task continual learning from well-labeled existing tasks to refine the foundation model and to learn transferable task-specific soft prompts, and (3) further improve zero-shot and few-shot ability using a linear combination of learned prompts.

By combining a strong EHR-centric foundation model with the flexibility and transferability of soft prompts, our proposed SptEHR has certain benefits.

– By pre-training a transformer-based foundation model on raw EHR data, we are able to build a solid foundation that captures the underlying structure and relationships within EHR data, which is crucial for effective downstream task performance.
– Our supervised multi-task continual training phase leverages well-labeled existing data to refine the foundation model and to learn transferable task-specific soft prompts, further enhancing its generalization capabilities and minimizing the risk of catastrophic forgetting.
– Lastly, the zero-shot generalization aspect of our method is particularly advantageous, as it allows the model to rapidly adapt to new unseen tasks without the need for extensive fine-tuning, saving both time and computational resources. By assigning a new prompt via a linear combination of multiple learned prompts from similar tasks, we effectively leverage the knowledge acquired during previous training and transfer it to the unseen task.

Thereby, our method contributes to the development of AI-driven healthcare systems that can quickly adapt to new tasks in few-shot or even zero-shot settings, facilitating enhanced patient care and better decision-making by healthcare professionals.

## 2    Related Work

### 2.1    Transformer-Based Model for EHR Data

Transformer-based models have a great ability to capture contextualized information and achieve great success in understanding complex healthcare data [5, 23–25]. Steinberg et al. [32] propose language model-based representations to learn structured data from EHRs. Si et al. [31] apply pre-training and finetuning strategy to learn general and transferable clinical language representations. Li et al. [17], inspired by BERT's [6] architecture, proposes BHERT, a transformer model for structured data in electronic health records. BHERT is scalable to perform well across a wide range of downstream predictions. Ren et al. [28] propose RAPT, a pre-training of time-aware transformer for learning robust healthcare representation.

### 2.2    Transfer Learning and Self-supervised Learning

One of the critical challenges of deep learning methods is data hungry [10]. These methods usually have a large number of parameters and thus are easy to overfit and have poor generalization ability without sufficient training data [2].

An important way to address this issue is via transfer learning [22,34]. To make use of previously learned knowledge to handle new problems, transfer learning formalizes a two-stage learning framework: a pre-training stage to capture knowledge from source task(s), and a fine-tuning stage to transfer the captured knowledge to target task(s). Due to the wealth of knowledge obtained in the pre-training stage, the fine-tuning stage can enable models to well perform target tasks with limited samples [10].

Self-supervised learning has become popular in recent years due to its ability to utilize large quantities of unlabelled data [10]. Large-scale self-supervised pre-trained models such as BERT [6] and GPT [27] have recently achieved great success and become a milestone in the field of artificial intelligence (AI). More specifically, BERT employs stacked Transformer Encoder blocks to achieve great performance on NLP tasks. Then, many variants of BERT, such as ALBERT [13], and StructBERT [38], have been proposed to achieve better performance. On the other hand, GPT-4, a transformer-based model pre-trained to predict the next token in a document, has exhibited human-level performance on various professional and academic benchmarks, including passing a simulated bar exam with a score around the top 10% of test takers [21]. It has been widely accepted in the AI community that all AI tasks should adopt a pre-trained model as the backbone to facilitate downstream tasks, rather than training models from scratch [10].

## 2.3   Prompt Learning

Although prompt learning has become more competitive when pre-trained models are getting bigger [20], the common finding is that prompt learning can reach the performance of traditional fine-tuning, and often outperform when there are no large-scale training samples [33].

Specifically, Brown et al. [3] propose a prompt design to achieve impressive few-shot performance, where their model is conditioned on a manual text prompt at inference time to perform different tasks. After that, several prompt-based learning approaches with carefully handcrafted prompts [11,29,30] have been proposed. However, Zhao et al. and Liu et al. [18,40] point out that hard prompts are sub-optimal and sensitive to the choice of the prompt. Therefore, learning soft prompts have become more popular in recent work [14,15,26].

On the other hand, Gu et al. [8] propose prompt transfer with hand-crafted pre-training tasks tailored to specific types of downstream tasks. In contrast, Vu et al. [36] use existing pre-training tasks as source tasks to learn soft prompts for target tasks, which significantly outperforms [14]'s work across many tasks. However, their work does not leverage the power of multitask learning and only focuses on NLP tasks, rather than EHR tasks.

## 2.4   Multitask Learning

Adding multitask learning to the pre-training stage has the potential to significantly enhance model performance during finetuning [4,9]. For instance, Aribandi et al. [1] have systematically studied the effect of scaling up the number of tasks during pre-training and analyzed co-training transfer amongst common families of tasks. The paper shows that multi-task scaling can largely improve models on its own. Xu et al. [39] propose a multitask pre-training approach called ZeroPrompt for zero-shot generalization. The paper shows that task scaling can be an efficient alternative to model scaling, i.e. the model size has less impact on performance while there is large number (i.e. 1000) of tasks for pre-training. Although applied both multitask training and soft prompt transfer, however, ZeroPrompt only focuses on task scaling and zero-shot prompting on NLP tasks.

## 3   SptEHR: Soft Prompt Transfer for EHR

In this section, we elaborate on our proposed pipeline toward Soft prompt transfer for Electronic Health Records (SptEHR). In particular, it begins with a self-supervised pre-training on raw EHR data for an EHR-centric foundation model (Stage One), followed by supervised multi-task continual learning from well-labeled existing tasks and learning corresponding transferable task-specific soft prompts (Stage Two). Lastly, the trained foundation model (pre-trained in Stage One and further enhanced in Stage Two), as well as the transfer from the soft prompts learned in the existing tasks, enhance zero-shot and few-shot ability for a new downstream task (Stage Three). An illustration of the proposed SptEHR is shown in Fig. 1.

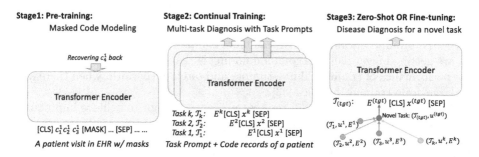

**Fig. 1.** An illustration of our proposed pipeline toward Soft prompt transfer for Electronic Health Records (SptEHR).

*Format of Electronic Health Records (EHR).* In this study, we focus specifically on the structured codes within Electronic Health Records (EHRs), including diagnosis, procedure, and medication codes, as our primary data source for model development. While EHR data is often heterogeneous, encompassing clinical notes, time series, and other data elements, we have chosen to concentrate on these structured codes due to their standardized nature and ease of interpretability. By doing so, we aim to create a robust and efficient model that is capable of extracting valuable insights from the structured EHR data, while also mitigating potential challenges and complexities associated with processing more diverse and unstructured data.

### 3.1   Self-supervised Pre-training on EHR

Self-supervised pre-training is the first and crucial component of our approach as it allows the model to learn meaningful representations from large volumes of unlabelled EHR data. This is valuable in the healthcare domain, as large-scale labeled samples in EHR are often inaccessible or very expensive in real clinical settings, such as in forecasting the progression of conditions for rare or novel diseases (such as Covid-19). Fortunately, self-supervised learning methods provide a way to enable the model to learn from the underlying patterns and structures presented in the EHR data, without the need for explicit labels. This self-supervised pre-training stage builds a strong foundation for the model and allows it to capture the nuances in EHR data, which in turn facilitates better performance and generalization capabilities when being fine-tuned for specific downstream tasks.

Specifically, inspired by Masked Language Modeling (MLM), one of the most popular self-supervised learning techniques in NLP, we have proposed our Masked Code Modeling (MCM) for EHR data. In MCM, the model is tasked with predicting masked EHR codes within the input data.

Formally, we first define the codes of a patient's visit as $\mathbb{C}^v = \{\mathbb{C}^{v,i}\}_i$, where $v$ denotes the patient's visit, and $i$ denotes the index of code categories, e.g., diagnosis, procedure, and medication. And, each $\mathbb{C}^{v,i} = \{c_1^{v,i}, \ldots, c_{N^i}^{v,i}\}$ denotes a set of International Classification of Diseases 9th edition (ICD-9) codes for diagnosis, procedure and Anatomical Therapeutic Chemical (ATC) codes for medication, where $N^i$ denotes the number of codes in $\mathbb{C}^{v,i}$. In the remainder, we omit the subscript, $v$, for clear writing if no confusion is caused.

Then, we apply random masking to $\forall c_j^i \in \mathbb{C}$ in a uniform distribution to obtain a masked set, $\bar{\mathbb{C}} = \{\bar{\mathbb{C}}^1, \bar{\mathbb{C}}^2, \dots\}$, where each $c_j^i$ is of $\alpha$ probability (e.g., 15% in our experiments) of being replaced by a special mask token, i.e., [MASK]. Next, a transformer [35] encoder is employed to produce contextualized representation for each code $\bar{c}_j^i$, i.e.,

$$\bar{H} = \{\bar{H}^i\}_i = \text{Transformer-Enc}(\,[\text{CLS}]\,\bar{\mathbb{C}}^1\,[\text{SEP}]\,\bar{\mathbb{C}}^2\,[\text{SEP}]\,\dots;\theta^{(\text{fm})}), \quad (1)$$

where $H^i \in \mathbb{R}^{d \times N^i}$ denotes the contextualized representations for all the codes in $\mathbb{C}^i$, $\theta^{(\text{fm})}$ parameterizes this transformer-based foundation model, and [CLS] and [SEP] are special tokens as in [7] to formulate model inputs.

Lastly, similar to MLM, we define the objective of our MCM in EHR domain as

$$L^{(\text{mcm})} = -\sum_{\mathbb{C}} \sum_i \sum_{j \in \mathbb{M}^i} \log P(c = c_i^j | \bar{\mathbb{C}}; \theta^{(\text{fm})}, \theta^{(\text{mcm})})$$
$$= -\sum_{\mathbb{C}} \sum_i \sum_{j \in \mathbb{M}^i} \log \text{softmax}(\text{MLP}(\bar{H}_j^i; \theta^{(\text{mcm})}))_{[c=c_i^j]}, \quad (2)$$

where $\mathbb{M}^i$ denotes a set of masked indices (i.e., the codes replaced by [MASK]) and the $\text{MLP}(\cdot; \theta^{(\text{mcm})})$ is used to map the masked contextualized representation into the original ICD code space.

To further refine the foundation model, apart from the MCM task, we also utilise another self surprised learning task (which is the prediction on length of stay for each visit). The detailed information on this task can be found in the Experiments section.

To sum up, the self-supervised pre-training allows the model to learn meaningful representations and acquire a comprehensive understanding of the EHR data, which serves as the basis for further supervised fine-tuning on specific tasks. In addition, the pre-trained foundation model provides a more robust starting point to tackle the issues of data sparsity, irregularity, and noisy labels commonly found in EHR data. By learning from the rich context within the EHR data during pre-training, the foundation model becomes better equipped to handle the challenges of missing or incomplete data and adapt to the variations in coding practices and subjective judgments. Ultimately, the self-supervised pre-training stage sets the stage for the further steps and enhances the model's generalization capabilities and performance on various downstream EHR tasks.

## 3.2 Multi-task Continual Learning from Existing Well-Labelled Tasks

Built upon the foundation model pre-trained in a self-supervised manner, we propose a supervised multi-task continual training to leverage well-labelled data

from other existing EHR tasks (e.g., patient data such as diagnosis information from common disease cohorts) to refine the foundation model and to learn
transferable task-specific soft prompts, so as to enhance model adaptability and
performance on EHR tasks.

Our motivation for this stage is to enhance the model's zero-shot and few-
shot abilities. By further training the model on multiple existing well-labeled
tasks, we expose it to a diverse range of related EHR scenarios, enhancing its
understanding of the EHR domain and fostering its ability to adapt to novel
tasks. Additionally, the learned task-specific soft prompts in this stage not only
improve the model performance on different EHR tasks but also can be transferred to new unseen tasks, making it possible to apply the knowledge learned
from existing tasks to future tasks.

By sharing the same foundation model, $\theta^{(fm)}$, task-specific soft prompts serve
as an efficient and flexible mechanism to condition the model for each specific
task, allowing it to adapt its behavior while leveraging its pre-trained knowledge. The multi-task learning part reduces the need for extensive fine-tuning
on downstream tasks and significantly cuts down on task-specific parameters,
thereby enabling more efficient transfer learning. For instance, for some rare
disease inference tasks, such as inferring whether a patient has HIV, where we
don't have enough sample patients, we can leverage well-labeled data from existing tasks such as sample patients from other diseases (e.g. immunocompromised
disease, chronic kidney disease, or even diabetes). And, "multi-task" here is
referred to the multiple inference tasks of the selected diseases that have sufficient sample patients. Thereby, we first suppose that we have a set of $M$ tasks,
i.e., $\mathbb{T} = \{\mathcal{T}_1, \ldots, \mathcal{T}_M\}$, each task is to diagnose whether a patient has a certain
disease, i.e., binary classification for yes or no. Formally, given a set of ICD codes,
$x^k$, (including diagnosis, procedure, and medication), the disease diagnosis task,
$\mathcal{T}_k$, is to find a model ($\mathcal{F}(\cdot)$) to map it to $\{0, 1\}$, i.e.,

$$\mathcal{F}(x^k) \rightarrow \{0, 1\}, \tag{3}$$

where 0 and 1 denote negative and positive, respectively. Then, we define a
soft prompt for each task, $\mathcal{T}_k$ as a list of $L$ virtual tokens, $[e_1^k, \cdots, e_L^k]$, and
they are assigned with a random initialized dense vector $\boldsymbol{E}^k = \{e_l^k\}_l \in \mathbb{R}^{d \times L}$.
They are called 'virtual' because these tokens $[e_1^k, \cdots, e_L^k]$ do not exist in the
standard ICD code but are newly designed for the specific task $\mathcal{T}_k$. Therefore,
the virtual codes are used to denote the task without any other meanings of
diagnosis, procedure, or medication codes. By following the paradigm of soft
prompt tuning [16, 19, 37, 39], the model defined in this work to solve the $\mathcal{T}_k$ can
be defined as

$$\boldsymbol{H} = \text{Transformer-Enc}([e_1^k, \cdots, e_L^k] \, [\texttt{CLS}] \, x^k \, [\texttt{SEP}] ; \theta^{(fm)}, \boldsymbol{E}^k), \tag{4}$$

$$\boldsymbol{v} = \text{CLS-Pool}(\boldsymbol{H}) = \boldsymbol{H}_{[\texttt{cls}]}, \tag{5}$$

$$\boldsymbol{o} = P(\text{b}|\theta^{(fm)}, \theta^{(out)}, \boldsymbol{E}^k) = \text{softmax}(\text{MLP}(\boldsymbol{v}; \theta^{(out)})) \in \mathbb{R}^2, \tag{6}$$

where CLS-Pool($\cdot$) denotes leveraging the contextualized embedding of the special token [CLS] to represent the whole sequence. Note that, as in other prompt-

tuning works, $\boldsymbol{E}^k$ is a learnable weight matrix specifically for the task $\mathcal{T}_k$, so it characterizes the unique knowledge of this task.

As such, we can define our multi-task supervised training as the combination of inference tasks for well-labeled common diseases, where all tasks share the same foundation model, (fm), but own task-specific prompts, $\boldsymbol{E}^k$. Thereby, our multi-task training object can be defined as a binary cross-entropy loss, i.e.,

$$L^{(\mathrm{mt})} = -\sum_{\mathcal{T}_k \in \mathbb{T}_k} \sum_{(x^k, y^*) \in \mathcal{T}} (y^* \cdot \log P(\mathrm{b} = 1 | \theta^{(\mathrm{fm})}, \theta^{(\mathrm{out})}, \boldsymbol{E}^k)$$
$$+ (1 - y^*) \cdot \log P(\mathrm{b} = 0 | \theta^{(\mathrm{fm})}, \theta^{(\mathrm{out})}, \boldsymbol{E}^k)). \qquad (7)$$

As a result, the multi-task continual training stage further enhances the model's zero-shot and few-shot ability, allowing the model to tackle new downstream EHR tasks with less training data.

### 3.3   Prompt Transfer to Enhance Zero-Shot and Few-Shot Ability

Our motivation for enhancing zero-shot and few-show ability for new downstream EHR tasks lies in the need for a scalable and adaptable solution in the ever-evolving medical domain. This prompt transfer stage empowers the model to effectively generalize to new EHR tasks, saving time, and computational resources, and making it possible to handle new tasks where there are no sufficient training samples.

Specifically, to enable the model to better leverage information from existing tasks, we apply a linear combination to the prompts learned for the existing tasks, to create a new prompt for each new task. The weights in the linear combination are derived from the similarities between the new task and the existing tasks (e.g. the new disease and the existing well-labeled diseases used in the multi-task continual training stage), which allows the model to capture useful information from existing tasks and harnesses it for the novel task.

More specifically, each disease has been assigned a set of diagnosis codes in ICD-9, as well as the corresponding code embeddings obtained in the self-supervised pre-training stage. Therefore, given a new disease inference task, $\mathcal{T}_{(\mathrm{tgt})}$, we denote its diagnosis codes as $[a_i^{(\mathrm{tgt})}]_{i=1}^n$, where $n$ is the number of diagnosis codes in the new task and the corresponding code embedding as $[\boldsymbol{u}_i^{(\mathrm{tgt})}]_{i=1}^n$. We employ mean pooling on $[\boldsymbol{u}_i^{(\mathrm{tgt})}]_{i=1}^n$ to obtain $\boldsymbol{u}^{(\mathrm{tgt})}$. Then, we can derive a new prompt for $\mathcal{T}_{(\mathrm{tgt})}$ by a linear combination among $\{\boldsymbol{E}^1, \ldots, \boldsymbol{E}^M\}$ based on

$$\boldsymbol{w}^{(\mathrm{tgt})} = \mathrm{softmax}([\boldsymbol{u}^1, \ldots, \boldsymbol{u}^M]^T \boldsymbol{u}^{(\mathrm{tgt})}), \qquad (8)$$

where $\boldsymbol{w}^{(tgt)} \in \mathbb{R}^M$ denotes the similarities between the new task and the existing tasks. Next, the prompt of the target task (i.e. the new task) can be formed by

$$\boldsymbol{E}^{(\mathrm{tgt})} = \sum_{k \in [1, M]} w_k^{(\mathrm{tgt})} \cdot \boldsymbol{E}^k. \qquad (9)$$

By assigning higher weights to the prompts learned for similar diseases, the model is encouraged to keep more relevant knowledge from similar tasks and transfer the knowledge to the new task. This approach allows our proposed SptEHR to generalize effectively to new downstream EHR tasks, achieving high performance without requiring extensive task-specific training. We can readily perform zero-shot inference for the new target task by

$$H = \text{Transformer-Enc}(\boldsymbol{E}^{(\text{tgt})}\,[\text{CLS}]\,x^{(\text{tgt})}\,[\text{SEP}]; \theta^{(\text{fm})}, \boldsymbol{E}^{(\text{tgt})}), \tag{10}$$

$$\boldsymbol{o} = P(\text{b}|\theta^{(\text{fm})}, \theta^{(\text{out})}, \boldsymbol{E}^{(\text{tgt})}) = \text{softmax}(\text{MLP}(\text{CLS-Pool}(\boldsymbol{H}); \theta^{(\text{out})})), \tag{11}$$

without the need for any fine-tuning or prompt-tuning. Certainly, the performance on the target task can be further boosted by further training using labeled data in the target task, i.e.,

$$L^{(\text{ft})} = - \sum_{(x^{(\text{tgt})}, y^*) \in \mathcal{T}_{(\text{tgt})}} (y^* \cdot \log P(\text{b} = 1|\theta^{(\text{fm})}, \theta^{(\text{out})}, \boldsymbol{E}^{(\text{tgt})})$$

$$+ (1 - y^*) \cdot \log P(\text{b} = 0|\theta^{(\text{fm})}, \theta^{(\text{out})}, \boldsymbol{E}^{(\text{tgt})})), \tag{12}$$

where $\boldsymbol{E}^{(\text{tgt})}$ is regarded as the prompt of the new task, $\mathcal{T}_{(\text{tgt})}$.

## 4   Experiments

In this section, we conduct experiments on a real-world medical claim dataset to evaluate the performance of the proposed SptEHR approach. Compared with the baseline models, The SptEHR yields better performance on different evaluation metrics, especially in few-shot settings.

### 4.1   Datasets

The MIMIC-III dataset [12] presents an open-source de-identified dataset containing the Electronic Healthcare Records (EHRs) of ICU patients between 2001 and 2012. The dataset adheres to the International Classification of Diseases 9th Edition (ICD-9) procedures and diagnoses codes standard. It contains information on 46,520 patients and 58,976 visits. For the purpose of our research, we utilize all visits to pre-train our foundation model on two tasks: the Masked Language Model and the prediction of the Length of Stay for each visit.

### 4.2   Data Pre-processing and Disease Selection

In MIMIC-III, the data contains records of patients with multiple visits, each of them consisting of various medical codes such as diagnosis codes, procedure codes, and medication codes. To be able to analyze this data accurately, multiple pre-processing steps are required to ensure the correctness and relevance of the data.

**Data Pre-processing.** The pre-training process of the MIMIC-III dataset involves several essential steps. First, we map the U.S. Food and Drug Administration (FDA) National Drug Codes (NDC) codes in the prescription table to the Anatomical Therapeutic Chemical (ATC) Level 3 to standardize the data. Second, we construct the input data format, represented as "<[CLS], diagnosis codes, [SEP], procedure codes, [SEP], medication codes, [SEP]>," where procedure and medication codes are optional. Third, we mask certain codes based on probabilities and generate labels for the length of stay, where the length of stay is labeled as 1 if it exceeds seven days and 0 if below or equal to seven days. With 58,976 pre-training samples, equal to the number of patient visits. For downstream tasks, we ensure consistency in the input format of pre-training samples by deriving data from patient records. We concatenate the diagnosis codes generated from all visits and apply the same method to the procedure and medication codes.

**Disease Selection.** Our downstream tasks involve inferring disease inside the selected six types of chronic conditions. It is noteworthy that each patient belongs to only one particular disease category. Negative samples in each category equal the number of patients in that category. These negative samples originate from patients outside of the selected six categories. See Table 1 for an overview of each disease category.

**Table 1.** Six selected diseases.

| Disease | ICD-9 Diagnosis Codes | Num of Patient |
|---|---|---|
| Cancer | 140–240 | 205 |
| Diabetes | 250.1–250.9 | 292 |
| COPD | 491, 492 | 446 |
| Chronic Kidney Disease (CKD) | 585 | 478 |
| HIV | 042 | 17 |
| Neurological conditions (NC) | 340, 345 | 135 |

### 4.3   Implementation Details

Our implementation of all the approaches used a BERT model from Huggingface Transformer with a hidden size of 768, 6 hidden layers, and 12 attention heads. For training models, we adopt AdamW, with a minibatch of 32 for pre-training and 16 for downstream tasks. We randomly split the data into training and validation sets, with a ratio of 95:5 for pre-training, 9:1 for source task training, and 6:2:2 for training, validation, and test sets in target task (i.e. new task) training. Soft prompts were assigned a size of 50, and a dropout rate of 0.1 was implemented using dropout strategies for all approaches.

### 4.4   Empirical Analysis

**Baselines.** We compare our proposed model to the following baselines:

- **Super-Train.** This is a traditionally supervised learning model without pre-training, multi-task continual training, and prompt learning. The model is a BERT model and trained from scratch only using the data samples from the specific downstream task. Therefore, the size of the training dataset is relatively small.
- **Prompt-Tune.** This approach contains a foundation model pre-trained on EHR and a soft prompt tuning model. The training data is the same as the input of Super-Train model. Unlike the original prompt tuning work in [14] which freezes the pre-trained models and only learns task-specific soft prompts, our pre-trained model is not frozen and will be tuned by the data sample from the downstream tasks.

**Target Task - Disease Inference.** In this task, we apply a leave-one-out strategy to evaluate the performance of the proposed SptEHR model against the baselines. Taking HIV as the target task (i.e. the new unseen task) as an example here: we first pre-train the SptEHR on unlabelled EHR. Then we continue to train the pre-trained model on the other five tasks (as existing well-labeled tasks) (i.e. Cancer, Diabetes, COPD, CKD, and NC). Finally, we apply soft prompt transfer and then fine-tune the entire model on the target task (i.e. HIV). The results are shown in Table 2.

The table summarizes the performance of the disease inference task of the three different approaches, namely Super-train, Prompt-tune, and SptEHR. With respect to the accuracy, SptEHR outperforms all other approaches, obtaining the highest result on all six disease tasks. Similarly, SptEHR achieves the highest F1 score on four out of six tasks, including Diabetes, CKD, HIV, and Cancer. Particularly, SptEHR shows the greatest advantage in HIV inference where there are only 17 training samples. These suggest that, compared with previous methods, SptEHR can achieve better performance on disease inference tasks, especially when a large amount of training data is not available. Overall, the results demonstrate the effectiveness of using the proposed model for disease inference, indicating its potential in clinical and medical applications, especially in few-shot settings.

**Source Prompts Transfer Strategy Analysis.** We conduct another set of experiments to explore the different prompt transfer strategies. Specifically, we compare three strategies to derive a new prompt for the target task, namely Random, Mean-Pooling and the proposed strategy SptEHR. For Random strategy, the new prompt for the target task is initialized randomly. For Mean-Pooling, the weights in the linear combination for each trained source prompt are the same. For SptEHR, the weights in the linear combination are derived from the similarities between the target task and the source tasks (e.g. the new disease

**Table 2.** Performance comparison for disease inference.

| Disease | Metric | Model (%) | | |
|---|---|---|---|---|
| | | Super-Train | Prompt-Tune | SptEHR |
| Cancer | Accuracy | 53.7 | 56.0 | **57.1** |
| | F1 Score | 55.7 | 55.6 | **56.9** |
| Diabetes | Accuracy | 55.7 | 59.1 | **61.9** |
| | F1 Score | 55.2 | 62.5 | **63.4** |
| COPD | Accuracy | 56.9 | 62.6 | **68.3** |
| | F1 Score | 60.7 | **67.1** | 65.5 |
| CKD | Accuracy | 59.6 | 62.0 | **67.6** |
| | F1 Score | 56.7 | 64.7 | **67.8** |
| HIV | Accuracy | 45.5 | 63.6 | **72.3** |
| | F1 Score | 57.1 | 66.7 | **72.3** |
| NC | Accuracy | 54.3 | 56.8 | **58.0** |
| | F1 Score | **58.4** | 52.1 | 56.4 |

and the existing well-labeled diseases used in the multi-task continual training stage). The results are shown in Table 3.

On average, the results show that the SptEHR strategy outperforms the other two strategies in terms of both performance metrics - achieving the highest accuracy in five out of six diseases (except Cancer) and the highest F1 score in four out of six diseases (i.e. Diabetes, COPD, CKD and HIV). Particularly, SptEHR

**Table 3.** Strategy comparison for prompt transfer.

| Disease | Metric | Strategy (%) | | |
|---|---|---|---|---|
| | | Random | Mean-Pooling | SptEHR |
| Cancer | Accuracy | 56.7 | **61.9** | 57.1 |
| | F1 Score | 57.7 | **62.5** | 56.9 |
| Diabetes | Accuracy | 58.0 | 59.7 | **61.9** |
| | F1 Score | 61.5 | 59.0 | **63.4** |
| COPD | Accuracy | 60.2 | 61.0 | **68.3** |
| | F1 Score | 57.4 | 63.6 | **65.5** |
| CKD | Accuracy | 66.7 | 64.1 | **67.6** |
| | F1 Score | 67.1 | 67.3 | **67.8** |
| HIV | Accuracy | 63.6 | 63.6 | **72.3** |
| | F1 Score | 63.6 | 63.6 | **72.3** |
| NC | Accuracy | 55.6 | 56.8 | **58.0** |
| | F1 Score | 55.0 | **57.8** | 56.4 |

transfer strategy shows the greatest advantage in HIV inference where there are only 17 training samples. Interestingly, the Mean-Pooling strategy shows the best performance for Cancer inference, which indicates Cancer may have similar connections to all other five selected diseases. Overall, the results suggest that the prompt transfer strategy in SptEHR is suitable to most cases especially when training data is not sufficient for the target task.

## 5      Conclusions

In conclusion, this paper proposes a novel pipeline, called Soft prompt transfer for Electronic Health Records (SptEHR), to address the challenges of developing efficient and adaptable models for EHR tasks, especially in few-shot settings. The proposed pipeline consists of three stages, including self-supervised pre-training, supervised multi-task continual training, and prompt transfer. The transformer-based foundation model which is learned from the first stage captures domain-specific knowledge. Then the multi-task continual training stage leverages well-labeled data from existing tasks to further refine the foundation model and to learn transferable task-specific soft prompts, improving model adaptability and performance on EHR tasks. Finally, the third stage leverages soft prompt transfer which is based on the similarity between the new and the existing tasks, to effectively address new tasks without requiring additional/extensive training. The effectiveness of the proposed method is demonstrated through experimentation on the MIMIC-III benchmark dataset, where the results indicate the superiority of the proposed SptEHR approach in comparison to traditional EHR models. Overall, the study contributes to the development of efficient and adaptable EHR models, which is critical in enabling the use of powerful deep learning models for various medical applications, especially in few-show settings such as in rare/novel disease inference.

## References

1. Aribandi, V., et al.: Ext5: towards extreme multi-task scaling for transfer learning. arXiv preprint arXiv:2111.10952 (2021)
2. Beltagy, I., Lo, K., Cohan, A.: SciBERT: a pretrained language model for scientific text. arXiv preprint arXiv:1903.10676 (2019)
3. Brown, T., et al.: Language models are few-shot learners. Adv. Neural. Inf. Process. Syst. **33**, 1877–1901 (2020)
4. Chen, T., Kornblith, S., Swersky, K., Norouzi, M., Hinton, G.E.: Big self-supervised models are strong semi-supervised learners. NeurIPS **33**, 22243–22255 (2020)
5. Choi, E., Xu, Z., Li, Y., Dusenberry, M., Flores, G., Xue, E., Dai, A.: Learning the graphical structure of electronic health records with graph convolutional transformer. In: Proceedings of the AAAI. vol. 34, pp. 606–613 (2020)
6. Devlin, J., Chang, M.W., Lee, K., Toutanova, K.: Bert: pre-training of deep bidirectional transformers for language understanding. arXiv preprint arXiv:1810.04805 (2018)

7. Devlin, J., Chang, M., Lee, K., Toutanova, K.: BERT: pre-training of deep bidirectional transformers for language understanding. In: NAACL-HLT, pp. 4171–4186. ACL (2019)

8. Gu, Y., Han, X., Liu, Z., Huang, M.: Ppt: Pre-trained prompt tuning for few-shot learning. arXiv preprint arXiv:2109.04332 (2021)

9. Gururangan, S., et al.: Don't stop pretraining: adapt language models to domains and tasks. arXiv preprint arXiv:2004.10964 (2020)

10. Han, X., et al.: Pre-trained models: past, present and future. AI Open **2**, 225–250 (2021)

11. Jiang, Z., Xu, F.F., Araki, J., Neubig, G.: How can we know what language models know? Trans. Assoc. Comput. Linguist. **8**, 423–438 (2020)

12. Johnson, A.E., et al.: Mimic-iii, a freely accessible critical care database. Sci. Data **3**, 160035 (2016)

13. Lan, Z., Chen, M., Goodman, S., Gimpel, K., Sharma, P., Soricut, R.: Albert: a lite BERT for self-supervised learning of language representations. arXiv preprint arXiv:1909.11942 (2019)

14. Lester, B., Al-Rfou, R., Constant, N.: The power of scale for parameter-efficient prompt tuning. arXiv preprint arXiv:2104.08691 (2021)

15. Li, X.L., Liang, P.: Prefix-tuning: Optimizing continuous prompts for generation. arXiv preprint arXiv:2101.00190 (2021)

16. Li, X.L., Liang, P.: Prefix-tuning: Optimizing continuous prompts for generation. In: Proceedings of ACL/IJCNLP 2021, (Volume 1: Long Papers), Virtual Event, August 1–6, 2021. pp. 4582–4597. ACL (2021)

17. Li, Y., et al.: BEHRT: transformer for electronic health records. Sci. Rep. **10**(1), 1–12 (2020)

18. Liu, P., Yuan, W., Fu, J., Jiang, Z., Hayashi, H., Neubig, G.: Pre-train, prompt, and predict: a systematic survey of prompting methods in natural language processing. ACM Comput. Surv. **55**(9), 1–35 (2023)

19. Liu, X., Ji, K., Fu, Y., Du, Z., Yang, Z., Tang, J.: P-tuning v2: prompt tuning can be comparable to fine-tuning universally across scales and tasks. CoRR abs/2110.07602 (2021)

20. Liu, X., et al.: P-tuning v2: prompt tuning can be comparable to fine-tuning universally across scales and tasks. arXiv preprint arXiv:2110.07602 (2021)

21. OpenAI: Gpt-4 technical report (2023)

22. Pan, S.J., Yang, Q.: A survey on transfer learning. IEEE Trans. Knowl. Data Eng. **22**(10), 1345–1359 (2010)

23. Peng, X., Long, G., Shen, T., Wang, S., Jiang, J.: Sequential diagnosis prediction with transformer and ontological representation. In: 2021 IEEE International Conference on Data Mining (ICDM), pp. 489–498. IEEE (2021)

24. Peng, X., Long, G., Shen, T., Wang, S., Jiang, J., Zhang, C.: Bitenet: bidirectional temporal encoder network to predict medical outcomes. In: 2020 IEEE International Conference on Data Mining (ICDM), pp. 412–421. IEEE (2020)

25. Peng, X., et al.: MIPO: mutual integration of patient journey and medical ontology for healthcare representation learning. arXiv preprint arXiv:2107.09288 (2021)

26. Qin, G., Eisner, J.: Learning how to ask: querying LMS with mixtures of soft prompts. arXiv preprint arXiv:2104.06599 (2021)

27. Radford, A., Narasimhan, K., Salimans, T., Sutskever, I., et al.: Improving language understanding by generative pre-training (2018)

28. Ren, H., Wang, J., Zhao, W.X., Wu, N.: Rapt: Pre-training of time-aware transformer for learning robust healthcare representation. In: Proceedings of the 27th ACM SIGKDD, pp. 3503–3511 (2021)

29. Schick, T., Schütze, H.: It's not just size that matters: Small language models are also few-shot learners. In: Proceedings of the 2021 Conference of the NAACL: Human Language Technologies, pp. 2339–2352 (2021)
30. Shin, T., Razeghi, Y., Logan IV, R.L., Wallace, E., Singh, S.: AutoPrompt: eliciting knowledge from language models with automatically generated prompts. arXiv preprint arXiv:2010.15980 (2020)
31. Si, Y., Bernstam, E.V., Roberts, K.: Generalized and transferable patient language representation for phenotyping with limited data. J. Biomed. Inform. **116**, 103726 (2021)
32. Steinberg, E., Jung, K., Fries, J.A., Corbin, C.K., Pfohl, S.R., Shah, N.H.: Language models are an effective representation learning technique for electronic health record data. J. Biomed. Inform. **113**, 103637 (2021)
33. Taylor, N., Zhang, Y., Joyce, D., Nevado-Holgado, A., Kormilitzin, A.: Clinical prompt learning with frozen language models. arXiv preprint arXiv:2205.05535 (2022)
34. Thrun, S., Pratt, L.: Learning to learn: Introduction and overview. learning to learn, pp. 3–17 (1998)
35. Vaswani, A., et al.: Attention is all you need. In: NeurIPS 2017, December 4–9, 2017, Long Beach, CA, USA, pp. 5998–6008 (2017)
36. Vu, T., Lester, B., Constant, N., Al-Rfou, R., Cer, D.: Spot: better frozen model adaptation through soft prompt transfer. arXiv preprint arXiv:2110.07904 (2021)
37. Vu, T., Lester, B., Constant, N., Al-Rfou', R., Cer, D.: Spot: better frozen model adaptation through soft prompt transfer. In: Proceedings of ACL, pp. 5039–5059. Association for Computational Linguistics (2022)
38. Wang, W., et al.: Structbert: Incorporating language structures into pre-training for deep language understanding. arXiv preprint arXiv:1908.04577 (2019)
39. Xu, H., Chen, Y., Du, Y., Shao, N., Wang, Y., Li, H., Yang, Z.: ZeroPrompt: scaling prompt-based pretraining to 1, 000 tasks improves zero-shot generalization. In: Findings of the Association for Computational Linguistics: EMNLP, pp. 4235–4252 (2022)
40. Zhao, Z., Wallace, E., Feng, S., Klein, D., Singh, S.: Calibrate before use: improving few-shot performance of language models. In: International Conference on Machine Learning, pp. 12697–12706. PMLR (2021)

# Graph Convolution Synthetic Transformer for Chronic Kidney Disease Onset Prediction

Di Zhu[1], Yi Liu[1], Weitong Chen[2], Yanda Wang[3], Yefan Huang[4], Xiaoli Wang[4], Ken Cai[5], and Bohan Li[1,6,7(✉)]

[1] Nanjing University of Aeronautics and Astronautics, Nanjing 211106, China
[2] Adelaide University, Adelaide SA 5005, Australia
[3] Nanjing Normal University, Nanjing 210023, China
[4] Xiamen University, Xiamen 361005, China
[5] Zhongkai University of Agriculture and Engineering, Guangzhou 510225, China
[6] Ministry Key Laboratory for Safety-Critical Software Development and Verification, Nanjing 211106, China
[7] National Engineering Laboratory for Integrated Aero-Space-Ground-Ocean Big Data Application Technology, Xian 710072, China
bhli@nuaa.edu.cn

**Abstract.** Effective disease prediction based on electronic health records (EHR) is an important topic in health informatics. The current methods usually use common deep-learning models for disease prediction. However, it is difficult to fully learn the graphic encounter structure of EHR to improve prediction performance. Moreover, in prediction tasks, chronic kidney disease (CKD) has a poor prognosis due to excessive risk factors and complex comorbidities. Therefore, we propose a CKD onset prediction model called Graph Convolution Synthetic Transformer (GCST) based on EHR, using Fusion Attention Mechanism to solve these challenges. By modifying Transformer, GCST uses Factorized Dense Attention and Medical Local Attention to learn global and local attention, generating Synthetic Attention to learn the potential encounter structure and meaningful medical knowledge of EHR. In addition, we also propose a transfer learning strategy based on sample weighted correction to guide the prediction of GCST in specific low-resource EHR. We conduct sufficient experiments on three datasets to test the performance of GCST. Experiments show that GCST has significant improvement over state-of-the-art models.

**Keywords:** Disease Prediction · Attention Mechanism · Transformer · Transfer Learning

## 1 Introduction

EHR system collects a large number of medical records (*e.g.* patients' basic information, clinical records, or drug prescriptions), providing great potential for risk prediction models to help doctors predict future diagnosis results more

X. Yang et al. (Eds.): ADMA 2023, LNAI 14178, pp. 33–47, 2023.
https://doi.org/10.1007/978-3-031-46671-7_3

accurately. Some previous work used autoencoders [2,19], recurrent neural network [4,21] and other deep learning methods to predict [8,16], achieved remarkable results.

However, some works do not make full use of the encounter structures of EHR to construct high-quality patient embeddings, which just treat each encounter as a group of disordered features, ignoring the potential graph structure of EHR. Moreover, current models usually map discrete medical concepts into continuous space and summarize interactions through complete search, which cannot follow internal medical knowledge of EHR to limit the search to meaningful attention distribution. In addition, when faced with low-resource EHR, existing models lack adaptability.

Consequently, we propose a CKD onset prediction model called GCST to solve the above limitations. GCST uses global attention to learn all potential graphic encounter structures in EHR, and uses local attention to learn medical knowledge to limit the search to meaningful attention distribution. Through the synthetic attention of global and local aspects, GCST finally provides richer patient-level information embedding. Additionally, we propose a transfer-learning module based on sample weighted correction to solve the problems of low-resource EHR. In note, We choose CKD as the prediction target, which is a general term for heterogeneous diseases that irreversibly change the structure of kidney or lead to the chronic decline of kidney function [12]. Most patients cannot feel symptoms or find diseases before the late stage of CKD, so it is very necessary to early predict the onset of CKD based on EHR.

The main contributions of this paper are summarized as follows:

- We propose a novel modification to Transformer, which uses the global and local attention mechanism to learn the potential encounter structure and meaningful medical knowledge in EHR respectively, guiding the synthetic self-attention to learn hospital-level patient embedding.
- We propose a transfer learning strategy based on sample weighted correction, which approximates the distribution of source and target domains with inconsistent data distribution, to solve the problem of low-resource EHR.
- We conduct experiments on open and synthetic datasets to verify the effectiveness of the proposed GCST for CKD onset prediction. The experiment results show that GCST is superior to all baseline models.

## 2   Related Work

### 2.1   EHR Data Mining

EHR data mining is a hot research topic in medical informatics. The analysis tasks include electronic genotyping [1,10], disease progression [15,18], drug detection [9,14], diagnosis prediction [4,5,16,17,20], etc.

The most common application in tasks is to predict disease through deep learning models. Deep neural networks play a limited role in early prediction tasks, requiring learning disease encounters from scratch. GRAM [4] uses the inherent multilevel structure of the medical ontology and PRIME [18] combines external medical knowledge to supplement EHR. Then, attention mechanism is introduced into risk prediction, Dipole [17] learns the weight of the visit and assigns weight to the diagnostic code in each visit. Retain [5] uses multiple attention mechanisms to model longitudinal EHR data. Currently, the fusion attention mechanism has been widely used to simulate the decision-making process of doctors during visits and achieve risk prediction on EHR. HiTANet [16] uses time information to build a hierarchical time-aware attention network. SETOR [20] captures the correlation between patient visits through a multi-layer transformer.

Deep learning models show great performance in large-scale data prediction. However, these methods are not fully applicable to small-scale, low-resource EHR datasets. The usual solution is to combine multiple source EHR datasets, but this requires complex data harmonization processes. Therefore, some methods like CONAN [8] use data augmentation techniques to deal with imbalanced positive sample datasets. G-BERT [24] uses pre-training and transfer learning methods to learn the representations of hospital visit records for risk prediction.

## 2.2   Medical Concept Representation

Medical concept representations in the medical field are mainly based on natural language processing (NLP), which encodes discrete medical concepts into vectors for learning embedding by focusing on patient [13], visit [25], medical code [3, 26].

When modeling medical concepts, the early NLP algorithms mainly focus on the feature dimension rather than the time dimension, ignoring the order of occurrence of medical concepts and the association between codes of EHR. Therefore, MIME [6] was proposed to use the proprietary structure of medical codes, mining multi-level embeddings of EHR, which requires that EHR contain complete structural information of diagnosis and treatment. GCT [7] was proposed to jointly learn the graphic structure of EHR, which proved that Transformer is an effective basic method for learning the potential encounter structures.

# 3   Method

## 3.1   Problem Statement

We formulate this problem as a binary prediction task on CKD onset. The EHR is in the form of sequences of hospital visits $S = \{V^1, V^2, \cdots, V^T\}$, $T$ is the number of visits. The $t$-th visit $V^t$ consists of diagnostic codes $D^t$, treatment codes $M^t$ and observation codes $O^t$. We want to predict whether patients will be diagnosed with CKD in subsequent visits $V^{t+1}$.

## 3.2   EHR as Graph Network

As shown in Fig. 1, in EHR, the $t$-th visit $V^t$ starts from the top visit node $v^t$. Under $v^t$ are the diagnosis codes $d_1^t, \cdots, d_{|d^t|}^t$, accompanied by treatment codes

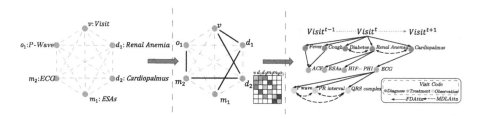

**Fig. 1.** Learn the potential graphic encounter structure of EHR. First, on the left, all nodes $c$ (including $d$, $m$, $o$) are considered fully connected. Then, in the middle, learn meaningful connections between nodes and construct attention matrix $B$. Finally, on the right, construct a connected graph network of EHR.

$m_1^t$, $\cdots$, $m_{|m^t|}^t$. The observation code $o_1^t$, $\cdots$, $o_{|o^t|}^t$ related to the treatment can be associated with a binary value (*e.g.* positive/negative reaction), where $|d|$, $|m|$, $|o|$ respectively represent the numbers of diagnostic codes, treatment codes, and observation codes in $V^t$.

Assuming that all features $d$, $m$, $o$ can be represented by collective term $c$ in the same potential space, $|c| = |d| + |m| + |o|$, each visits $V$ (we called encounter) can be viewed as a graph composed of an adjacency matrix $A$ and $c$ nodes, where $A$ represents the connection relationship between $c$. Therefore, given $c$ and $A$, we can use EHR as a graph network and generate the visit representation $v$ for CKD onset predicted tasks.

However, it is thoughtlessly to simply consider $V$ as a bag of nodes $c$, and EHR usually lack obvious $A$ because doctors should follow medical rules when making medical decisions. Motivated by [6,7], assuming that all $c$ are fully connected, Transformer can learn the underlying encounter structure in EHR as a special graph embedding algorithm, which can be formulated as:

$$C^j = MLP^j \left( \text{softmax} \left( \frac{Q^j K^{j\top}}{\sqrt{d_{\text{attn}}}} \right) V^j \right) \tag{1}$$

where $Q^j = C^{j-1} W_Q^j$, $K^j = C^{j-1} W_K^j$, $V^j = C^{j-1} W_V^j$, and $d_{\text{attn}}$ is the dimension of attention mechanism, $W_K^j, W_Q^j, W_V^j$ are linear variable matrix of the $j$-th Transformer self-attention block. Because of the unordered features in fully connected encounters, this method does not use sine and cosine functions for location coding.

This is an effective method: using Transformer as a basic method for understanding the underlying structures in electronic health records (EHRs) and representing them as graph networks,

Therefore, Transformer is an effective basic algorithm for understanding the potential encounter structures in EHR and representing them as graph networks. However, the existing models usually rely entirely on pairwise interaction between tokens, do not make full use of prior knowledge to modularize a pre-trained network. Meanwhile, single directed self-attention cannot effectively learn global and local attention in EHR. Therefore, we propose GCST to learn the potential encounter structure of EHR more completely and efficiently.

## 3.3 Graph Convolutional Synthesizer Transformer

In this subsection, as shown in Fig. 2, we introduce the proposed GCST in detail, including the synthetic attention based on factorized dense attention and medical local attention, and the low-resource transfer learning strategy.

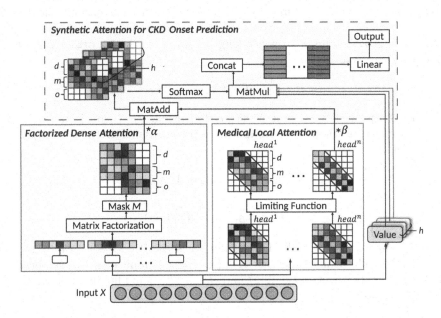

**Fig. 2.** Overview of the proposed GCST model.

**Factorized Dense Attention for Potential Encounter Structure.** FDAttn removes the notion of query-key-values in the self-attention module of Transformer and directly generates the attention weights by searching the entire attention space. In practice, the autocoder receives a series of EHR data as $\{x'_i\}_{t=1}^{T}$, where $x'_i = [x_i; d_i] \in \mathbb{R}^{n \times 1}$, and generates the hospital visit matrix $X_\ell \in \mathbb{R}^{n \times d}$, where $l$ represents the length of sequence, and $d$ represents the dimension of model. FDAttn projects $x'_i$ from $d$ dimension to $n$ dimension through the function $F(\cdot)$, which adopts two Feed-Forward Layers with $ReLU$ activation. Hence, each token can learn the weights of each token in the input sequence, which is formulated as:

$$B_{FDAttn} = F(X) = \sigma_R (XW_1) W_2 \tag{2}$$

$$V = C(X) = XW_3 \tag{3}$$

where $W_1 \in \mathbb{R}^{d \times d}$, $W_2 \in \mathbb{R}^{d \times \ell}$, $W_3 \in \mathbb{R}^{d \times d}$ are all learnable weights, $C(\cdot)$ is a parameterized function of $X$, similar to Value $V(\cdot)$ in Transformer.

Consequently, we eliminate the overfocus between dot products by changing the $QK$ projection in Transformer. However, the input sequence length $l$ of EHR is usually larger than the model dimension $d$, which makes it necessary to think more about the input sequence. Specifically, we use 1) a diagnose-observation mask $M_{FDAttn}$ and 2) a low-rank matrix factorization of sequence, reducing the parameter costs. Firstly, considering that diagnosis codes can only connect to treatment codes, but not directly to observation codes, we design Mask $M_{FDAttn}$ to block the unacceptable connection $B_M \in \mathbb{R}^{|d| \times |o|}$ during attention generation. Then, we reduce the parameter size by low rank matrix factorization of the sequence length. Assuming sequence length $l = a \times b$, the factorized variant of dense attention is formulated as:

$$B_1, B_2 = F_{B_1}(X), F_{B_2}(X) \tag{4}$$

where $F_{B_1}(\cdot)$ projects the input $X$ into $a$ dimension, $F_{B_2}(\cdot)$ projects the input $X$ into $b$ dimension. Hence, the computation processes of $FDAttn$ with $M_{FDAttn}$ can be factorized as:

$$Y_{FDAttn} = \mathrm{softmax}\left(\tilde{B}_1 \odot \tilde{B}_2 + M_{FDAttn}\right) C(X)$$

$$\tilde{B}_1, \tilde{B}_2 = H_{B_1}(B_1), H_{B_2}(B_2) \tag{5}$$

where $B^{h_i}$ in Eq 2 is factorized by element-wise multiplication $\odot$ into $\tilde{B}_1^{h_i} \odot \tilde{B}_2^{h_i} \in \mathbb{R}^{n \times n}$. The tiling functions $H_{B_1}(\cdot)$, $H_{B_2}(\cdot)$ duplicates the vector $a$, $b$ times respectively to get $\tilde{B}_1^{h_i} \in \mathbb{R}^{a \times b}$ and $\tilde{B}_2^{h_i} \in \mathbb{R}^{b \times a}$, and finally multiply them bit by bit.

**Medical Local Attention for Meaningful Medical Knowledge.** Considering the fact that there might be a potential connection in the medical codes of same types, such as complications in diagnostic codes or adverse drug reactions in treatment codes, and the multi-head mechanism allows capturing relationships from multiple aspects to synthesize multi-view information. We propose MDLAttn to restrict the search to the attention space that may contain meaningful medical knowledge.

First, We block the visit matrix $X$ into $M_{diag} = A_D \oplus A_M \oplus A_O$ to achieve a fine-grained feature extraction, where $\oplus$ is the direct sum operator, $M_{diag}$ is a diagonal conditional matrix, block matrices $A_D \in \mathbb{R}^{|d| \times |d|}$, $A_M \in \mathbb{R}^{|m| \times |m|}$, $A_O \in \mathbb{R}^{|o| \times |o|}$ and $|d| + |m| + |o| = l$. Note that $M_{diag}$ has the same size as the FDAttn matrix $B_{FDAttn}$.

Then, we define function $\Gamma(\cdot)$, which limits the current token to interacting only with adjacent tokens within an appropriate range during the training stage. By assigning attention weight only to the current token and its adjacent tokens, the attention weight of other tokens out of width is set to 0, which can be expressed as:

$$\Gamma(t, c, i) = \min \left\{ \max \left\{ \frac{1}{r} \left( r + \mu_\theta - |t + c - i| \right) \right\}, 1 \right\} \tag{6}$$

where hyperparameter $r$ determines the change gradient of the attention width. The appropriate value of $\mu_\theta \in [0, \mu]$ is determined during the training stage, which controls the maximum attention span and range of backpropagation allowed. Parameter $c \in [0, \max\{d, m, o\} - \min\{d, m, o\}]$ learns the maximum width of local attention because the long-range interaction of tokens is out of MDLAttn consideration. $|c + t - i|$ defines the neighborhood distance at timestep $t$, which limits the attention weight of MDLAttn close to the center. Therefore, the single-headed MDLAttn can be represented as:

$$Y_{MDLAttn}^{h_i} = \text{softmax}(\bar{B}_{MDLAttn}^{h_i}) C^{h_i}(X) \tag{7}$$

$$\bar{B}_{t,i} = \frac{\Gamma_{t,c,i} exp(B_{t,i})}{\sum_j \Gamma_{t,c,j} exp(B_{t,j})} \tag{8}$$

where each attention weight matrix $B$ of is multiplied by function $\Gamma(\cdot)$ to get the final attention weight $\bar{B}_{MDLAttn}^{h_i}$, and the calculation of $C^{h_i}(\cdot)$ is similar to FDAttn.

**Synthetic Attention for CKD Onset Prediction.** Given FDAttn and MDLAttn, we can derive SYNAttn to guide GCST in learning the true graph structure of EHR for CKD onset prediction. According to Eq 5, and Eq 7, we can derive the following formula:

$$Y_{SYNAttn}^{h_i} = \text{softmax} \left( \alpha(\tilde{B}_1 \odot \tilde{B}_2 + M_{FDAttn}) + \beta \bar{B}_3^{h_i} \right) C^{h_i}(X) \tag{9}$$

where hyperparameters $\alpha$ and $\beta$ determine the attention matrix weight of FDAttn $(\tilde{B}_1^{h_i} \odot \tilde{B}_2^{h_i} + M)$ and MDLAttn $(\Gamma_{t,c,i} B_3^{h_i})$ in the attention matrix of SYNAttn. Therefore, the calculation of multi-head SYNAttn with $n$ attention heads is as follows:

$$SYNAttn(X) = Concat \left( Y_{SYNAttn}^{h_1}, \ldots, Y_{SYNAttn}^{h_n} \right) W^O \tag{10}$$

where $W^O \in \mathbb{R}^{(h_n \times d) \times d}$ is the $SYNAttn$ final multihead output matrix, $Concat$ is the concatenate operation. Therefore, the binary label $y$ of CKD onset prediction can be expressed as:

$$\mathcal{L} = -\frac{1}{N} \sum_{j=1}^{N} (y_j \log \hat{y}_j + (1 - y_j) \log (1 - \hat{y}_j)) \tag{11}$$

$$\hat{y} = \sigma \left( w_y^\top c + b_y \right) \tag{12}$$

where $\hat{y}$ is the prediction score of CKD onset. $N$ is the number of patients. The training objective is to minimize binary cross-entropy loss using predicted score $\hat{y}$ and true label $y$.

### 3.4 Transfer Learning Strategy for Low-Resource EHR

In practice, we find through experiments that due to the interference of patient groups, recording devices and external environment in different hospitals, EHR from different sources can not satisfy the same distribution conditions, so the classification model trained on the old dataset can not classify the new data with different distribution efficiently and accurately. In addition, the number of samples is usually limited, Transformer model is not effective for specific low-resource EHR. Therefore, it is important to adopt a transfer learning strategy to transfer the basic knowledge from the original EHR to low-resource data to improve the adaptability of GCST.

Specifically, we propose a GCST modification based on a sample weighted correction mechanism, called GCST-TL, to solve the problem of GCST processing low-resource EHR. It approximates the distribution of inconsistent data in the source and target domains. The pre-trained EHR is used as the source domain, and the low-resource EHR for learning and testing is used as the target domain. All input is $X$, the output is $Y$, and the data distribution of the source domain and target domain is different, where $P_{\text{target}}(x) \neq P_{\text{source}}(x)$, but the conditional distribution is similar, where $P_{\text{target}}(y \mid x) \approx P_{\text{source}}(y \mid x)$.

First, we initialize the weights of the two domains, finding the sample in the source domain most similar to samples in the target domain. Then, we merge samples from the two domains $X' = X_{\text{source}} \cup X_{\text{target}}$, and conduct training. While samples in the source domain recorded as negative, and samples in the target domain recorded as positive. Because $n_{sourc} \gg n_{target}$, the positive samples of the loss function have a higher weight, so we can get the initial normalized positive sample strange rate as $\beta = normalize(P(y = 1 \mid x_{\text{source}}))$.

In the preliminary experiment, we noticed that there are some samples in the source domain very different from the target domain, which affected the speed and effect of model training. Therefore, we have set the exclusion policy of the lowest threshold value $w_{min}$ during the normalization process as follows:

$$X'_{source} = X_s - \{x_i \mid w_i \in W_s, x_i \in X_s, w_i \leq P_q\} \tag{13}$$

where $X_s = \{x_1, \ldots, x_n\}$ is source sample, $W_s = \{w_1, \ldots, w_n\}$ is the normalized weight corresponding to the source sample, and $P_q$ is the $q$ - th quantile of $X_{source}$. The low-weight samples are removed based on the weight quantile, and only the samples with strong correlation with the target domain are retained. Therefore, we can combine the data again $T = X'_{source} \cup X_{\text{target}}$, and normalize the weight $\beta$ of $X'_{\text{source}}$ to get $\beta'$, and assign the weights of the two domains as:

$$w = \begin{cases} \beta', i = 1, \ldots, n_s \\ \max(\beta'), i = n_s + 1, \ldots, n_s + m_t \end{cases} \tag{14}$$

where $n_s$ and $m_t$ is the number of samples in the source domain and target domain. In the $t$-th iteration, the unlabeled dataset and the combined training set $T$ with normalized weight vector $w$ are used to perform GCST to generate a weak classifier $GCST_t$. Therefore, we calculate the error rate of $GCST_t$ on target sample $X_{target}$ as follows:

$$\varepsilon_t = \sum_{i=n+1}^{n+m} \frac{w_i^t \left| GCST_t \left( x_i - y_i \right) \right|}{\sum w_i^t} \tag{15}$$

set $\rho = 1/(1 + \sqrt{2\ln n/N})$ and $\beta_t = \varepsilon_t/(1 - \varepsilon_t)$. Then, the weight value is updated as:

$$w_i^{t+1} = \begin{cases} w_i^t \rho^{|h_t(x_i - y_i)|} & , i = 1, \ldots, n_s \\ w_i^t \beta_t^{-|h_t(x_i - y_i)|} & , i = n_s + 1, \ldots, n_s + m_t \end{cases} \tag{16}$$

in the iteration process, according to the sample exclusion strategy of the lowest threshold, normalize $w_i^{t+1}$ and exclude the sample $x_i$ with weight $w_i^{t+1} < w_{min}$ in the source domain, the final prediction model $h(\cdot)$ of GCST-TL calculated as follows:

$$GCST(x) = \begin{cases} 1, \sum_{t=0}^{N} \ln (1/\beta_t) h_t(x) \geq \frac{1}{2} \sum_{t=0}^{N} \ln (1/\beta_t) \\ 0, \text{ others} \end{cases} \tag{17}$$

## 4  Experiments

In this section, we evaluate the performance of the proposed GCST by conducting extensive experiments on datasets and answering the following four questions:

- RQ1: How does the prediction performance of GCST compare with the most advanced methods?
- RQ2: Can we improve the prediction performance of GCST by learning the potential encounter structure of EHR at local and global levels?
- RQ3: Is it effective to solve the problem of low-resource EHR through the transfer learning strategy of sample weighting correction?
- RQ4: What is the effect of different hyper-parameter settings on GCST's performance?

### 4.1  Data Description

**Datasets.** To our knowledge, public EHR datasets often lack structural information. To evaluate the predictive and adaptive ability of GCST to learn EHR structure, we reference two real EHR datasets: MIMIC-III and eICU, and generate a Synthetic dataset. Note that the Synthetic datasets are only used to test GCST's transfer-learning ability. We split and modify the datasets, which makes the distribution of MIMIC-III, eICU, and Synthetic datasets different from each other to simulate the requirements of transfer learning on low-resource EHR. We extract kidney disease sequences from the datasets and treat CKD onset prediction as a binary classification problem to predict whether patients will be diagnosed with CKD.

Table 1 summarizes the details of datasets. Since most patients have less than 30 visits, we reserve 30 recent visits for each patient to improve scalability. The datasets are divided into training, validation and test sets at the ratio of 75/10/15.

**Table 1.** Dataset Details.

| Dataset | MIMIC-III | eICU | Synthetic |
|---|---|---|---|
| Total # of patients | 29,304 | 50,950 | 15,000 |
| # of cases (positive) | 9,767 | 16,966 | 5,000 |
| # of controls (negative) | 19,537 | 67,968 | 10,000 |
| # of patients for training | 21,978 | 38,212 | 11,250 |
| # of patients for validation | 2,931 | 5.095 | 1,500 |
| # of patients for testing | 4,395 | 7,643 | 2,250 |

### 4.2  Baseline Methods

To demonstrate the performance of our proposed GCST, we compare it with eight classical and advanced prediction models.

- **Logistic regression (LR)** [23]: It serves as the fundamental foot-stone for comparison. At the last time of patient input, it count the existence of medical codes and summarized the observation results.
- **Multi-layer perceptron (MLP)** [22]: It is also the basic baseline. We use a method similar to LR and add a fully connected layer activated by relu.
- **RNN & BiRNN** [11]: They are the basic framework for most prediction models. We use the full connection layer coding input, and propagate vectors to the forward/bidirectional RNN layer, the last state predicted with logical regression.
- **Retain** [5]: It learns the embedding of medical concepts and performs heart failure prediction through a reverse RNN with attention mechanism.
- **Dipole** [17]: It uses a bidirectional RNN layer and three attention mechanisms to predict the patient's visit information.
- **HiTANet** [16]: It uses the dynamically fused attention of these two time-aware attention weights to predict risk.
- **SETOR** [20]: It uses an end-to-end transformer to combine medical ontology with medical information for risk prediction.

**Evaluation Metrics.** We choose the area under the precision-recall curve (AUPRC) to evaluate the model performance on the unbalanced dataset. AUPRC can effectively estimate the proportion of true positive in the positive prediction. In addition, we also calculate the negative logarithmic likelihood (NLL) based on Eq. 11 to measure the model loss on the testset.

## 4.3   Overall Performance (RQ1)

We evaluate GCST and all baselines on two testsets, Table 2 shows AUPRC and NLL scores. The best model in each column is shown in bold. We note that:

**Table 2.** Performance comparisons of all models on the two datasets.

| Method | | Classical Methods | | Plain RNNs | | Attention-based Models | | | | Ours0 |
|---|---|---|---|---|---|---|---|---|---|---|
| | | LR | MLP | RNN | BiRNN | RETAIN | Dipole | HiTANet | SETOR | GCST |
| MIMIC-III | AUPRC | 0.4179 | 0.4392 | 0.4934 | 0.4949 | 0.5274 | 0.5325 | 0.5638 | 0.5692 | **0.5768** |
| | NLL | 0.3573 | 0.3115 | 0.2987 | 0.2977 | 0.2969 | 0.2962 | 0.2947 | 0.2935 | **0.2927** |
| eICU | AUPRC | 0.4981 | 0.5132 | 0.5538 | 0.5611 | 0.5942 | 0.6064 | 0.6283 | 0.6323 | **0.6412** |
| | NLL | 0.3432 | 0.3082 | 0.2877 | 0.2864 | 0.2853 | 0.2846 | 0.2838 | 0.2811 | **0.2793** |

• **Plain RNNs (RNNs and BiRNNs) are superior to classical methods (LR and MLP)**, because sequential models capture the underlying patterns of disease encounters more effectively. Classical methods of non-sequential model lack the ability of sequence modeling, and cannot distinguish each access, can only learn from the aggregated information of EHR. Intelligible, in practical medical settings, it's not possible for doctors to diagnose CKD based on isolated symptoms. Rather, they must meticulously examine the patient's medical records and complement them with extensive medical evaluations to arrive at a conclusive diagnosis.

• **All attention-based models show better performance in predictions.** Obviously, predictive disease benefits from learning the underlying encounter structure. Retain and Dipole use attention mechanisms to learn sequence information from scratch to assign scores to each visit, which is a good choice for original disease prediction tasks. HiTANet and SETOR consider the influence of time on prediction in feature aggregation, making the fused attention mechanism a better scheme. However, the above methods do not fully consider the influence of pathological knowledge on disease prediction, because important risk factors may be interwoven. In addition, making significant predictions based solely on the existence of multiple risk factors is challenging without comprehensive prior knowledge.

• **Our proposed GCST reaches the most advanced score in indicators.** By introducing the Synthetic attention of FDAttn and CondAttn, GCST effectively learns all the potential encounter structures of EHR, limits the search to meaningful attention distribution according to medical knowledge, produces a high-quality patient embedding, compresses the information of the input sequence, so that our model can better identify the relationship between medical diagnosis and historical medical records. Compared with the most effective baseline model (SETOR), GCST increased by 1.34% and 1.41% in the AUPRC of the two datasets respectively.

## 4.4   Ablation Study

**Impact of Synthetic Attention (RQ2).** We propose global attention FDAttn to obtain all the potential encounter structures in EHR and conditional attention CondAttn to limit the search to meaningful attention distributions. Now we study whether the Synthetic attention is necessary. As shown in Table 3, the Fusion Attention Mechanism demonstrates its superiority in encoding medical concepts. Reasonably, with the addition of joint attention to pathological knowledge, the information increased as expected but with limited intensity. The performance is improved in each case, which demonstrates the validity of GCST and the significance of learning patient embedding at both local and global levels.

**Table 3.** Performance comparison of the models using different attentions.

| Dataset | MIMIC-III | | | eICU | | |
|---|---|---|---|---|---|---|
| Model | w/o FDAttn | w/o CondAttn | w/ SYNAttn | w/o FDAttn | w/o CondAttn | w/ SYNAttn |
| AUPRC | 0.5264 | 0.5532 (+5.09%) | 0.5768 (+9.57%) | 0.6011 | 0.6279 (+4.46%) | 0.6412 (+6.67%) |
| NLL | 0.2970 | 0.2951 | 0.2927 | 0.2849 | 0.2841 | 0.2793 |

**Impact of Transfer Learning Module (RQ3).** We propose a transfer-learning strategy based on sample weighted correction to solve the problem of low-resource EHR, which is called GCST-TL. Now we study whether the transfer learning strategy is effective. We take GCST as the basic classifier and compare it in three cases: the auxiliary training set $T_d$ (MIMIC-III or eICU), the source training set $T_s$ (Synthetic dataset) and the combined training set $T_t$ ($T_s \cup T_d$), where $T_s \ll T_d$. In Fig. 3, the ratio between the same-distribution and different-distribution training examples increases from 0.05 to 0.30. The results show that GCST-TL has good prediction performance when processing small datasets with inconsistent data distribution (*i.e.* ratio $< 0.20$). However, when the amount of target dataset reaches a certain level, the improvement effect of GCST-TL on GCST is no longer significant, because the different distribution of training numbers can generate noise and affect prediction. Therefore, this proves that GCST-TL makes GCST effective when dealing with low-resource EHR.

## 4.5   Hyper-parameter Study (RQ4)

**Effectiveness of embedding size.** The size of patient embedding is related to the selection of data set and can be evaluated by specific tasks. Figure 4(a) shows that the performance is poor in the case of low dimensions, while excessive high dimensions may cause over-fitting problems. When the embedded size is 64, the performance is best

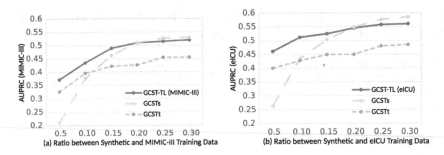

**Fig. 3.** Performance comparison of GCST using different processing strategies of low-resource datasets.

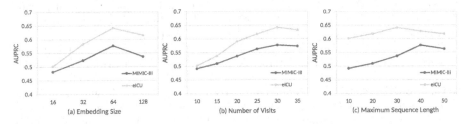

**Fig. 4.** Performance comparison of GCST models using different parameter settings.

**Effectiveness of the number of visits.** Figure 4(b) shows that the far times-tamps obtain worse performance, and the performance is best in the first 25–30 visits before the diagnosis. One possible explanation is that symptoms such as hyperglycemia, hypertension, and obesity may not be direct causes of the disease, but they significantly elevate the risk of developing it over time.

**Effectiveness of the maximum sequence length.** We set a limit on the sequence length to enhance the processing of sequence data. As illustrated in Fig. 4(c), the data characteristics of each dataset are different, requiring us to choose different treatment methods. Therefore, we chose the appropriate length for each dataset to achieve optimal results in the comparative experiment.

## 5    Conclusion

Learning effective potential information of EHR is an important step to improve the performance of disease prediction tasks. In this paper, we propose GCST, a CKD onset prediction model based on EHR. By modifying Transformer, GCST learns the potential encounter structure and meaningful medical knowledge of EHR from the global and local levels. In addition, with the proposed transfer learning strategy based on sample weighted correction, GCST can be applied to the prediction task of low-resource EHR. Experiments show that GCST is superior to all baselines in CKD onset prediction. In the future, we plan to extend GCST to general disease prediction tasks. We will also use more advanced attention algorithms to propose lightweight models.

**Acknowledgment.** This work is supported by the Natural Science Foundation of China (62172351), 14th Five-Year Plan Civil Aerospace Pre-research Project of China (D020101), and the Fund of Prospective Layout of Scientific Research for Nanjing University of Aeronautics and Astronautics.

# References

1. Bastarache, L.: Using phecodes for research with the electronic health record: from PheWAS to PheRS. Ann. Rev. Biomed. Data Sci. **4**, 1–19 (2021)
2. Che, Z., Kale, D., Li, W., Bahadori, M.T., Liu, Y.: Deep computational phenotyping. In: Proceedings of the 21th ACM SIGKDD International Conference on Knowledge Discovery and Data Mining, pp. 507–516 (2015)
3. Chen, Y., Ma, T., Yang, X., Wang, J., Song, B., Zeng, X.: Muffin: multi-scale feature fusion for drug-drug interaction prediction. Bioinformatics **37**(17), 2651–2658 (2021)
4. Choi, E., Bahadori, M.T., Song, L., Stewart, W.F., Sun, J.: Gram: graph-based attention model for healthcare representation learning. In: Proceedings of the 23rd ACM SIGKDD International Conference on Knowledge Discovery and Data Mining, pp. 787–795 (2017)
5. Choi, E., Bahadori, M.T., Sun, J., Kulas, J., Schuetz, A., Stewart, W.: Retain: an interpretable predictive model for healthcare using reverse time attention mechanism. In: Advances in Neural Information Processing Systems 29 (2016)
6. Choi, E., Xiao, C., Stewart, W., Sun, J.: Mime: multilevel medical embedding of electronic health records for predictive healthcare. In: Advances in Neural Information Processing Systems 31 (2018)
7. Choi, E., et al.: Learning the graphical structure of electronic health records with graph convolutional transformer. In: Proceedings of the AAAI Conference on Artificial Intelligence, vol. 34, pp. 606–613 (2020)
8. Cui, L., Biswal, S., Glass, L.M., Lever, G., Sun, J., Xiao, C.: Conan: complementary pattern augmentation for rare disease detection. In: Proceedings of the AAAI Conference on Artificial Intelligence, vol. 34, pp. 614–621 (2020)
9. Ding, Y., Tang, J., Guo, F.: Identification of drug-side effect association via multiple information integration with centered kernel alignment. Neurocomputing **325**, 211–224 (2019)
10. Ghassemi, M., Naumann, T., Schulam, P., Beam, A.L., Chen, I.Y., Ranganath, R.: A review of challenges and opportunities in machine learning for health. AMIA Summits Transl. Sci. Proc. **2020**, 191 (2020)
11. Jagannatha, A.N., Yu, H.: Bidirectional RNN for medical event detection in electronic health records. In: Proceedings of the Conference. Association for Computational Linguistics. North American Chapter. Meeting, vol. 2016, p. 473. NIH Public Access (2016)
12. Kalantar-Zadeh, K., Jafar, T.H., Nitsch, D., Neuen, B.L., Perkovic, V.: Chronic kidney disease. Lancet **398**(10302), 786–802 (2021)
13. Khope, S.R., Elias, S.: Critical correlation of predictors for an efficient risk prediction framework of ICU patient using correlation and transformation of MIMIC-III dataset. Data Sci. Eng. **7**(1), 71–86 (2022). https://doi.org/10.1007/s41019-022-00176-6

14. Lee, C.Y., Chen, Y.P.P.: Prediction of drug adverse events using deep learning in pharmaceutical discovery. Brief. Bioinform. **22**(2), 1884–1901 (2021)
15. Li, J., Wu, B., Sun, X., Wang, Y.: Causal hidden markov model for time series disease forecasting. In: Proceedings of the IEEE/CVF Conference on Computer Vision and Pattern Recognition, pp. 12105–12114 (2021)
16. Luo, J., Ye, M., Xiao, C., Ma, F.: HitaNet: hierarchical time-aware attention networks for risk prediction on electronic health records. In: Proceedings of the 26th ACM SIGKDD International Conference on Knowledge Discovery & Data Mining, pp. 647–656 (2020)
17. Ma, F., Chitta, R., Zhou, J., You, Q., Sun, T., Gao, J.: Dipole: diagnosis prediction in healthcare via attention-based bidirectional recurrent neural networks. In: Proceedings of the 23rd ACM SIGKDD International Conference on Knowledge Discovery and Data Mining, pp. 1903–1911 (2017)
18. Ma, F., Gao, J., Suo, Q., You, Q., Zhou, J., Zhang, A.: Risk prediction on electronic health records with prior medical knowledge. In: Proceedings of the 24th ACM SIGKDD International Conference on Knowledge Discovery & Data Mining, pp. 1910–1919 (2018)
19. Miotto, R., Li, L., Kidd, B.A., Dudley, J.T.: Deep patient: an unsupervised representation to predict the future of patients from the electronic health records. Sci. Rep. **6**(1), 1–10 (2016)
20. Peng, X., Long, G., Shen, T., Wang, S., Jiang, J.: Sequential diagnosis prediction with transformer and ontological representation. In: 2021 IEEE International Conference on Data Mining (ICDM), pp. 489–498. IEEE (2021)
21. Pham, T., Tran, T., Phung, D., Venkatesh, S.: DeepCare: a deep dynamic memory model for predictive medicine. In: Bailey, J., Khan, L., Washio, T., Dobbie, G., Huang, J.Z., Wang, R. (eds.) Advances in Knowledge Discovery and Data Mining: 20th Pacific-Asia Conference, PAKDD 2016, Auckland, New Zealand, April 19-22, 2016, Proceedings, Part II, pp. 30–41. Springer, Cham (2016). https://doi.org/10.1007/978-3-319-31750-2_3
22. Ramchoun, H., Ghanou, Y., Ettaouil, M., Janati Idrissi, M.A.: Multilayer perceptron: architecture optimization and training (2016)
23. Seber, G.A., Lee, A.J.: Linear Regression Analysis, vol. 330. John Wiley & Sons (2003)
24. Shang, J., Ma, T., Xiao, C., Sun, J.: Pre-training of graph augmented transformers for medication recommendation. arXiv preprint arXiv:1906.00346 (2019)
25. Si, Y., et al.: Deep representation learning of patient data from electronic health records (EHR): a systematic review. J. Biomed. Inform. **115**, 103671 (2021)
26. Zhu, D.: A survey of advanced information fusion system: from model-driven to knowledge-enabled. Data Sci. Eng. **8**(2), 85–97 (2023). https://doi.org/10.1007/s41019-023-00209-8

# csl-MTFL: Multi-task Feature Learning with Joint Correlation Structure Learning for Alzheimer's Disease Cognitive Performance Prediction

Wei Liang[1,2], Kai Zhang[1,2], Peng Cao[1,2,3(✉)], Xiaoli Liu[4], Jinzhu Yang[1,2,3(✉)], and Osmar R. Zaiane[5]

[1] Computer Science and Engineering, Northeastern University, Shenyang, China
[2] Key Laboratory of Intelligent Computing in Medical Image of Ministry of Education, Northeastern University, Shenyang, China
caopeng@mail.neu.edu.cn, yangjinzhu@cse.neu.edu.cn
[3] National Frontiers Science Center for Industrial Intelligence and Systems Optimization, Shenyang 110819, China
[4] DAMO Academy, Alibaba Group, Hangzhou, China
[5] Alberta Machine Intelligence Institute, University of Alberta, Edmonton, AB, Canada

**Abstract.** Alzheimer's disease (AD) is a common chronic neurodegenerative disease and the accurate prediction of the clinical cognitive performance is important for diagnosis and treatment. Recently, multi-task feature learning (MTFL) methods with sparsity-inducing regularization have been widely investigated on cognitive performance prediction tasks. Although they have proved to achieve improved performance compared with single-task learning, the major challenges are still not fully resolved. They involve how to capture the non-linear correlation among the tasks or features, and how to introduce the learned correlation for guiding the MTFL learning. To resolve these challenges, we introduce a correlation structure learning method through self-attention learning and sequence learning for jointly capturing the complicated but more flexible relationship for features and tasks, respectively. Moreover, we develop a dual graph regularization to encode the inherent correlation and an efficient optimization algorithm for solving the nonsmooth objective function. Extensive results on the ADNI dataset demonstrate that the proposed joint training framework outperforms existing methods and achieves state-of-the-art prediction performance of AD. Specifically, the proposed algorithm achieves an nMSE (normalized Mean Squared Error)/wR (weighted R-value) of 3.808/0.438, obtaining a relative improvement of 3.84%/7.35% compared with the MTFL method.

**Keywords:** Alzheimer's disease · Multi-task learning · Sparse learning · Structure learning · Biomarker identification

## 1 Introduction

Alzheimer's disease (AD), characterized by progressive cognitive dysfunction and impaired daily living abilities, is a common chronic neurodegenerative brain disease,

X. Yang et al. (Eds.): ADMA 2023, LNAI 14178, pp. 48–62, 2023.
https://doi.org/10.1007/978-3-031-46671-7_4

which ultimately results in a heavy socio-economic burden. Although the progression of AD is irreversible, the alleviation of the disease symptoms is possible through timely intervention and treatment at the early stages [1,2]. Magnetic resonance imaging (MRI) is an important neuroimage tool in the early diagnosis of AD, since it provides more stable and sensitive biomarkers [3]. Recently, rather than predicting categorical variables as a classification task, several studies begin to estimate continuous clinical variables (e.g. cognitive performance) from brain images. Accurate cognitive performance prediction of AD is key to the development, assessment and monitoring of new treatments for AD. Various Neuropsychological tests can be conducted to get the cognitive or clinical performance of subjects, which can reflect their cognitive functioning. The widely used cognitive performance involve ADAS (Alzheimer's Disease Assessment Scale cognitive total score), MMSE (the MiniMental State Exam score) and RAVLT (the Rey Auditory Verbal Learning Test). Predicting subjects' cognitive performance from MRI and identifying relevant imaging biomarkers are considered to be significant research directions in Alzheimer's disease research [5, 16, 21].

Multi-task feature learning (MTFL) [5] has been successfully studied and achieves better performance by exploring the common features that are important for all the tasks with an $\ell_{2,1}$-norm regularization. However, the assumption of MTFL is too restrictive since it equally treats all the cognitive predictive tasks and MRI features, ignoring the underlying correlation among the cognitive tasks and among the MRI features at the same time. Consideration of the interaction between features and the correlation among tasks is vital in MTFL. In real-world applications, such a priori information on task correlation or feature correlation may be not easy to obtain. FTSMTFL [25] constructed the task correlation graph and feature correlation graph with Pearson Correlation Coefficient (PCC) and proposed graph correlation regularization to model the interdependencies among the tasks or features to guide the MTFL. The major limitation of PCC is the linear association and its sensitivity to the range of observations [17]. Furthermore, some data driven methods are proposed to learn the task relationship structure or the feature interaction. Lin et al. proposed a multi-task feature interaction learning (MTIL) [9] to exploit the task relatedness from high-order feature interactions via shared representations of feature interactions. Goncalves et al. proposed multi-task sparse structure learning (MSSL) method for learning the structure of task relationships as well as parameters for individual tasks [8]. Nonetheless, the linear assumption in both methods usually does not hold due to the inherently complex relationships among the features or the tasks.

Driven by this important issue, a question arises: can we seek an approach that is capable of considering the complicated relationship among the MRI markers or among the cognitive performance prediction tasks? Modeling the correlation among cognitive performance or the correlation among neuroimage measures with nonlinear functions, such as kernel-based methods, may provide enhanced flexibility and the potential ability to better capture the complex relationship. In kernel-based methods, the nonlinear relation is based on the high-order features induced by the kernel function, which determines the mapping between the input space and the feature space. However, the most important limitation is that inappropriate kernels can not accurately capture the correlation structure in the features or tasks [22]. Compared with kernel-based methods, the expressive power of deep learning to extract the underlying complex patterns from data has been well recognized. The power of deep learning lies in automatically learning relevant and powerful patterns or correlations, which are learned through end-to-end architectures. Due to the high nonlinearity and complexity of the MRI data and the predictive tasks, we consider exploiting the nonlinear correlations in tasks or features by deep learning methods. More specifically, we propose a joint self-attention feature interaction learning and sequence task correlation learning approach to better capture the highly non-linear structure. We conclude our major contributions as follows:

1. We propose a pairwise self-attention mechanism for exploiting the nonlinear feature correlation and a sequence learning model for nonlinearly modeling the task correlation.
2. Our model remains interpretability property through the combination of the linear graph regularized MTFL and the deep learning based correlation learning. This is a desired property in many practical applications, where high-quality features are the key to predictive performance.
3. We propose an efficient optimization strategy for jointly estimating the weight parameters in the proposed method.

## 2    Multi-task Feature Learning with Joint Correlation Structure Learning, csl-MTFL

### 2.1    The Formulation of Multi-task Feature Learning

In this paper, we focus on the problem of predicting the cognitive performance using MRI data. Here, instead of independent prediction, we aim to predict multiple predictive tasks at the same time. Assume that it is a multi-task learning (MTL) setting with $T$ tasks. The dataset are $D = \{(\boldsymbol{x}_1, \boldsymbol{y}_1), \ldots, (\boldsymbol{x}_N, \boldsymbol{y}_N)\}$, where $\boldsymbol{x}_i \in \mathbb{R}^C$ indicates the MRI features of $i$-th subject (e.g. the volume of hippocampus), $\boldsymbol{y}_i \in \mathbb{R}^T$ indicates the prediction scores of $i$-th subject, $C$ and $N$ are the number of the feature dimensionality and the training instances, respectively. The model parameter is denoted as $\Theta = [\boldsymbol{\theta}^1, \ldots, \boldsymbol{\theta}^T] \in \mathbb{R}^{C \times T}$. It is noteworthy that for $\boldsymbol{x}, \boldsymbol{y}$ and $\boldsymbol{\theta}$, the superscript and the subscript represent a column and a row of them respectively. The multi-task regression problem with the task correlation and the feature structure is formulated as:

$$\min_{\Theta}  L(X, Y, \Theta) + \lambda_f R_f(\Theta) + \lambda_t R_t(\Theta) + \lambda_e R_e(\Theta) \,, \tag{1}$$

where $X = [\boldsymbol{x}_1, \ldots, \boldsymbol{x}_N]^\top \in \mathbb{R}^{N \times C}$ and $Y = [\boldsymbol{y}_1, \ldots, \boldsymbol{y}_N]^\top \in \mathbb{R}^{N \times T}$ denote the features and cognitive performance of training samples. $L(X, Y, \Theta) = \frac{1}{2}\|Y - X\Theta\|_F^2 = \sqrt{\sum_{i=1}^{N}\sum_{t=1}^{T} |\, y_i^t - \boldsymbol{x}_i \boldsymbol{\theta}^t \,|^2}$ is a least square loss function. $R_e(\cdot)$ is a regularization for constraining $\Theta$ and the commonly used regularization is $\ell_{2,1}$-norm, the assumption of which is if one feature is important for one task, it is also important for the other tasks. $\ell_{2,1}$-norm is formulated as: $\|\Theta\|_{2,1} = \sum_{c=1}^{C} \|\boldsymbol{\theta}_c\|_2$. $R_t(\cdot)$ and $R_f(\cdot)$ are regularizations for encoding the relationship of tasks and features through incorporating the prior knowledge. $\lambda_e, \lambda_f$ and $\lambda_t$ are the regularization parameters to control the balance among the different regularizations.

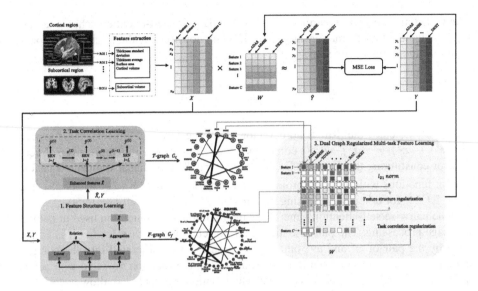

**Fig. 1.** An illustration of the csl-MTFL framework. **Step1**: With the input MRI features $X$, $F$-graph is constructed by a pairwise self-attention scheme to perform the feature transformation and aggregation. **Step2**: With the enhanced features obtained by **Step1**, $T$-graph is learned through the sequence learning with the memory mechanism of RNN layer. **Step3**: The learned $F$-graph and $T$-graph are integrated to the multi-task learning model to guide the learning and improve the performance

## 2.2 Overview

The major limitation of the traditional MTFL is the lack of considering the correlation among features or tasks. To solve this limitation, we propose a MTFL formulation with joint correlation structure learning (named csl-MTFL), involving $(i)$ feature structure learning with a self-attention scheme, $(ii)$ task correlation learning with a sequence learning scheme and $(iii)$ multi-task feature learning with graph correlation regularizations for guiding the learning process. The overall framework is shown in Fig. 1.

## 2.3 Nonlinear Graph Correlation Learning

To model the nonlinear relationships, we construct $F$-graph $\mathbb{G}_f = (V_f, E_f, G_f)$ and $T$-graph $\mathbb{G}_t = (V_t, E_t, G_t)$ to represent the feature-feature relations and task-task relations. $V_f$ and $V_t$ represent the set of nodes in $\mathbb{G}_f$ and $\mathbb{G}_t$. Each vertex of $V_f$ denotes a feature of a brain region (e.g. SV of L.Hippocampus, TA of R.Entorhinal). Each vertex of $V_t$ denotes a task (e.g. ADAS, MMSE). $E_f$ and $E_t$ represent the edges in $\mathbb{G}_f$ and $\mathbb{G}_t$. $G_f \in \mathbb{R}^{C \times C}$ and $G_t \in \mathbb{R}^{T \times T}$ denote the corresponding correlation matrix of $\mathbb{G}_f$ and $\mathbb{G}_t$.

**The Construction of $F$-Graph.** We propose a pairwise self-attention scheme to perform the feature transformation and aggregation for constructing the $F$-graph. The pairwise self-attention is defined as:

$$\hat{x}^i = \sum_{j=1}^{C} \delta(\gamma(\boldsymbol{x}^i), \psi(\boldsymbol{x}^j))\beta(\boldsymbol{x}^j) , \tag{2}$$

where $x^j$ indicates the $j$-th feature vector of all the training subjects, the function $\delta(\gamma(\boldsymbol{x}^i), \psi(\boldsymbol{x}^j)) = \gamma(\boldsymbol{x}^i)^\top \psi(\boldsymbol{x}^j)$ computes the weights of the $i$-th feature and $j$-th feature for constituting the $F$-graph $\mathbb{G}_f$. $\gamma(\cdot)$, $\psi(\cdot)$ and $\beta(\cdot)$ are linear trainable transformation functions with different projection matrices $U_\gamma \in \mathbb{R}^{N \times C^l}$, $U_\psi \in \mathbb{R}^{N \times C^l}$ and $U_\beta \in \mathbb{R}^{N \times N}$ for efficient processing. In our study, $C^l$ is empirically set to 50.

The pairwise self-attention generalizes the traditional self-attention used in natural language processing. The pairwise means the correlation of $\delta(\gamma(\boldsymbol{x}^i), \psi(\boldsymbol{x}^j))$ is calculated by the pair of $\boldsymbol{x}^i$ and $\boldsymbol{x}^j$, other locations are not incorporated into the computation. Similar to the self-attention mechanism [24], the function $\gamma(\cdot)$ and $\psi(\cdot)$ act as query and key, while $\beta(\cdot)$ acts as a value. The function $\gamma(\cdot)$ and $\psi(\cdot)$ allow us to introduce trainable transformations into the construction of the feature correlation. Hence, the part of $\delta(\gamma(\boldsymbol{x}^i), \psi(\boldsymbol{x}^j))$ enables us to explore correlation among the features, and the output value constitutes the $F$-graph by $G_f(i, j) = \delta(\gamma(\boldsymbol{x}^i), \psi(\boldsymbol{x}^j))$ for representing the feature correlation of each feature pair $\boldsymbol{x}^i$ and $\boldsymbol{x}^j$. The function $\beta(\cdot)$ produces the transformed feature vectors, which are aggregated by the correlation weight obtained by $\delta(\gamma(\boldsymbol{x}^i), \psi(\boldsymbol{x}^j))$ to generate an enhanced feature representation. These different transformation functions enhance the generalization performance by projecting the features into different representation subspaces. Each enhanced feature $\hat{x}^i$ can be obtained through the Eq. (2) in parallel. Then, these enhanced features are utilized to be fed into the following construction of $T$-graph.

**The Construction of $T$-Graph.** Recurrent Neural Network (RNN) is designed to solve sequence prediction problems through an internal memory. Specifically, it performs the same function for every input of data while the output of the current input depends on the past computation. With the $\hat{X}$ transformed by pairwise self-attention, we model task correlation through the sequence learning with the memory mechanism of the RNN layer. In our study, we employ a simple RNN (SRN) to model the task correlation, which is a basic variation of RNN. Each iteration in the SRN layer produces an updated prediction considering the task correlation. We formulate the task correlation learning as a sequence prediction problem as follows:

$$Y^{(1)} = \sigma(\hat{X}W_r) , \tag{3}$$

$$Y^{(l)} = \sigma(\hat{X}W_r + Y^{(l-1)}G_t) , \tag{4}$$

where $\hat{X}$ represents the enhanced feature obtained by Eq. (2), $Y^{(l)}$ denotes the output vector. The feature transformation $W_r \in \mathbb{R}^{C \times T}$ takes in the enhanced feature descriptor and transforms it to an output space. The memory transformation $G_t \in \mathbb{R}^{T \times T}$ takes in the output from the previous iteration and transforms it to the same output space.

We assume that there are $L$ iterations in total and an output is produced for each iteration. Equation (3) indicates that the sequence learning produces a prediction output based on the $\hat{X}W_r$ in the first iteration, which predicts each task independently without considering the correlation of other tasks. From the second iteration, our sequence learning starts to consider the outputs from the previous iteration to produce better predictions $Y^{(2)}, Y^{(3)}, \ldots, Y^{(L)}$. Specifically, Eq. (4) involves two components, the input term $\hat{X}W_r$ indicating the output without considering the task correlation in each iteration, and the memory term $Y^{(l-1)}G_t$ serving as the model for task correlation by taking in the previous output and transforming it to the same output space. Combined $\hat{X}W_r$ with $Y^{(l-1)}G_t$, $Y^{(L)}$ is used as the final prediction. Due to the linear property of the feature transformation $W_r$ and memory transformation $G_t$ in SRN, the task correlation can be captured. That is the reason why we chose SRN as the model.

## 2.4 Dual Graph Regularized Multi-task Feature Learning

To capture the feature structure and task correlation, we propose a graph correlation regularization for penalizing large deviations of high correlations. The graph correlation regularization term can be expressed as:

$$\|\hat{G}_f \Theta\|_1 = \sum_{e(p,q) \in E_f} |\hat{g}_f^{(p,q)}| |\hat{\theta}_p - \text{sign}(\hat{g}_f^{(p,q)})\hat{\theta}_q| , \tag{5}$$

$$\|\Theta \hat{G}_t\|_1 = \sum_{e(m,n) \in E_t} |\hat{g}_t^{(m,n)}| |\hat{\theta}^m - \text{sign}(\hat{g}_t^{(m,n)})\hat{\theta}^n| , \tag{6}$$

where $\hat{G}_t$ and $\hat{G}_f$ are normalized matrices of $G_t$ and $G_f$. $|\hat{g}_f^{(p,q)}|$ and $|\hat{g}_t^{(m,n)}|$ indicate the strength weights of the correlation between two features or two tasks connected by an edge. If the two features/tasks are highly correlated, the difference between the two corresponding regression coefficients $\hat{\theta}_p$ and $\hat{\theta}_q$, or $\hat{\theta}^m$ and $\hat{\theta}^n$ will be penalized more than differences for other pairs of features/tasks with weaker correlation. Due to the same procedure, we show the procedure of $\hat{g}_f^{p,q}$ as Eq. (7). In Eq. (7), $K$ denotes the amount of edges. Therefore, the proposed graph correlation regularization encourages the two terms to take the same value and can yield a solution that has sparsity in the differences of $\hat{\theta}_p$ and $\text{sign}(\hat{g}_f^{(p,q)})\hat{\theta}_q$, or $\hat{\theta}^m$ and $\text{sign}(\hat{g}_t^{(m,n)})\hat{\theta}^n$.

$$\hat{g}_f^{(p,q)} = \begin{cases} -\dfrac{g_f^{(p,q)}}{K} & (p,q) \in E, p \neq q \\ \dfrac{\sum_{p=1,q\neq p}^C |g_f^{(p,q)}|}{K} & (p,q) \in E, p = q \\ 0 & (p,q) \notin E \end{cases} \tag{7}$$

Then, we integrate both graph correlation regularizations into the multi-task feature learning model as follows:

$$\min_{\Theta} \frac{1}{2}\|Y - X\Theta\|_F^2 + \lambda_e\|\Theta\|_{2,1} + \lambda_f\|\hat{G}_f\Theta\|_1 + \lambda_t\|\Theta\hat{G}_t\|_1 , \tag{8}$$

## 2.5 Optimization

The objective in Eq. (8) can be considered as a combination of smooth terms and non-smooth terms. The proposed formulation is challenging to solve due to 1) multiple regularizations and 2) the non-smooth penalties including the $\ell_{2,1}$-norm and graph correlation regularizations. To solve these issues, we optimize the objective function in Eq. (8) using an alternating direction method of multipliers (ADMM) [20] combined with proximal gradient method [19]. According to the ADMM-based optimization procedure, Eq. (8) is equivalent to the following constrained optimization problem by introducing auxiliary variables $B$, $M$ and $V$:

$$\min_{\Theta,B,M,V} \frac{1}{2}L(\|Y - X\Theta\|_F^2) + \lambda_e\|B\|_{2,1} + \lambda_f\|M\|_1 + \lambda_t\|V\|_1 \tag{9}$$
$$\text{s.t. } \Theta - B = 0, \ \hat{G}_f\Theta - M = 0, \ \Theta\hat{G}_t - V = 0 .$$

The augmented Lagrangian of Eq. (9) is:

$$L_\rho(\Theta, B, M, V) = \frac{1}{2}\|Y - X\Theta\|_F^2 + \lambda_e\|B\|_{2,1} + \lambda_f\|M\|_1 + \lambda_t\|V\|_1 + <Z_1, \Theta - B>$$
$$+ \frac{\rho}{2}\|\Theta - B\|^2 + <Z_2, \hat{G}_f\Theta - M> + \frac{\rho}{2}\|\hat{G}_f\Theta - M\|^2 + <Z_3, \Theta\hat{G}_t - V> + \frac{\rho}{2}\|\Theta\hat{G}_t - V\|^2 , \tag{10}$$

where $\langle\cdot,\cdot\rangle$ indicates the matrix scalar product, e.g. $\langle A, B\rangle = \sum_{i=1}^{I}\sum_{j=1}^{J}(a_{ij}b_{ij}) = \text{Tr}(A^\top B)$, $Z_1$, $Z_2$ and $Z_3$ are augmented lagrangian multipliers, $\rho$ is a nonnegative penalty parameter.

**Update $\Theta$:** The update function of $\Theta$ at the $(t + 1)$-th iteration is a non-constrained smooth convex optimization problem which can be expressed as Eq. (11):

$$\Theta^{(t+1)} = \min \frac{1}{2}\|Y - X\Theta\|_F^2 + <Z_1^{(t)}, \Theta - B^{(t)}> + \frac{\rho}{2}\|\Theta - B^{(t)}\|^2 + <Z_2^{(t)}, \hat{G}_f\Theta - M^{(t)}>$$
$$+ \frac{\rho}{2}\|\hat{G}_f\Theta - M^{(t)}\|^2 + <Z_3^{(t)}, \Theta\hat{G}_t - V^{(t)}> + \frac{\rho}{2}\|\Theta\hat{G}_t - V^{(t)}\|^2, \tag{11}$$

Following [18], the closed form solution of Eq. (11) can be solved by setting its derivative to zero and employed with Cholesky factorization:

$$\Theta^{(t+1)} = F^{-1}B^{(t)} , \tag{12}$$

where $F = X^\top X + \rho I + \rho\hat{G}_f\hat{G}_f + \rho\hat{G}_t\hat{G}_t$ and $B^{(t)} = X^\top Y - Z_1^{(t)} + \rho B^{(t)} - \hat{G}_fZ_2^{(t)} - \hat{G}_tZ_3^{(t)} + \rho\hat{G}_fM^{(t)} + \rho\hat{G}_tS^{(t)}$.

**Update $B$:** According to Eq. (10), the update of $B$ can be considered as a standard $l_{2,1}$-norm regularization problem:

$$B^{(t+1)} = \arg\min_B \frac{1}{2}\|B - \Lambda_1^{(t+1)}\|^2 + \frac{\lambda_e}{\rho}\|B\|_{2,1} , \tag{13}$$

where $\Lambda_1^{(t+1)} = \Theta^{(t+1)} + \frac{Z_1^{(t)}}{\rho}$. $B^{(t+1)}$ is updated according to the proximal gradient algorithm [5]. Specifically, the Eq. (13) can be decoupled into:

$$b_i^{(t+1)} = \arg\min_{b_i} \frac{1}{2}\|b_i - \alpha_{1(i)}^{(t+1)}\|^2 + \frac{\lambda_e}{\rho}\|b_i\|_1 , \tag{14}$$

where $\alpha_{1(i)}$ and $b_i$ are the $i$-th row of $\Lambda_1^{(t+1)}$ and $B^{(t+1)}$. For any $\lambda_e \geq 0$, we can calculate Eq. (14) by the following:

$$b_i^{t+1} = \frac{\max\left(\|\alpha_{1(i)}^{(t+1)}\|_2 - \frac{\lambda_e}{\rho}, 0\right)}{\|\alpha_{1(i)}^{(t+1)}\|_2}\alpha_{1(i)}^{(t+1)} . \tag{15}$$

**Update $M$:** Updating $M$ equals solve the following problem:

$$M^{(t+1)} = \arg\min_M \frac{1}{2}\|M - \Lambda_2^{(t+1)}\|^2 + \frac{\lambda_f}{\rho}\|M\|_1 , \tag{16}$$

where $\Lambda_2^{(t+1)} = \hat{G}_f\Theta^{(t+1)} + \frac{Z_2^{(t)}}{\rho}$. According to [23], for any $\lambda_f \geq 0$, we can solve Eq. (16) according to the following:

$$m_{i,j}^{(t+1)} = \text{sign}(\alpha_{2(i,j)}^{(t+1)})\max|(\alpha_{2(i,j)}^{(t+1)}| - \frac{\lambda_f}{\rho}, 0) , \tag{17}$$

where $m_{i,j}^{(t+1)}$ and $\alpha_{2(i,j)}^{(t+1)}$ represent the element of $M^{(t+1)}$ and $\Lambda_2^{(t+1)}$.

**Update $V$:** The optimization of $V$ is the same as the procedure of $M$.

**Update $Z_1$, $Z_2$ and $Z_3$:** the updates of augmented lagrangian multipliers are obtained according to the standard ADMM.

$$\begin{aligned} Z_1^{(t+1)} &= Z_{(1)}^{(t)} + \rho(\Theta^{(t+1)} - B^{(t+1)}) \\ Z_2^{(t+1)} &= Z_{(2)}^{(t)} + \rho(S\Theta^{(t+1)} - M^{(t+1)}) \\ Z_3^{(t+1)} &= Z_{(3)}^{(t)} + \rho(\Theta^{(t+1)}Z - V^{(t+1)}) \end{aligned} \tag{18}$$

Generally, the optimization process of csl-MTFL can be summarized as Algorithm 1. For our proposed two-stage csl-MTFL, we introduce three strategies to optimize the models. 1) Two-step training. Both nonlinear graph correlation learning and dual graph regularized MTFL are supervised with their own regression loss function independently. 2) Jointly training. We optimize nonlinear graph correlation learning and dual graph regularized multi-task feature learning together with a single loss in end-to-end manner. 3) Pre-training. We first optimize the nonlinear graph correlation learning with the loss function, then the overall csl-MTFL is further trained on the same data under the supervision, which is based on the nonlinear graph correlation learning with the trained weights.

**Algorithm 1.** The optimization of csl-MTFL

---
**Require:** Feature matrix $X$, Target matrix $Y$, $\lambda_e$, $\lambda_f$, $\lambda_t$
**Ensure:** $\Theta$
  **repeat**
    Update $U_\gamma, U_\psi$ and $U_\beta$ to obtain the enhanced features and feature correlation matrix $G_f$
    Update $W_r$ to obtain the task correlation matrix $G_t$ with the enhanced features
  **until** Convergence.
  Normalize the matrices $G_f$ and $G_t$ into $\hat{G}_f$ and $\hat{G}_t$
  Incorporate the learned $\hat{G}_f$ and $\hat{G}_t$ into the formulation of Eq. (8)
  **repeat**
    Update $\Theta^{(t+1)}$ by Eq. (11)
    Update $B^{(t+1)}$ Eq. (13)
    Update $M^{(t+1)}$ and $V^{(t+1)}$ by Eq. (16)
    Update $Z_1^{(t+1)}, Z_2^{(t+1)}, Z_3^{(t+1)}$ by Eq. (18)
  **until** Convergence
---

## 3 Experiment

### 3.1 Settings

The dataset used in this study was obtained from Alzheimer's Disease Neuroimaging Initiative (ADNI) [4]. 788 subjects with 319 features are involved in our study. All the subjects include three groups: Alzheimer's Disease (AD), Mild Cognitive Impairment (MCI) and Normal Control (NC). The summary of subject information is shown in Table 1. In our experiments, correlation coefficient (CC) and root mean squared error (rMSE) are chosen to evaluate the performance on each task. nMSE and wR are applied to evaluate the comprehensive performance of the proposed model on all the tasks. The measures are formulated as Eq. (19) to Eq. (22). $y_t$ and $\hat{y}_t$ are the ground truth value and predicted value of task $t$. $Y$ and $\hat{Y}$ are the ground truth and predicted values of all tasks. $n_t$ is the number of subjects in task $t$. A smaller (higher) value of nMSE and rMSE (CC and wR) represents better regression performance. We use 10-fold cross-validation to evaluate our model and conduct the comparison, each trials of which involves a training set (90%) and test set (10%). In each of the 10 trials, a 5-fold nested cross-validation procedure is employed to tune the regularization parameters. The scope of the regularization parameters $\lambda_e, \lambda_f, \lambda_t$ in our model is [0.1,1,10,100,1000]. With the best hyperparameter chosen on the training set, the performance on the test set is evaluated and the mean score over all 10 folds is reported as the overall performance.

$$\text{CC}\,(y_t, \hat{y}_t) = \frac{cov(y_t, \hat{y}_t)}{\sigma(y_t)\sigma(\hat{y}_t)} \tag{19}$$

$$\text{rMSE}\,(y_t, \hat{y}_t) = \frac{\|y_t - \hat{y}_t\|_2^2}{N} \tag{20}$$

$$\text{nMSE}\left(Y, \hat{Y}\right) = \frac{\sum_{t=1}^T \frac{\|y_t - \hat{y}_t\|_2^2}{\sigma(y_t)}}{\sum_{t=1}^T n_t} \tag{21}$$

$$\text{wR}\left(Y, \hat{Y}\right) = \frac{\sum_{t=1}^{T} \text{CC}(\boldsymbol{y}_t, \hat{\boldsymbol{y}}_t) n_t}{\sum_{t=1}^{T} n_t} \tag{22}$$

Table 1. Summary of subject information.

| Category | CN | MCI | AD |
|---|---|---|---|
| Number | 225 | 390 | 173 |
| Gender (M/F) | 116/109 | 252/138 | 88/85 |
| Age (ave $\pm$ std) | $75.87 \pm 5.04$ | $74.75 \pm 7.39$ | $75.42 \pm 7.25$ |
| Educatio (ave $\pm$ std) | $16.03 \pm 2.85$ | $15.67 \pm 2.95$ | $14.65 \pm 3.17$ |

* F: female, M: male, ave: average, std: standard deviation

## 3.2 The Performance Comparison

To validate the effectiveness of the proposed method, we compare the proposed method with the previous studies, including multi-task multi-kernel learning with sparsity ($\ell_{2,1}$-$\ell_1$ MKMTL) [6], multi-task feature learning with sparse group lasso (SGL-MTFL) [21], Robust Multi-Task Feature Learning (rMTFL) [14], multi-task feature interaction learning by exploiting the task relatedness from high-order feature interactions(MTIL) [9], multi-task relationship learning with Gaussian graphical model (MSSL) [8] and Group-Sparse Multi-task Regression and Feature Selection (G-SMuRFS) [15]. The formulation of the comparable methods is shown in Table 2. The experiment results are shown in Tables 3 and 4. Both the task graph and feature graph learned are visually shown in Fig. 2.

From Tables 3 and 4, we can observe that our proposed csl-MTFL can consistently outperform the existing multi-task learning and single-task learning methods. Specifically, the proposed algorithm achieves an nMSE/wR of 3.808/0.438, resulting in a 3.84%/7.35% increase compared with the MTFL. In addition, csl-MTFL, SGL-MTFL, G-SMuRFS and $\ell_{2,1} - \ell_1$ MKMTL perform better than single-task learning methods, while MTIL and MSSL perform worse than single-task learning methods. This justifies the motivation of learning multiple tasks simultaneously, indicating that multiple cognitive performance prediction tasks are not independent, introducing the appropriate regularization consistent with prior knowledge can significantly improve multi-task learning performance, and vice versa. The proposed model with learned correlation provides more flexibility in predicting multiple cognitive performances simultaneously.

## 3.3 Ablation Study

To verify the impact of each component we proposed, we conduct an ablation study. The experimental results in terms of the nMSE and wR are reported in Table 5. In Table 5, we can observe that: 1) The dual graph regularization is better than any single graph regularization. For instance, the proposed dual graph guided joint learning method csl-MTFL outperforms the feature/task graph guided joint learning method

**Table 2.** The formulation of the comparable methods

| Learning method | Name | Formulation |
|---|---|---|
| Single-task learning methods | Ridge | $\min_{\boldsymbol{\theta}_i} \frac{1}{2}\|\boldsymbol{y}_i - X\boldsymbol{\theta}_i\|_F^2 + \lambda\|\boldsymbol{\theta}_i\|_2$ |
| | Lasso | $\min_{\boldsymbol{\theta}_i} \frac{1}{2}\|\boldsymbol{y}_i - X\boldsymbol{\theta}_i\|_F^2 + \lambda\|\boldsymbol{\theta}_i\|_1$ |
| Multi-task learning methods | MTFL | $\min_{\Theta} \frac{1}{2}\|Y - X\Theta\|_F^2 + \lambda\|\Theta\|_{2,1}$ |
| | rMTFL [14] | $\min_{\Theta} \frac{1}{2}\|Y - X(L+S)\|_F^2 + \lambda_1\|L\|_{2,1} + \lambda_2\|S^T\|_{2,1}$ |
| | SGL-MTFL [21] | $\min_{\Theta} \frac{1}{2}\|Y - X\Theta\|_F^2 + \lambda_1\|\Theta\|_{2,1} + \lambda_2\|\Theta\|_1$ |
| | G-SMuRFS [15] | $\min_{\Theta} \frac{1}{2}\|Y - X\Theta\|_F^2 + \lambda_1\|\Theta\|_{G_{2,1}} + \lambda_2\|\Theta\|_{2,1}$ |
| | MTIL [9] | $\min_{\Theta,Q} \sum_{i=1}^{n} \frac{1}{2}\left\|\mathbf{x}_i^T\Theta + \mathbf{x}_i^T Q\mathbf{x}_i - y_i\right\|_2^2 + \frac{\lambda}{2}\|\Theta\|_2^2 + \mu\|\mathbf{Q}\|_{1,1}$ |
| | MSSL [8] | $\min_{\Theta,\Omega \succ 0} \quad L(X,Y,\Theta) - \frac{k}{2}\log\Omega + \mathrm{Tr}\left(\Theta\Omega\Theta^T\right) + \lambda_1\|\Omega\|_1 + \lambda_2\|\Theta\|_1$ |
| | $\ell_{2,1}$-$\ell_1$MKMTL | $\min_{\Theta,\boldsymbol{\xi}} \frac{1}{2}\left(\sum_{j=1}^{k}\left(\sum_{h=1}^{t}\|\hat{\boldsymbol{\theta}}_{j,h}\|_2^2\right)^{\frac{1}{2}}\right)^2 + \frac{\lambda}{2}\sum_{h=1}^{t}\sum_{i=1}^{n_h}\xi_{ti}^2$ |

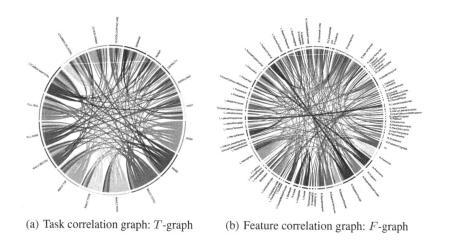

(a) Task correlation graph: $T$-graph          (b) Feature correlation graph: $F$-graph

**Fig. 2.** The learned correlation graphs.

csl-MTFL-feaGraph/csl-MTFL-taskGraph with gains of 1.50%/1.14%, 3.55%/4.53% regarding nMSE and wR respectively. This demonstrates that considering both the task correlation and the feature correlation simultaneously resulted in a better solution. An integration of two graph regularizations can facilitate the prediction performance of multi-task learning compared with only one structure regularization. 2) csl-MTFL with a joint training scheme outperforms csl-MTFL with a Two-step training scheme and csl-MTFL with a Pre-training scheme. In addition, csl-MTFL-feaGraph/ csl-MTFL-taskGraph outperform MTFL-feagraph/MTFL-taskGraph with gain of 2.79%/2.53%, 2.95%/2.17% regarding nMSE and wR respectively. These results demonstrate that the scheme of joint training is more effective for multi-task learning performance. The nonlinear regularized model has the potential to better capture the complex relationship between brain structure and cognitive performance. It also validates our motivation to build a nonlinear correlation learning model. The major limitation of the traditional MTFL model with Pearson correlation is that correlation estimation and learning are treated as two separate tasks in these methods, so potential inconsistency between the correlation estimated by Pearson correlation and MTFL degrades the predictive

**Table 3.** Performance comparison of various methods in terms of rMSE and nMSE on eighteen cognitive performance. A student's t-test (with the significance level of 0.05) on the nMSE is performed by our method and each competing method.

| Method | ADAS | MMSE | RAVLT | | | | | DSPAN | |
|---|---|---|---|---|---|---|---|---|---|
| | | | TOTAL | TOT6 | TOTB | T30 | RECOG | For | BAC |
| Ridge | 7.433±0.477 | 2.783±0.179 | 11.18±0.788 | 3.859±0.380 | 1.984±0.117 | 4.018±0.298 | 4.283±0.427 | 2.405±0.207 | 2.571±0.188 |
| Lasso | 6.936±0.670 | 2.258±0.169 | 10.43±0.767 | 3.422±0.303 | 1.731±0.199 | 3.517±0.210 | 3.776±0.281 | 2.072±0.235 | 2.192±0.186 |
| MTFL | 6.881±0.489 | 2.248±0.105 | 9.715±0.776 | 3.339±0.255 | **1.651±0.162** | 3.471±0.270 | **3.608±0.181** | 2.004±0.151 | **2.117±0.183** |
| SGL-MTFL [21] | 6.689±0.466 | **2.191±0.104** | 9.815±0.707 | 3.317±0.281 | 1.664±0.163 | 3.434±0.280 | 3.618±0.229 | 2.002±0.151 | 2.131±0.189 |
| rMTFL [14] | 6.991±0.443 | 2.375±0.235 | 10.79±0.686 | 3.468±0.330 | 1.695±0.155 | 3.602±0.253 | 3.836±0.401 | 2.036±0.730 | 2.167±0.208 |
| G-SMuRFS [15] | 6.899±0.533 | 2.258±0.102 | **9.673±0.794** | 3.324±0.256 | 1.654±0.158 | 3.442±0.296 | **3.608±0.202** | 2.010±0.154 | 2.123±0.190 |
| MTIL [9] | 6.885±0.551 | 2.932±0.132 | 10.55±0.777 | 3.481±0.298 | 1.729±0.136 | 3.619±0.242 | 3.748±0.278 | 2.124±0.126 | 2.214±0.206 |
| MSSL [8] | 7.048±0.473 | 2.813±0.390 | 10.93±0.751 | 3.594±0.372 | 1.782±0.140 | 3.727±0.293 | 3.929±0.420 | 2.174±0.150 | 2.266±0.199 |
| $\ell_{2,1}$-$\ell_1$ MKMTL [6] | 6.696±0.461 | 2.210±0.102 | 9.725±0.743 | 3.319±0.273 | 1.674±0.165 | 3.434±0.283 | 3.617±0.210 | 2.006±0.156 | 2.141±0.192 |
| csl-MTFL | **6.562±0.413** | 2.203±0.090 | 9.694±0.644 | **3.316±0.263** | 1.680±0.168 | **3.427±0.291** | 3.616±0.223 | **1.999±0.155** | 2.133±0.193 |

| Method | FLU | | LOGMEM | | CLOCK | | BOSNAM | ANART | DIGIT | nMSE |
|---|---|---|---|---|---|---|---|---|---|---|
| | ANIM | VEG | IMMTOTAL | DELTOTAL | DRAW | COPYSCORE | RECOG | | | |
| Ridge | 6.312±0.603 | 4.284±0.194 | 4.673±0.399 | 5.211±0.542 | 1.155±0.104 | 0.779±0.041 | 4.675±0.423 | 11.21±0.731 | 12.76±1.305 | 5.354±0.325* |
| Lasso | 5.554±0.434 | 3.755±0.181 | 4.382±0.424 | 4.778±0.514 | 1.022±0.093 | 0.665±0.079 | 4.113±0.553 | 10.39±1.233 | 12.26±1.524 | 4.419±0.530* |
| MTFL | 5.251±0.492 | 3.729±0.237 | 4.142±0.377 | 4.560±0.509 | **0.971±0.110** | **0.648±0.882** | 4.044±0.501 | 9.434±0.698 | 11.58±1.275 | 3.991±0.229* |
| SGL-MTFL [21] | 5.264±0.505 | 3.681±0.206 | 4.162±0.358 | 4.549±0.494 | 0.988±0.108 | 0.658±0.086 | **3.945±0.462** | 9.500±0.680 | 11.43±1.296 | 3.960±0.223* |
| rMTFL [14] | 5.599±0.493 | 3.846±0.281 | 4.299±0.307 | 4.768±0.491 | 1.004±0.154 | 0.688±0.156 | 4.211±0.511 | 10.39±0.730 | 12.44±1.169 | 4.512±0.278* |
| G-SMuRFS [15] | 5.245±0.482 | 3.717±0.232 | 4.162±0.374 | 4.565±0.523 | 0.973±0.108 | 0.649±0.086 | 4.044±0.526 | 9.425±0.694 | 11.57±1.297 | 3.984±0.216* |
| MTIL [9] | 5.532±0.531 | 3.873±0.281 | 4.334±0.376 | 4.747±0.422 | 1.013±0.113 | 0.716±0.072 | 4.382±0.364 | 10.01±0.666 | 12.00±1.306 | 4.485±0.250* |
| MSSL [8] | 5.861±0.605 | 3.993±0.283 | 4.442±0.366 | 4.897±0.507 | 1.054±0.091 | 0.775±0.120 | 4.484±0.386 | 10.51±0.696 | 12.59±1.219 | 4.815±0.318* |
| $\ell_{2,1}$-$\ell_1$ MKMTL [6] | 5.249±0.505 | 3.686±0.218 | 4.143±0.372 | 4.521±0.523 | 1.009±0.108 | 0.673±0.088 | 3.960±0.473 | 9.451±0.679 | 11.38±1.263 | 3.945±0.221* |
| csl-MTFL | **5.236±0.468** | **3.679±0.205** | **4.139±0.372** | **4.519±0.540** | 1.004±0.095 | 0.668±0.074 | 3.953±0.490 | **9.421±0.680** | **11.28±1.273** | **3.808±0.194** |

**Table 4.** Performance comparison of various methods in terms of CC and wR on eighteen cognitive scores. A student's t-test (with the significance level of 0.05) on the wR is performed by our method and each competing method.

| Method | ADAS | MMSE | RAVLT | | | | | DSPAN | |
|---|---|---|---|---|---|---|---|---|---|
| | | | TOTAL | TOT6 | TOTB | T30 | RECOG | For | BAC |
| Ridge | 0.601±0.053 | 0.421±0.067 | 0.407±0.124 | 0.362±0.133 | 0.141±0.090 | 0.375±0.135 | 0.268±0.112 | 0.011±0.060 | 0.031±0.118 |
| Lasso | 0.638±0.071 | 0.510±0.057 | 0.455±0.104 | 0.466±0.111 | 0.271±0.124 | 0.489±0.100 | 0.375±0.131 | 0.025±0.073 | 0.129±0.094 |
| MTFL | 0.638±0.077 | 0.541±0.066 | 0.512±0.107 | 0.488±0.123 | **0.331±0.087** | 0.495±0.109 | **0.419±0.124** | 0.027±0.075 | 0.210±0.129 |
| SGL-MTFL [21] | 0.665±0.065 | **0.548±0.068** | 0.504±0.094 | 0.500±0.122 | 0.320±0.081 | 0.512±0.111 | 0.415±0.126 | 0.050±0.117 | 0.180±0.130 |
| rMTFL [14] | 0.636±0.051 | 0.506±0.056 | 0.429±0.122 | 0.443±0.122 | 0.275±0.082 | 0.455±0.117 | 0.343±0.115 | 0.092±0.092 | 0.157±0.122 |
| G-SMuRFS [15] | 0.638±0.077 | 0.542±0.065 | **0.522±0.097** | 0.497±0.118 | 0.327±0.080 | 0.511±0.104 | **0.419±0.127** | 0.001±0.062 | 0.189±0.140 |
| MTIL [9] | 0.629±0.052 | 0.408±0.067 | 0.421±0.131 | 0.414±0.145 | 0.209±0.104 | 0.434±0.124 | 0.330±0.114 | 0.021±0.066 | 0.121±0.106 |
| MSSL [8] | 0.645±0.054 | 0.366±0.086 | 0.443±0.128 | 0.444±0.135 | 0.244±0.119 | 0.454±0.116 | 0.373±0.122 | 0.032±0.110 | 0.138±0.086 |
| $\ell_{2,1}$-$\ell_1$ MKMTL [6] | 0.664±0.066 | 0.542±0.071 | 0.515±0.093 | 0.500±0.120 | 0.324±0.088 | 0.513±0.110 | 0.417±0.128 | 0.057±0.126 | 0.183±0.139 |
| csl-MTFL | **0.671±0.066** | 0.537±0.067 | 0.515±0.094 | **0.503±0.119** | 0.327±0.056 | **0.519±0.115** | 0.418±0.131 | **0.097±0.118** | **0.213±0.130** |

| Method | FLU | | LOGMEM | | CLOCK | | BOSNAM | ANART | DIGIT | wR |
|---|---|---|---|---|---|---|---|---|---|---|
| | ANIM | VEG | IMMTOTAL | DELTOTAL | DRAW | COPYSCORE | RECOG | | | |
| Ridge | 0.201±0.130 | 0.389±0.128 | 0.418±0.111 | 0.433±0.121 | 0.227±0.107 | 0.133±0.095 | 0.363±0.145 | 0.049±0.083 | 0.390±0.045 | 0.290±0.055* |
| Lasso | 0.315±0.097 | 0.495±0.076 | 0.473±0.115 | 0.507±0.123 | 0.334±0.055 | 0.068±0.115 | 0.444±0.101 | 0.100±0.087 | 0.402±0.077 | 0.361±0.051* |
| MTFL | 0.395±0.084 | 0.490±0.091 | **0.511±0.084** | 0.531±0.094 | **0.389±0.085** | 0.223±0.097 | 0.465±0.103 | 0.160±0.121 | 0.429±0.114 | 0.403±0.063* |
| SGL-MTFL [21] | 0.384±0.100 | 0.509±0.080 | 0.503±0.091 | 0.535±0.102 | 0.380±0.076 | 0.232±0.100 | **0.484±0.093** | 0.161±0.100 | 0.460±0.064 | 0.408±0.055* |
| rMTFL [14] | 0.299±0.126 | 0.473±0.104 | 0.474±0.103 | 0.488±0.120 | 0.373±0.090 | 0.230±0.097 | 0.424±0.138 | 0.085±0.071 | 0.402±0.043 | 0.366±0.056* |
| G-SMuRFS [15] | 0.396±0.07 | 0.498±0.086 | 0.508±0.087 | 0.534±0.092 | 0.379±0.081 | 0.224±0.113 | 0.458±0.082 | 0.161±0.116 | 0.432±0.107 | 0.402±0.061* |
| MTIL [9] | 0.265±0.129 | 0.441±0.107 | 0.455±0.100 | 0.475±0.112 | 0.340±0.104 | 0.166±0.088 | 0.386±0.129 | 0.075±0.073 | 0.401±0.066 | 0.333±0.064* |
| MSSL [8] | 0.318±0.112 | 0.456±0.089 | 0.467±0.094 | 0.496±0.092 | 0.333±0.111 | 0.165±0.094 | 0.378±0.124 | 0.098±0.071 | 0.418±0.049 | 0.345±0.065* |
| $\ell_{2,1}$-$\ell_1$ MKMTL [6] | 0.389±0.096 | 0.509±0.079 | 0.511±0.090 | **0.543±0.098** | 0.378±0.078 | 0.231±0.103 | 0.481±0.086 | 0.165±0.107 | 0.463±0.068 | 0.410±0.057* |
| csl-MTFL | **0.405±0.090** | **0.517±0.081** | 0.511±0.092 | 0.541±0.101 | 0.388±0.073 | **0.239±0.105** | 0.479±0.087 | **0.170±0.105** | **0.478±0.065** | **0.438±0.056** |

performance. 3) It is interesting that csl-MTFL-taskGraph outperforms csl-MTFL-feaGraph. The reason is that only 5% of features correlation is bigger than 0.5 for feature correlation, whereas more than 18% of task correlation is bigger than 0.5 in the task correlation. This observation suggests that the task correlation has more impact on the final performance. Overall, the results demonstrate that the correlation structure learning methods can enable more accurate capacity estimations as compared to the traditional estimation methods with Pearson correlation.

**Table 5.** Ablation study on the comparable methods with different regularizations and learning strategies

| Method | learning strategy | nMSE | wR |
|---|---|---|---|
| csl-MTFL | Jointly training | **3.808** | **0.438** |
| csl-MTFL | Two-step training | 3.822 | 0.422 |
| csl-MTFL | Pre-training | 3.848 | 0.428 |
| csl-MTFL-feaGraph | Jointly training | 3.866 | 0.419 |
| csl-MTFL-taskGraph | Jointly training | 3.852 | 0.423 |
| graph regularized MTFL-feaGraph | Pearson correlation | 3.977 | 0.407 |
| graph regularized MTFL-taskGraph | Pearson correlation | 3.952 | 0.414 |
| two graphs regularized MTFL [16] | Pearson correlation | 3.923 | 0.422 |

**Table 6.** The identified important features

| MTFL | FTS-MTFL [16] | csl-MTFL |
|---|---|---|
| SV of L.Hippocampus | SV of L.Hippocampus | SV of L.Hippocampus |
| TA of L.MidTemporal | CV of R.Entorhinal | CV of L.MidTemporal |
| CV of R.Entorhinal | TA of L.MidTemporal | SA of L.SupFrontal |
| SV of L.LateralVentricle | SV of L.LateralVentricle | CV of R.Entorhinal |
| TA of R.Entorhinal | CV of L.MidTemporal | TA of R.Fusiform |
| TA of R.IsthmusCingulate | TA of R.Entorhinal | SA of L.Entorhinal |
| TA of L.Parahippocampal | SV of CCAnterior | TA of L.Parahippocampal |
| TS of L.SuperiorFrontal | SA of L.RostralAnterior | TA of L.InferiorParietal |
| TA of R.InferiorParietal | SA of L.SuperiorParietal | SV of L.LateralVentricle |
| TA of L.InferiorTemporal | TA of L.Precuneus | SV of CCAnterior |

## 3.4   The Biomarker Identification

Besides the evaluation regarding the predictive performance, we also focus on the identification of the MRI biomarkers. To investigate the biomarker patterns identified by our model, we show the top 10 features identified by MTFL, FTS-MTFL and csl-MTFL in Table 6. It can be found that some important brain regions including the hippocampus, middle temporal and entorhinal are identified by both methods, demonstrating that our method has the ability to identify the important features. Moreover, we find our model identifies some important features that are missed by MTFL, such as SuperiorParietal and RostralAnterior, which are important brain regions for the AD diagnosis [11,12]. The discovery of these important features benefits from the introduction of the feature graph and task graph guided regularization. Due to the SV of L.Hippocampus has stronger correlations with the SA of L.SuperiorParietal and L.Rostral, feature graph guided regularization enforces them to share similar weights. With the help of feature graph guided regularization, both the potentially important features missed by MTFL can be identified by csl-MTFL.

## 4   Conclusion

We presented a method for multi-task feature learning for AD prediction problems that is capable of learning the inherent feature dependence structure and the inherent task correlation. Such correlations are incorporated into the multi-task feature learning formulation. Moreover, an optimization method is developed for jointly learning the proposed generalized multi-task feature learning objective function. The experimental results indicate that the proposed method not only improves the predictive performance of multi-task feature learning but also benefits the interpretation of identifying stable biomarkers.

**Acknowledgment.** This research was supported by the 111 Project (B16009), National Natural Science Foundation of China (No. 62076059) and the Science Project of Liaoning Province (2021-MS-105).

## References

1. Yang, Y., Li, X., Wang, P., Xia, Y., Ye, Q.: Multi-source transfer learning via ensemble approach for initial diagnosis of Alzheimer's disease. IEEE J. Transl. Eng. Health Med. **8**, 1–10 (2020)
2. Fritzsche, K.H., Stieltjes, B., Schlindwein, S., Van Bruggen, T., Essig, M., Meinzer, H.P.: Automated MR morphometry to predict Alzheimer's disease in mild cognitive impairment. Int. J. Comput. Assist. Radiol. Surg. **5**(6), 623–632 (2010)
3. Pan, Y., Liu, M., Xia, Y., Shen, D.: Disease-image-specific learning for diagnosis-oriented neuroimage synthesis with incomplete multi-modality data. IEEE Trans. Pattern Anal. Mach. Intell. **44**, 6839–6853 (2021)
4. Marinescu, R.V., et al.: TADPOLE challenge: prediction of longitudinal evolution in Alzheimer's disease (2018)
5. Liu, J., Ji, S., Ye, J.: Multi-task feature learning via efficient $\ell_{2,1}$-norm minimization. In: Proceedings of the Twenty-Fifth Conference on Uncertainty in Artificial Intelligence (2009)
6. Cao, P., Liu, X., Yang, J., Zhao, D., Huang, M., Zaiane, O.: $\ell_{2,1}$-$\ell_1$ regularized nonlinear multi-task representation learning based cognitive performance prediction of Alzheimer's disease. Pattern Recogn. **79**, 195–215 (2018)
7. Zhou, J., Liu, J., Narayan, V.A., et al.: Modeling disease progression via multi-task learning. Neuroimage **78**, 233–248 (2013)
8. Gonçalves, A.R., Von Zuben, F.J., Banerjee, A.: Multi-task sparse structure learning with Gaussian copula models. J. Mach. Learn. Res. **17**, 1205–1234 (2016)
9. Lin, K., Xu, J., Baytas, I.M., Ji, S., Zhou, J.: Multi-task feature interaction learning. In: The 22nd SIGKDD Conference, pp. 1735–1744 (2016)
10. Zhou, J., Liu, J., Narayan, V.A., Ye, J., Alzheimer's Disease Neuroimaging Initiative: Modeling disease progression via multi-task learning. NeuroImage **78**, 233–248 (2013)
11. Prawiroharjo, P., et al.: Disconnection of the right superior parietal lobule from the precuneus is associated with memory impairment in oldest-old Alzheimer's disease patients. Heliyon **6**(7), e04516 (2020)

12. Koch, G., et al.: Transcranial magnetic stimulation of the precuneus enhances memory and neural activity in prodromal Alzheimer's disease. Neuroimage **169**, 302–311 (2018)
13. Wang, H., et al.: Sparse multi-task regression and feature selection to identify brain imaging predictors for memory performance. In: 2011 International Conference on Computer Vision, pp. 557–562 (2011)
14. Gong, P., Ye, J., Zhang, C.: Robust multi-task feature learning. In: Proceedings of the 18th ACM SIGKDD International Conference on Knowledge Discovery and Data Mining, pp. 895–903 (2012)
15. Yan, J., et al.: Cortical surface biomarkers for predicting cognitive outcomes using group $\ell_{2,1}$-norm. Neurobiol. Aging **36**, S185–S193 (2015)
16. Cao, P., Liang, W., Zhang, K., Tang, S., Yang, J.: Joint feature and task aware multi-task feature learning for Alzheimer's disease diagnosis. In: 2021 IEEE International Conference on Bioinformatics and Biomedicine (BIBM), pp. 2643–265 (2015)
17. Janse, R.J., et al.: Conducting correlation analysis: important limitations and pitfalls. Clin. Kidney J. **14**(11), 2332–2337 (2021)
18. Cao, P., et al.: Generalized fused group lasso regularized multi-task feature learning for predicting cognitive outcomes in Alzheimers disease. Comput. Methods Programs Biomed. **1**(162), 19–45 (2018)
19. Tanabe, H., Fukuda, E.H., Yamashita, N.: Proximal gradient methods for multiobjective optimization and their applications. Comput. Optim. Appl. **72**(2), 339–61 (2019)
20. Boyd, S., Parikh, N., Chu, E., Peleato, B., Eckstein, J.: Distributed optimization and statistical learning via the alternating direction method of multipliers. Found. Trends® Mach. Learn. **3**(1), 1–22 (2011)
21. Wang, H., et al.: Sparse multi-task regression and feature selection to identify brain imaging predictors for memory performance. In: International Conference on Computer Vision, pp. 557–562 (2011)
22. Cao, P., Liu, X., Yang, J., Zhao, D., Zaiane, O.: Sparse multi-kernel based multi-task learning for joint prediction of clinical scores and biomarker identification in Alzheimer's disease. In: International Conference on Medical Image Computing and Computer-Assisted Intervention, pp. 195–202 (2017)
23. Nesterov, Y.: Introductory Lectures on Convex Optimization: A Basic Course. Springer, New York (2003). https://doi.org/10.1007/978-1-4419-8853-9
24. Vaswani, A., et al.: Attention is all you need. Adv. Neural Inf. Process. Syst. **30** (2017)
25. Cao, P., Liang, W., Zhang, K., Tang, S., Yang, J.: Joint feature and task aware multi-task feature learning for Alzheimer's disease diagnosis. In: 2021 IEEE International Conference on Bioinformatics and Biomedicine (BIBM), pp. 2643–2650 (2021)

# Multi-level Transformer for Cancer Outcome Prediction in Large-Scale Claims Data

Leah Gerrard[1,2]([✉]), Xueping Peng[1], Allison Clarke[2], and Guodong Long[1]

[1] Australian AI Institute, Faculty of Engineering and IT, University of Technology Sydney, Sydney, Australia
leah.gerrard@student.uts.edu.au, {xueping.peng,guodong.long}@uts.edu.au
[2] Health Economics and Research Division, Australian Government Department of Health and Aged Care, Canberra, Australia
{leah.gerrard,allison.clarke}@health.gov.au

**Abstract.** Predicting outcomes for cancer patients initiating chemotherapy is essential for care planning and offers potential to support clinical and health policy decision-making. Existing models leveraging deep learning with longitudinal healthcare data have demonstrated the benefits of Transformer-based approaches to learning temporal relationships among medical codes (e.g., diagnoses, medications, procedures). Recent applications have also recognised the benefit of including patient information such as demographics to improve predictions. However, much of the existing work has focused on Electronic Health Record (EHR) data, and applications to administrative claims data, which has a differing temporal structure to EHR, are limited. Furthermore, it is still unclear how to best encode medical data from both EHR and claims data and model it collectively in Transformer models. Motivated by the above, this work proposes a Multi-Level Transformer specifically designed for claims data (Claims-MLT) to enhance cancer outcome prediction. The model uses a dual-level structure to learn effective patient representations by considering the low-level claims item relationships and sequential patterns in patient claim histories. We also integrate patient demographic and clinical features to provide additional information to the model. We evaluate our approach on two tasks from a real-world cancer dataset containing breast and colorectal cancer patients, and demonstrate the proposed model outperforms comparative baselines.

**Keywords:** Cancer · Transformer · BERT · Claims data · EHR

## 1 Introduction

Anticipating likely outcomes for patients initiating chemotherapy is crucial for providing optimal cancer care. Chemotherapy is a frequently used cancer treatment and can offer benefits such as lowering risk of cancer recurrence and improving survival outcomes [5]. However, chemotherapy also contributes to the range of symptoms and side effects patients often experience during treatment, such

© The Author(s), under exclusive license to Springer Nature Switzerland AG 2023
X. Yang et al. (Eds.): ADMA 2023, LNAI 14178, pp. 63–78, 2023.
https://doi.org/10.1007/978-3-031-46671-7_5

as complications and adverse events. Two important considerations for cancer patients are likely survival outcomes and cardiovascular (heart) disease risk. Existing research has indicated there are few tools for mortality risk prior to chemotherapy and accurate predictions could be useful to clinicians and patients to inform discussions and decisions [5]. In addition, evidence suggests that patients may be at increased risk of cardiovascular diseases following cancer treatment, and models to predict this risk are important for supporting treatment plans and preventive interventions [1]. Prediction models also offer the potential to inform cancer-related health policy through patient risk stratification.

Longitudinal health data sources, such as Electronic Health Records (EHRs) and administrative claims data, contain patient information relating to diagnoses, medications, and treatments, and therefore have detailed information on cancer care and outcomes. However, the characteristics of this data, notably the high-dimensionality, heterogeneity, and temporality [2,18,19], have presented challenges for traditional machine learning approaches, and thus deep learning is being increasingly adopted for prediction models.

In recent years, Transformer-based models have emerged as state-of-the-art deep learning approaches for EHR data. Much of this implementation has focused on the Bidirectional Encoder Representations from Transformers (BERT), which has demonstrated capability in modelling temporal relations in patient medical codes (e.g., diagnoses, medications and procedures) and has outperformed other sequential methods such as recurrent neural networks (RNNs) [11,12,21]. However, existing research has primarily focused on the use of EHR data, with only a limited number of applications to claims data [28,29]. This is critical due to the different temporal structure of claims data, which does not naturally fall into a pattern of visits like EHR. Furthermore, there are still open questions on how to best encode the structure of medical data and collectively model it in BERT-based approaches [21,28].

In addition to the above, there is a growing trend of leveraging patient information outside of medical codes for developing health-specific Transformer-based models. This has largely included patient demographic information, such as age and gender, and some clinical features (e.g., lab tests) [7,8,10,15]. While there is evidence to suggest inclusion of additional patient features offers improvements for model predictions, many existing approaches do not evaluate the benefit of including such features, and often only incorporate a small number of features. Further exploration and evaluation of patient features with Transformer-based models are required.

To address the indicated challenges, we propose a **Multi-Level T**ransformer that leverages **Claims** data (**Claims-MLT**) and information from a cancer registry to predict outcomes for breast and colorectal cancer patients. We evaluate our approach on a real-world cancer dataset from Australia and demonstrate the benefit of our approach compared to baselines. To summarise, the main contributions of this paper are: 1) An end-to-end multi-level Transformer that is tailored to claims data and predicts survival and heart disease diagnosis for cancer patients at the point of chemotherapy initiation. 2) We propose the use of a dual feature encoder block, where each block consists of a Transformer encoder and attention pooling, to learn context-aware vector representations for low-level claims item relationships within a month and patient-level relationships from sequential claims patterns. 3) Fusion of patient features from a cancer registry with the patients' claim representation, to capture the important static demographic and clinical information at cancer diagnosis. 4) An experimental study on real-world cancer data, demonstrating Claims-MLT outperforms all comparative methods. This is, to the best of our knowledge, the first Transformer-based model to predict cancer outcomes using temporal claims and static patient demographic and clinical data.

The remainder of this paper is organised as follows: Sect. 2 briefly reviews the related work on Transformers for EHR and claims data. Section 3 presents the model framework and approach. Section 4 describes the data and experimental results. Section 5 concludes the paper.

## 2   Related Work

The success of models based on BERT naturally led to interest in such approaches for EHR data. One of the first was BEHRT (BERT for EHR) [12], designed to learn diagnosis code and patient age relationships to predict future diagnoses. This model was extended in BEHRT-HF [20], which added medication information and calendar year to predict heart failure. In a cancer-related example, the authors in [21] developed Med-BERT, a BERT-based model to predict pancreatic cancer and heart failure for diabetes patients. These models, although demonstrating the advantages of learning contextualised medical relations for prediction tasks, were all built for EHR data, and did not explore claims data.

Of the Transformer-based models for claims data, the most relevant to this work are Claim-PT [29] and the Transformer-based Multimodal AutoEncoder (TMAE) [28]. Developed by the same authors, these approaches obtain medical visit representations using a max-pooling layer to capture the most important features. The pooled outputs are then fed into a Transformer encoder to provide a patient representation, which is then used for prediction tasks, such as the survival and asthma exacerbation predictions with Claim-PT. While these applications demonstrate the superiority of the Transformer compared to RNN models, we note a number of existing limitations and challenges. First, as indicated earlier, approaches need to consider the differing temporal structure of claims data compared to EHR, and it is unclear how to develop an optimal data representation for claims data. Claim-PT and TMAE define medical visits as a claim type (inpatient, outpatient, pharmacy) comprised of medical codes, and appear to aggregate medical visits by service date. However, this separates healthcare services that may occur closely in time. Approaches using time windows to split patient sequences are common for deep learning in healthcare [17], and use of longer time periods (i.e., month) has demonstrated benefits in predicting patient outcomes [16]. Second, attention pooling may be a more appropriate pooling approach for maintaining important features than max-pooling [6]. Third, like other EHR models (i.e., BEHRT, Med-BERT), Claim-PT and TMAE employ only a single Transformer encoder to learn the relationships between medical information. A recent extension to BEHRT, called Hi-BEHRT [11], has demonstrated benefits in using a two-level Transformer to capture associations in longer patient EHR sequences. Our approach leverages the ideas of attention pooling and multi-level Transformers to enhance claims data modelling. We also represent a patient's claim history as a sequence of months to model clinical events in close proximity while maintaining temporal information.

Finally, our work also relates to Transformer models that have integrated additional patient features such as demographics into predictions. [8] evaluated several data types for predicting recurrence of colorectal cancer, including tabular (demographics, tumour characteristics, and treatment parameters) and time-series (test results) data. They found the use of multiple data types offered improved model performance, however, did not include medical codes as features. [15] proposed a Transformer model to predict depression for patients with breast cancer. This model included medical codes, patient age and gender, and clinical notes, however, only clinical notes were evaluated in terms of impact on predictions. The authors in [10] extended Med-BERT to include age, gender, medications, clinical measures, and state information in a model called ExMed-BERT. Other work has also explored inclusion of demographics such as age and gender, patient geographical information, and/or temporal observations (lab tests, vitals), in addition to medical codes [7,15,23]. Much of this work, however, fails to evaluate the impact of the additional patient features, making their benefit to model performance unclear. In this work, we provide an ablation study to determine the effect of patient demographic and clinical information in predicting cancer outcomes.

## 3  Methodology

This section describes the methodology of the proposed approach. It firstly provides an overview of the model and then details the individual model components.

### 3.1  Model Overview

An overview of the proposed Claims-MLT model is shown in Fig. 1, which is trained end-to-end and can be viewed as four parts. In the first part, the *claims input embedding*, an embedding layer is used to encode all claims items to dense numerical vectors. Next, sequences of claims items within a month travel through a *feature encoder block*, which contains a Transformer encoder (based on BERT) and attention pooling. The output of the Transformer encoder is a hidden vector for each claims item in the input sequence, which are then compressed via the attention pooling layer into a fixed context-aware representation for each month. To differentiate between months of claims data for each patient, position embeddings are added to the month representations, and a second feature encoder block is used to learn a fixed-size representation of a patient's claim history. Static patient features are then fused with the patient claims representation to provide demographic and clinical information, which forms the final *patient representation*. Finally, fully connected and sigmoid layers enable the *cancer outcome prediction* tasks, which are binary classification tasks and trained independently.

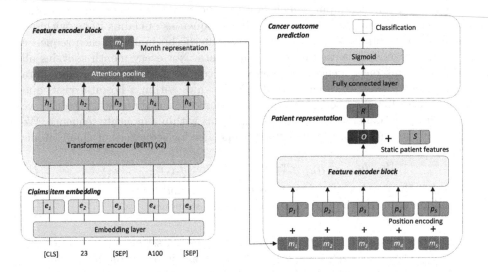

**Fig. 1.** Overview of the proposed Claims-MLT model.

## 3.2    Claims Item Embedding

A patient's claims history can be represented as a sequence of months $[M^1, M^2, ..., M^T]$ where $T$ is the number of months in the patient's history and each month is temporally related. A month $M^t$ contains a subset of claims codes $X = [x_1, x_2, ..., x_n]$, where $n$ is the number of unique claims codes in the dataset, and includes medical services, prescriptions, and diagnoses, and $x_i$ is a one-hot vector of dimension size $n$. An embedding layer transforms discrete claims in $X$ to low-dimensional dense vector representations $E = [e_1, e_2, ..., e_n]$. This can be formally written as $E = W^{(e)}X$, where the claim code embedding weight matrix $W^{(e)}$ is fine-tuned during training.

## 3.3    Feature Encoder Block

To capture the inherent dependencies between claims items in patient histories, we adopt a multi-level Transformer. This includes dual feature encoder blocks, which each contain a Transformer encoder and attention pooling.

**Transformer Encoder.** Implementation of the Transformer encoder is based on BERT [4], which contains two sub-layers: a multi-head attention mechanism, which relies on self-attention, and a position-wise fully connected feed-forward network. Residual connections and layer normalisation surround each sub-layer. We refer readers to the original papers on BERT [4] and the Transformer [22] for additional detail.

The Transformer encoder takes as input sequences of items and outputs a hidden vector for each input, which captures the context of the input within the sequence. In the first feature encoder block, sequences of claims items within a month serve as input. We use a [CLS] token at the beginning of each sequence and [SEP] tokens after each type of claims data. Not every month contains all types of claims data, hence a sequence can contain one or more claims data types. Due to no strict ordering of diagnosis codes within a hospital visit, and the presence of medical services and prescriptions occurring on the same date, we do not use position embeddings for claims items within a monthly sequence. We represent the output of the Transformer encoder for sequences of claims items as:

$$[h_1, h_2, \ldots, h_k] = \texttt{TransformerEnc}([e_1, e_2 \ldots, e_k]) \tag{1}$$

where $k$ is the number of claims items in a month, and $h_i (1 \leq i \leq k)$ represents the hidden vector of the $i$-th claims item in the sequence.

**Attention Pooling.** Attention pooling compresses a set of inputs into a fixed vector representation, retaining the most important information during pooling to capture the input context [6]. For claims items within a month sequence, we take the hidden representation output $H^t = [h_1^t, h_2^t, \ldots, h_k^t]$ of the $t$-th month as an example:

$$g(h_i^t) = w^T \sigma(W^{(1)} h_i^t + b^{(1)}) + b, \tag{2}$$

where $h_i^t$ $(1 \leq i \leq k)$ is the $i$-th row of $H^t$, $\sigma$ is ReLU function and $w$, $W^{(1)}$, $b^{(1)}$, $b$ are learnable parameters. The probability distribution can be represented as:

$$\alpha_t = \texttt{softmax}([g(h_i^t)]_{i=1}^k). \tag{3}$$

The final output $m_t$ of the attention pooling is the weighted average of sampling a claims item based on its contribution to the input sequence, i.e.,

$$m_t = \sum_{i=1}^k \alpha_t \odot [h_i^t]_{i=1}^k, \tag{4}$$

where $m_t \in \mathbb{R}^d$ $(1 \leq t \leq T)$ represents the $t$-th month in the patient claims history, which in the low-dimensional dense vector representations is denoted as $M = [m_1, \ldots, m_t, \ldots, m_T]$.

### 3.4   Patient Representation

The patient representation aims to capture the temporal relationships in a patient's claim history and fuse the output with additional features to include demographic and clinical information. This involves the use of position encoding, a second feature encoder bock and fusion of static patient features as described below.

**Position Encoding.** Position encoding incorporates information on the order of each month in a patient's claim history. A position embedding matrix $P_e$, which embeds each temporal position, is fused with the patient claims history using an Addition and Normalisation layer. The updated vector representation is denoted as:

$$J = \texttt{LayerNorm}(M + P_e). \tag{5}$$

**Second Feature Encoder Block.** The updated vector representations containing position information are used as input to the second feature encoder block. This is equivalent to the process described above, in which the Transformer encoder outputs hidden vector representations of the inputs, i.e., $[o_1, o_2, \ldots, o_k] = \texttt{TransformerEnc}([j_1, j_2 \ldots, j_k]$, and attention pooling is used to obtain a fixed-length vector representation for the patient's claim history. This can be represented as $O = \texttt{AttentionPooling}([o_1, o_2 \ldots, o_k])$, where $O$ signifies the pooled patient claims history representation.

**Static Patient Features.** We extract tabular static patient features to include demographics and clinical information in the proposed model. To transform the static features into the same dimension as the patient claim representation, we pass the static features through a two-layer neural network, with ReLU activation function after each layer. The output is then fused with the patient's claim history to provide the final patient representation (containing temporal and static features) used for predictions. The final patient representation, where $S$ represents the static patient features, is denoted as:

$$R = O + S. \tag{6}$$

### 3.5   Cancer Outcome Prediction

Given a patient's final representation $R$, the multi-level Transformer is trained end-to-end to learn the survival and heart disease prediction tasks independently.

**Survival Prediction.** The objective of this task is to predict patient mortality/death within one year (365 days) of chemotherapy initiation, determined using time between death date and chemotherapy initiation date. If no death date is present, a patient is considered to have survived the study period.

**Heart Disease Prediction.** The objective of this task is to predict whether a patient will have a heart disease diagnosis recorded in hospital within one year (365 days) of chemotherapy initiation, determined using time between hospital admission date and chemotherapy initiation date. Heart disease-related admissions were identified using the following International Classification of Diseases Tenth Revision (ICD-10) codes (I10-I11, I13, I15, I20-I25, I27, I34-I36, I42, I482, I50, U821-823).

Both prediction tasks are binary classification tasks, with average cross-entropy loss computed as follows:

$$\mathcal{L}_{\mathcal{R}} = -\frac{1}{N} \sum_{t=1}^{N} \left( \boldsymbol{y_R}^{\mathrm{T}} \log \hat{\boldsymbol{y}}_R + (1 - \boldsymbol{y_R})^{\mathrm{T}} \log \left( 1 - \hat{\boldsymbol{y}}_R \right) \right), \tag{7}$$

where $N$ is the number of data points and $\boldsymbol{y_R}$ is the ground truth of the prediction task (i.e. survival or heart disease diagnosis) and takes the value 0 or 1, and $\hat{\boldsymbol{y}}_R$ is the probability for the $N$ data point.

## 4   Experiments

In this section, we introduce the cancer dataset and describe the data preparation and cancer cohort criteria. We then present the experimental setup and results.

### 4.1   Dataset

We conduct experiments on a real-world cancer dataset that contains data on individuals diagnosed with breast and/or colorectal cancer (ICD-10 codes C50 and C18-21, respectively) in Victoria, Australia, between 2008 and 2019[1]. The data contains linked Commonwealth and Victorian state datasets including the Medicare Benefits Schedule (MBS), Pharmaceutical Benefits Scheme (PBS), Victorian Admitted Episodes Dataset (VAED) and Victorian Cancer Registry (VCR), which captures patient-level medical services, medicine prescriptions, hospital diagnoses, and demographics and clinical information, respectively.

### 4.2   Data Preparation

**Claims Data.** Claims data from 1 July 2012 onwards was included for analysis[2] and prepared as follows:

- MBS items were mapped to current items using publicly available mapping data[3]. Bulk-billing incentive and modifier items were excluded from analysis.

---

[1] Use of the data has approval from the Australian Institute of Health and Welfare (AIHW) Ethics Committee (EO2015/4/219).

[2] From this date there was increased capture of medicine dispensing in the PBS through inclusion of under co-payment data and chemotherapy medicines supplied in public and private hospitals as part of the Efficient Funding of Chemotherapy arrangements [14]. For consistency, this date range was also used for MBS and hospital admission data.

[3] http://www.mbsonline.gov.au/.

MBS subgroup codes were used to reduce the number of categories. The number of services was derived by summing each patient's service count by subgroup code and date of service. Services with a sum of zero or less were excluded.

- PBS items were mapped to chemical substance (Anatomical Therapeutic Chemical fifth level (ATC5)) codes, with items unable to be mapped excluded from analysis.
- Diagnosis codes from hospital admissions were refined to three-digit codes to reduce the number of categories. Diagnoses in the same hospital visit (identified using admission date and maximum separation date) were combined. Admissions without corresponding diagnosis information were excluded from analysis.
- MBS subgroup, ATC codes, and diagnoses were aggregated by month and combined to produce the input data. To reduce the length of the input sequences, only the first code was kept for claims items that appeared multiple times in the same month.

**Cancer Registry Data.** A summary of the static patient features included from the cancer registry data can be found in Table 1. Data was one-hot encoded prior to use in the analysis pipeline.

**Table 1.** Patient static features.

| Feature | Description |
| --- | --- |
| Age | Age group at diagnosis (<40, 10-year age groups to 90+). |
| Gender | As recorded in the cancer registry at diagnosis. |
| Tumour grade | Describes grade of tumour and level of differentiation. |
| Cancer stage | Derived cancer stage (stage I–IV, or missing). Later stage cancers are more advanced. |
| Region | Victorian region where the patient usually resided at diagnosis. |
| SEIFA quintile | Quintile of Patient Socio-Economic Indexes for Areas (SEIFA) Index of Relative Socioeconomic Disadvantage (IRSD) based on usual residence at diagnosis. |
| Screen detected | Whether the cancer was detected through national cancer screening programs. Only partial data for colorectal cancer. Does not include cancers diagnosed in private or other screening. Missing data filled with integer. |
| Country of birth | Born in Australia, overseas, or unallocated. |
| Diagnosis year | Calendar year of cancer diagnosis. |

### 4.3   Cohort Definitions

Patients with breast and/or colorectal cancer between 1 July 2008 and 31 December 2019 formed the basis of the cancer cohort. Following exclusions[4], chemotherapy initiation date was identified using a patient's first supply of an *antineoplastic agent* (ATC code *L01*). Cohorts were split by cancer types and prediction tasks. Patients with a death date within the study period were excluded from the heart disease diagnosis task but included in survival predictions. Table 2 provides the summary statistics of the cancer cohorts for the prediction tasks.

**Table 2.** Statistics of the cancer data.

| Detail | Survival | | Heart Disease | |
|---|---|---|---|---|
| | Breast | Colorectal | Breast | Colorectal |
| # of unique patients | 15,211 | 10,746 | 13,759 | 8,709 |
| # of unique medical services | 250 | 233 | 230 | 217 |
| # of unique prescriptions | 663 | 690 | 646 | 658 |
| # of unique diagnoses | 1249 | 1271 | 1203 | 1200 |
| Total # of unique claims codes | 2162 | 2194 | 2079 | 2075 |
| Proportion positive class | 0.032 | 0.128 | 0.168 | 0.332 |

### 4.4   Experimental Setup

**Baselines.** We compare our proposed model to the following baselines:

- **Single level Transformer (Claims-SLT)**. This is equivalent to Claims-MLT but does not include the Transformer encoder in the first feature encoder block. It uses only attention pooling on the embedded claims inputs to obtain month representations. This is similar to the existing single-level Transformer encoders used for claims [29] and EHR data [11,12].
- **Shallow machine learning models**. We use the **Decision Tree (DT)** and **Random Forest (RF)** for shallow machine learning approaches, as they have been frequently used for predicting cancer-related outcomes, including survival [9,25,30], cancer prognosis and prediction [13,24], and treatment outcomes [3]. These models are trained on claims items and static patient features, however do not consider the temporal relationships between claims items. For these approaches, claims data is transformed to a single vector of size $n$ (number of unique claims codes in the dataset), with values being 1 if a claims item is present in the patient's claim history, else 0. The claims data is then concatenated with the static patient features.

---

[4] The following exclusions applied: individuals with linkage errors or data quality issues (defined as identifiers linked to multiple dates of birth); records where a cancer diagnosis was reported by death certificate only; secondary cancer diagnoses; and cancer diagnoses prior to 1 July 2012 (for consistency with claims data). Patients who did not initiate chemotherapy or had a death date on or prior to chemotherapy initiation were excluded.

– **Claims- and static-only models.** To investigate the impact of static patient information, we also implement a claims-only multi-level Transformer (Claims-only-MLT) and claims- and static-only DT models.

**Evaluation Metrics.** The following performance metrics are recorded for models: precision, recall, F1 score, and accuracy. We use the **F1 score as the primary evaluation measure.** The F1 score is the harmonic mean of precision and recall, where precision equals true positives/(true positives+false positives); and recall equals true positives/(true positives+false negatives).

**Implementation Detail.** We implement all coding and analysis pipelines with Python 3.7.6 and run models on an Intel(R) Xeon(R) CPU E5-4650 v2 @ 2.40 GHz with 65 GB RAM and 16 CPUs. We use Pytorch 1.10.2+cu102 for Transformer models and Scikit-learn is used for the training of DT and RF models.

For model training, we randomly split data into training, validation, and test sets in ratios of 80%, 10% and 10% respectively. To avoid data leakage, only data in the months prior to the chemotherapy initiation date was used as input for model training, which was limited to the most recent 18 months of claims data for each patient. We use up-sampling of the minority (positive) class to obtain balanced training sets. For RF models, we set the number of trees to 40 for model training. All models were run three times to calculate mean and standard deviation of performance metrics on the test set. For reproducibility, seeds 7, 32, and 42 were used for experiments.

For Transformer models, we use Adadelta [27] for gradient descent optimisation with a minibatch of 32 patients and a learning rate of 0.01. We set the hidden dimension to 128. For each Transformer encoder, we used two stacked encoder layers with 4 attention heads. To avoid overfitting, drop-out strategies are used for all approaches (dropout rate = 0.1). We use 10 training iterations (epochs) for all tasks. Experimentation was also done with various batch sizes (4, 32, 256), learning rates (0.0001, 0.001, 0.01, 0.1), and optimisation methods (Adam, Adadelta), with parameters selected based on F1 scores from the validation dataset.

## 4.5   Results

**Survival Prediction Results.** To evaluate the proposed model on survival prediction, we compare performance metrics in mean and standard deviation (in brackets) against the baselines (Table 3). As demonstrated by the F1 scores, the proposed Claims-MLT model outperforms shallow machine learning approaches (i.e., DT and RF) and the single-level Transformer model (Claims-SLT). This indicates the benefits of Claims-MLT to model temporal claims data leveraging dual Transformer encoder blocks. We also observe that the way in which claims items in patient histories are represented is critical for deep learning models, as the Claims-SLT model is also outperformed by the DT in predicting survival for colorectal cancer. Finally, while we note the higher accuracy scores of the shallow machine learning approaches, the imbalanced data makes this an improper measure of model performance [26]. The class imbalance also contributes to the difficulty of the prediction task, and exploration of approaches to deal with imbalanced data, or small datasets (such as transfer learning), may offer improvements. Further work training the survival model on the total cancer cohort rather than only those that undergo chemotherapy could also improve predictions.

**Heart Disease Prediction Results.** We also compare model performance of Claims-MLT with baselines for the heart disease prediction task (Table 4). Like the survival task, Claims-MLT provides better predictions than shallow machine learning and single-level Transformer baselines. This further supports the advantages of leveraging month- and patient-level claims item relationships with dual Transformer encoders for predicting cancer outcomes. Compared to survival prediction, we see better model performance (in terms of F1 scores) for the heart disease task, indicating the cancer data may be better suited for prediction of future disease. This may be due to the historical patient context available within claims data, including past diagnoses and medications that may be related to cardiovascular or other associated conditions. Given the noted link between cancer treatment and future heart disease [1], this finding may be relevant for the development of cardiovascular-specific risk prediction models for cancer patients.

**Table 3.** Model performance for survival prediction.

|  | Precision | Recall | F1 | Accuracy |
|---|---|---|---|---|
| *Breast Cancer* | | | | |
| Claims-MLT | 0.120 (0.015) | 0.583 (0.021) | **0.208** (0.019) | 0.858 (0.019) |
| Claims-SLT | 0.119 (0.008) | 0.604 (0.042) | 0.199 (0.013) | 0.845 (0.006) |
| DT | 0.167 (0.014) | 0.208 (0.021) | 0.185 (0.017) | 0.942 (0.001) |
| RF | 0.333 (0.577) | 0.007 (0.012) | 0.014 (0.024) | 0.968 (0.001) |
| *Colorectal Cancer* | | | | |
| Claims-MLT | 0.234 (0.054) | 0.587 (0.081) | **0.329** (0.045) | 0.680 (0.115) |
| Claims-SLT | 0.210 (0.028) | 0.681 (0.099) | 0.319 (0.030) | 0.624 (0.067) |
| DT | 0.300 (0.020) | 0.346 (0.044) | 0.321 (0.031) | 0.813 (0.003) |
| RF | 0.618 (0.071) | 0.097 (0.008) | 0.168 (0.014) | 0.876 (0.003) |

**Table 4.** Model performance for heart disease prediction.

|  | Precision | Recall | F1 | Accuracy |
|---|---|---|---|---|
| *Breast Cancer* | | | | |
| Claims-MLT | 0.474 (0.012) | 0.807 (0.020) | **0.597** (0.004) | 0.816 (0.008) |
| Claims-SLT | 0.463 (0.006) | 0.815 (0.021) | 0.591 (0.002) | 0.810 (0.004) |
| DT | 0.437 (0.010) | 0.476 (0.018) | 0.456 (0.009) | 0.808 (0.005) |
| RF | 0.668 (0.034) | 0.368 (0.028) | 0.474 (0.030) | 0.862 (0.007) |
| *Colorectal Cancer* | | | | |
| Claims-MLT | 0.600 (0.014) | 0.746 (0.004) | **0.666** (0.007) | 0.750 (0.009) |
| Claims-SLT | 0.600 (0.016) | 0.762 (0.036) | 0.656 (0.007) | 0.747 (0.007) |
| DT | 0.554 (0.020) | 0.564 (0.009) | 0.559 (0.015) | 0.703 (0.013) |
| RF | 0.712 (0.016) | 0.642 (0.043) | 0.658 (0.005) | 0.788 (0.006) |

**Effect of Static Patient Information.** To explore the impact of patient demographic and clinical information, we conduct an ablation study and compute performance metrics for models with claims- and static-only data (Table 5). When comparing the claims-only Transformer to Claims-MLT (Tables 3 and 4), we find inclusion of static patient information provides better model performance for both tasks and cancer types. Similar results are also seen across cancers in the survival task for DT models. However, for the heart disease prediction task, the DT model with claims and static patient information outperforms the claims-only DT for colorectal cancer, but not breast cancer. This difference from the Transformer models may reflect the distinct static feature integration approaches, where fusion is used for Claims-MLT compared to the concatenation approach for DTs. The result could also be due to cancer-specific differences in prediction labels for the patient demographic and clinical features. For example, colorectal cancer has a higher proportion of patients with a future heart disease diagnosis for all age groups and cancer stages, and hence these features may be less useful in making predictions for breast cancer patients with the DT. Future work exploring individual demographic and clinical features would be helpful to further understand their impacts on performance, however we note the broad benefit of including these static patient features, particularly with the Transformer model.

**Table 5.** Model F1 scores for claims- and static-only data.

|                  | Survival          | Heart Disease     |
|------------------|-------------------|-------------------|
| *Breast Cancer*  |                   |                   |
| Claims-only-MLT  | 0.182 (0.011)     | 0.592 (0.004)     |
| DT (claims only) | 0.154 (0.016)     | 0.480 (0.022)     |
| DT (static only) | 0.105 (0.001)     | 0.285 (0.011)     |
| *Colorectal Cancer* |                |                   |
| Claims-only-MLT  | 0.304 (0.036)     | 0.656 (0.003)     |
| DT (claims only) | 0.241 (0.012)     | 0.535 (0.004)     |
| DT (static only) | 0.292 (0.009)     | 0.442 (0.008)     |

## 5    Conclusion

In this paper, we present a multi-level Transformer designed to model complex relationships between claims data and integrate additional demographic and clinical features to predict cancer outcomes. Compared to a single-level Transformer and traditional machine learning approaches, results indicate the proposed approach provides improved survival and heart disease diagnosis predictions for breast and colorectal cancer patients initiating chemotherapy. There are several avenues for future research, including alternative claims data processing approaches, consideration of machine learning techniques such as transfer learning to further leverage claims item relationships, additional exploration of patient features, and expansion of models to alternative data and prediction tasks. This work demonstrates the potential of leveraging patients' claim histories and static features with Transformers to predict cancer outcomes, and offers future opportunities to improve predictive models for cancer, and more broadly in healthcare.

**Acknowledgements.** The authors acknowledge the Australian Government Department of Health and Aged Care for supporting this research. We also acknowledge the data sharing collaboration with the Victorian Department of Health, as well as the AIHW and Centre for Victorian Data Linkage (CVDL) for data linkage. This research is supported by an Australian Government Research Training Program Scholarship.

## References

1. Altena, R., Hubbert, L., Kiani, N., Wengström, Y., Bergh, J., Hedayati, E.: Evidence-based prediction and prevention of cardiovascular morbidity in adults treated for cancer. Cardio-Oncology **7**, 20 (2021)
2. Ayala Solares, J.R., et al.: Deep learning for electronic health records: a comparative review of multiple deep neural architectures. J. Biomed. Inform. **101**, 103337 (2020)
3. Chu, C., Lee, N., Adeoye, J., homson, P., Choi, S.W.: Machine learning and treatment outcome prediction for oral cancer. J. Oral Pathol. Med. **49**, 977–985 (2020)

4. Devlin, J., Chang, M.W., Lee, K., Toutanova, K.: BERT: pre-training of deep bidirectional transformers for language understanding. arXiv preprint arXiv:1810.04805 (2018)
5. Elfiky, A.A., Pany, M.J., Parikh, R.B., Obermeyer, Z.: Development and application of a machine learning approach to assess short-term mortality risk among patients with cancer starting chemotherapy. JAMA Netw. Open 1(3), e180926–e180926 (2018)
6. Er, M.J., Zhang, Y., Wang, N., Pratama, M.: Attention pooling-based convolutional neural network for sentence modelling. Inf. Sci. 373, 388–403 (2016)
7. Fouladvand, S., et al.: Predicting opioid use disorder from longitudinal healthcare data using multi-stream transformer. CoRR abs/2103.08800 (2021)
8. Ho, D., Tan, I.B.H., Motani, M.: Predictive models for colorectal cancer recurrence using multi-modal healthcare data. In: Proceedings of CHIL, pp. 204–213. ACM (2021)
9. Jung, J.O., et al.: Machine learning for optimized individual survival prediction in resectable upper gastrointestinal cancer. J. Cancer Res. Clin. Oncol. 149, 1691–1702 (2022)
10. Lentzen, M., et al.: A transformer-based model trained on large scale claims data for prediction of severe COVID-19 disease progression. medRxiv (2022)
11. Li, Y., et al.: Hi-BEHRT: hierarchical transformer-based model for accurate prediction of clinical events using multimodal longitudinal electronic health records. CoRR abs/2106.11360 (2021)
12. Li, Y., et al.: BEHRT: transformer for electronic health records. CoRR abs/1907.09538 (2019). http://arxiv.org/abs/1907.09538
13. Manikandan, P., Durga, U., Ponnuraja, C.: An integrative machine learning framework for classifying seer breast cancer. Sci. Rep. 13, 5362 (2023)
14. Mellish, L., et al.: The Australian pharmaceutical benefits scheme data collection: a practical guide for researchers. BMC. Res. Notes 8, 1–13 (2015)
15. Meng, Y., Speier, W., Ong, M.K., Arnold, C.W.: Bidirectional representation learning from transformers using multimodal electronic health record data for chronic to predict depression. CoRR abs/2009.12656 (2020)
16. Min, X., Yu, B., Wang, F.: Predictive modeling of the hospital readmission risk from patients' claims data using machine learning: a case study on COPD. Sci. Rep. 9, 2362 (2019)
17. Morid, M.A., Sheng, O.R.L., Dunbar, J.: Time series prediction using deep learning methods in healthcare. ACM Trans. Manage. Inf. Syst. 14(1), 1–29 (2023)
18. Peng, X., Long, G., Shen, T., Wang, S., Jiang, J.: Sequential diagnosis prediction with transformer and ontological representation. In: 2021 IEEE International Conference on Data Mining (ICDM), pp. 489–498. IEEE (2021)
19. Peng, X., Long, G., Shen, T., Wang, S., Jiang, J., Zhang, C.: BiteNet: bidirectional temporal encoder network to predict medical outcomes. In: 2020 IEEE International Conference on Data Mining (ICDM), pp. 412–421. IEEE (2020)
20. Rao, S., et al.: BEHRT-HF: an interpretable transformer-based, deep learning model for prediction of incident heart failure. Eur. Heart J. 41, ehaa946-3553 (2020)
21. Rasmy, L., Xiang, Y., Xie, Z., Tao, C., Zhi, D.: Med-BERT: pretrained contextualized embeddings on large-scale structured electronic health records for disease prediction. NPJ Digit. Med. 4(1), 1–13 (2021)
22. Vaswani, A., et al.: Attention is all you need. In: NeurIPS, pp. 5998–6008 (2017)

23. Wang, Y., Guan, Z., Hou, W., Wang, F.: TRACE: early detection of chronic kidney disease onset with transformer-enhanced feature embedding. CoRR abs/2012.03729 (2020)
24. Xu, C., Wang, J., Zheng, T., Cao, Y., Fan, Y.: Prediction of prognosis and survival of patients with gastric cancer by weighted improved random forest model. Arch. Med. Sci. **18**, 1208 (2021)
25. Yang, Y., Xu, L., Sun, L., Zhang, P., Farid, S.S.: Machine learning application in personalised lung cancer recurrence and survivability prediction. Comput. Struct. Biotechnol. J. **20**, 1811–1820 (2022)
26. Yanminsun, S., Wong, A., Kamel, M.S.: Classification of imbalanced data: a review. Int. J. Pattern Recognit Artif Intell. **23**, 687–719 (2011)
27. Zeiler, M.D.: ADADELTA: an adaptive learning rate method. arXiv preprint arXiv:1212.5701 (2012)
28. Zeng, X., Lin, S., Liu, C.: Transformer-based unsupervised patient representation learning based on medical claims for risk stratification and analysis (2021)
29. Zeng, X., Lin, S.M., Liu, C.: Pre-training transformer-based framework on large-scale pediatric claims data for downstream population-specific tasks. CoRR abs/2106.13095 (2021)
30. Zhang, I., Hart, G., Qin, B., Deng, J.: Long-term survival and second malignant tumor prediction in pediatric, adolescent, and young adult cancer survivors using random survival forests: a seer analysis. Sci. Rep. **13**, 1911 (2023)

# Individual Functional Network Abnormalities Mapping via Graph Representation-Based Neural Architecture Search

Qing Li[1], Haixing Dai[2], Jinglei Lv[3], Lin Zhao[2], Zhengliang Liu[2], Zihao Wu[2], Xia Wu[1], Claire Coles[4(✉)], Xiaoping Hu[4(✉)], Tianming Liu[2(✉)], and Dajiang Zhu[5(✉)]

[1] Beijing Normal University, Beijing 100875, China
{liqing,wuxia}@bnu.edu.cn
[2] University of Georgia, Athens, GA 30602, USA
{hd54134,lin.zhao,zl18864,zw63397,tliu}@uga.edu
[3] The University of Sydney, Sydney 2006, Australia
jinglei.lv@sydney.edu.au
[4] Emory University, Atlanta, GA 30322, USA
ccoles@emory.edu, xhu@engr.ucr.edu
[5] University of Texas at Arlington, Arlington, TX 76013, USA
dajiang.zhu@uta.edu

**Abstract.** Prenatal alcohol exposure (PAE) has garnered increasing attention due to its detrimental effects on both neonates and expectant mothers. Recent research indicates that spatio-temporal functional brain networks (FBNs), derived from functional magnetic resonance imaging (fMRI), have the potential to reveal changes in PAE and Non-dysmorphic PAE (Non-Dys PAE) groups compared with healthy controls. However, current deep learning approaches for decomposing the FBNs are still limited to hand-crafted neural network architectures, which may not lead to optimal performance in identifying FBNs that better reveal differences between PAE and healthy controls. In this paper, we utilize a novel graph representation-based neural architecture search (GR-NAS) model to optimize the inner cell architecture of recurrent neural network (RNN) for decomposing the spatio-temporal FBNs and identifying the neuroimaging biomarkers of subtypes of PAE. Our optimized RNN cells with the GR-NAS model revealed that the functional activation decreased from healthy controls to Non-Dys PAE then to PAE groups. Our model provides a novel computational tool for the diagnosis of PAE, and uncovers the brain's functional mechanism in PAE.

**Keywords:** Prenatal Alcohol Exposure · Graph Representation-based Neural Architecture Search · Brain Network Decomposition · fMRI

Q. Li and H. Dai—Equal contribution.

X. Yang et al. (Eds.): ADMA 2023, LNAI 14178, pp. 79–91, 2023.
https://doi.org/10.1007/978-3-031-46671-7_6

# 1  Introduction

Prenatal alcohol exposure (PAE) can induce adverse outcomes among young mothers [1, 2]. Though the PAE-related abnormalities include functional cognitive behavioral impairment have been reported [3, 4], the adverse effects on the health of young mothers are often overlooked. Based on functional magnetic resonance (fMRI), researchers have identified altered brain network organization in individuals exposed to ethanol [5, 6], resulting in a significant decrease of small-worldness in spatio-temporal functional brain networks (FBNs). Notably, the spatio-temporal FBNs are fundamental components of brain activities that reflect transformations in brain function. Therefore, analyzing variations in brain function from the perspective of spatio-temporal FBNs could potentially reveal the effect across the different subtypes of PAE [6].

The spatio-temporal FBNs have been extensively investigated in the neuroimaging community using deep learning approaches [7–12]. For example, a deep sparse recurrent autoencoder (DSRAE) [8, 13] was developed to simultaneously decompose FBNs and demonstrated the effectiveness of recurrent neural network (RNN) models in extracting meaningful spatio-temporal networks from 4D fMRI data. Generally, designing appropriate or optimal neural network architectures manually as aforementioned approaches is a challenging and time-consuming process that heavily relies on domain knowledge and experience. To address these challenges, neural architecture search (NAS) related approaches have been proposed for identifying the optimal architectures for FBNs analysis [14–18]. Among them, differentiable neural architecture search (DARTS) [19] has improved the efficiency of searching optimal RNN architectures while maintaining comparable performances, which has been adopted in spatio-temporal FBNs decomposition, called spatio-temporal DARTS (ST-DARTS) [20].

Despite the efficiency of the DARTS framework that relaxed the operations on inner nodes with the maximum approximation for the optimization process, DARTS-based algorithms are still limited by the discrete space among the inner nodes in RNN cells. Additionally, these algorithms do not consider the topological information among inner nodes within RNN cells [19], and may be trapped in local optimum that further degrade performance. Considering that PAE-related FBNs are suggested to be associated with functional connectivity among brain regions, which is a kind of topology of the human brain [21], applying the DARTS-based methods directly on assessing the brain functions with PAE may be negatively affected due to the lack of topological information.

In this work, we use a novel graph representation-based neural architecture search (GR-NAS) to optimize the RNN cell architecture for decomposing the spatio-temporal FBNs [22] and identifying the neuroimaging biomarkers of subtypes of PAE. Specifically, to optimize the DARTS's discrete searching process, we represent the RNN cell architectures as graphs and embed them into a latent continuous search space via graph isomorphism network (GIN) [23] encoder that is the graph representation process. Then, the GR-NAS can utilize the embedded graph to search the optimal RNN cells in a continuous space and preserve the topological information. In this paper, we employed reinforce learning (RL) as the main search strategy engine for optimizing RNN cells on the PAE task fMRI dataset [24], which includes the normal controls, exposed Non-dysmorphic PAE (Non-Dys PAE) and exposed dysmorphic PAE participants. The results demonstrate the robustness and the reliability of the identified biomarkers of PAE in both group-wise and individual manner. To our best knowledge, this paper is one of the earliest contributions to PAE FBN analysis with NAS-based deep models, providing a new perspective for the abnormal brain early diagnosis with fMRI data.

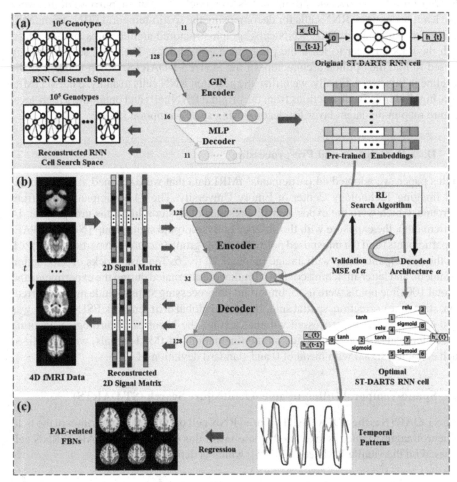

**Fig. 1.** Our framework of GR-NAS. (a) A GIN encoder and a multiple layer perceptron (MLP) [17] decoder are used to embed the RNN cell architectures to obtain pre-trained embeddings. (b) RL search strategy is used on the pre-trained embeddings to search for the optimal architecture based on the architecture performance estimation. (c) Decomposed spatio-temporal functional network learning with the learned optimal cell architecture.

## 2  Materials and Method

### 2.1  Overview

Figure 1 presents the overview of the GR-NAS model. The general purpose of this work is to learn the optimal RNN cells for decomposing the spatio-temporal FBNs from fMRI data for the subtypes of PAE. RNN cells' graph-structured architectures are embedded with the GIN encoder to learn the pre-trained embeddings. Then, RL search strategy is used to find the optimal architecture of the SOTA ST-DARTS, which is taken as the baseline in this paper. Finally, we utilize the acquired RNN cells to analyze the 4D fMRI data, highlighting the distinctions from the original RNN cells and obtaining a high-level feature map in the latent layer for decomposing spatio-temporal FBNs.

### 2.2  Data Description and Pre-processing

In this paper, we adopted 44 participants' fMRI data that were scanned at the Biomedical Imaging Technology Center of Emory University. The 44 participants were from 3 groups, which were the exposure with presence of dysmorphic signs group (PAE, 14 participants), the exposure with the absence of dysmorphic signs group (Non-Dys PAE, 14 participants) and the unexposed normal controls group (Control, 16 participants) [24]. All the participants were with an age range of 20 to 26. Ten task blocks of subtraction arithmetic and letter-matching control stimuli were alternated during the experiment, and in total 100 time points were used. Important preprocessing steps include motion correction, slice time correction, spatial smoothing, and global drift removal. FSL-FLIRT[1] was used to register the preprocessed volumes against the Montreal Neurological Institute (MNI) template. In order to focus on the fluctuations of fMRI signals, we normalized each extracted signal with mean of 0 and standard deviation of 1.

### 2.3  Spatio-Temporal Differentiable Architecture Search (ST-DARTS)

The ST-DARTS is the SOTA DARTS-based RNN cell optimization algorithm in the field of neuroimaging, which is taken as the baseline in this paper. The ST-DARTS RNN cell is based on the vanilla RNN cell [20, 25], which is defined as:

$$h_t = tanh(W_{xh}x_t + U_{hh}h_{t-1} + b_h) \tag{1}$$

in which, the $h_t$ is the RNN cell's hidden state that maintains the sequence memory of the temporal information of brain dynamics, $x_t$ is the input of the fMRI signal matrix, $W_{xh}$ and $U_{hh}$ are the weights of the current input and the previous hidden state, respectively, $b_h$ is the bias, and $tanh$ $(\cdot)$ is the activation function in RNN cell that implements the non-linearity to squash the activations to the range $[-1, 1]$. Then the RNN cell's output $y_t$ could be defined based on the hidden state $h_t$ and the weights of output $V_{yh}$:

$$y_t = V_{yh}h_t \tag{2}$$

---

[1] https://fsl.fmrib.ox.ac.uk/fsl/fslwiki.

Inherited from the vanilla RNN cell, for each ST-DARTS RNN cell, there are two inputs and a single output, which are the current step input $x_t$ (the volume sample on the $t$-th time point), the hidden state from the previous step $h_{t-1}$, and the concatenation of all the intermediate nodes $y_t$. Each ST-DARTS RNN cell is a directed acyclic graph consisting of an ordered sequence of $N$ nodes. The candidate operation choices on nodes are relaxed by *softmax* to make the search space continuous as follows:

$$\bar{o}^{(i,j)}(node) = \sum_{o \in \mathcal{O}} \frac{\exp(\alpha_o^{(i,j)})}{\sum_{o'} \exp(\alpha_{o'}^{(i,j)})} o(node) \tag{3}$$

where the function $o(\cdot)$ indicates the operation that is applied on the inner node $node^{(i)}$ of the cell. For more specifically, $\bar{o}^{(i,j)}$ represents the mixed operation from $node^{(i)}$ to $node^{(j)}$, and $o'$ denotes the one-step forward model's operation. And $node$ denotes the collect of the inner nodes of such ST-DARTS RNN cell. $node^{(i)}$ Represents the $i^{\text{th}}$ node in the cell architecture that is a latent representation. $\alpha_o^{(i,j)}$ is the operation mixing weight of the given operation $o(\cdot)$ from $node^{(i)}$ to $node^{(j)}$.

After the jointly learning process of the cell architecture parameter $\alpha$ and the entire ST-DARTS RNN architecture weights $w$, the discrete architecture can be obtained by replacing the mixed operation $\bar{o}^{(i,j)}$ with the most likely operation.

Though the ST-DARTS RNN cell is searched after relaxing the discrete operations into a continuous space, such cell only focuses on the operations between the inner nodes and ignores the topological information within RNN cells. In other words, the current ST-DARTS RNN cell makes the topological information actually in the discrete space, instead of in the continuous space, which would be improved with a future NAS method to convert the whole acyclic graph into a continuous space to promote spatio-temporal FBN decomposition.

## 2.4 Graph Representation Neural Architecture Search

In this work, we use the novel GR-NAS model to search for the optimal RNN cell architecture, and then apply it to the PAE-related FBNs decomposition. GR-NAS employs a variational graph isomorphism autoencoder to embed topological graph of RNN cell into a continuous searching space; then RL strategy is used to explore the optimal architecture in this searching space.

**Variational Graph Isomorphism Autoencoder.** During the embedding process, each RNN cell was represented as a directed acyclic graph (DAG). We denote the DAG as $DAG = (V, E)$, where $V$ is the set of nodes and $E$ is the set of edges. Each node of the DAG is associated with one of four predefined activation operations, including *tanh*, *identity* (indicating there is a connection between two nodes without activation operation), *sigmoid* and *ReLU*. Therefore, the $DAG$ can be represented by two matrices: one upper triangular adjacency matrix $A \in \mathbb{R}^{N \times N}$ that encodes the connections among $N$ nodes and one-hot operation matrix $X \in \mathbb{R}^{N \times K}$ that records the operation of each node.

To preserve the topological information of RNN cell, GIN was used to encode the graph-structured architectures into embedding space $Z \in \mathbb{R}^{N \times K}$ as follows:

$$q(Z|X, A) = \prod_{i=1}^{N} q(z_i|X, A), \text{ with } q(z_i|X, A) = \mathcal{N}(z_i|\mu_i, \text{diag}(\sigma_i^2)) \qquad (4)$$

where a one-layer GIN encoder $q(\cdot)$ is used to embed the adjacency matrix $A$ and operation matrix $X$ into the embedding vector $Z$. $z_i$ represents the $i^{th}$ value of the latency embedding vector $Z$. $\mu, \sigma$ are the mean and variance of approximation $q(z_i|X, A)$, respectively. $\mathcal{N}(\cdot)$ indicates Gaussian distribution. Then the embedding matrix $H$ was as:

$$H = MLP((1 + \epsilon) \cdot X + A \times X) \qquad (5)$$

where $H$ is the output node embedding matrix, $\epsilon$ is a trainable bias, $MLP$ denotes a multi-layer perceptron, in which each layer is a linear-batchnorm-ReLU triplet. With Eq. (5), based on $H$, the mean $\mu$ and the variance $\sigma$ could be obtained from $H$ as the Gaussian distribution parameters to approximate $q(Z|X, A)$.

We use generative model $p(\cdot)$ to obtain the reconstructed connection $\widehat{A}$ and one-hot operation matrix $\widehat{X}$ from the latent variable $Z$. The one-layer MLP decoder is:

$$p\left(\widehat{A}|Z\right) = \prod_{i=1}^{N} \prod_{j=1}^{N} p(\widehat{A}_{ij}|z_i, z_j), \text{ with } p(\widehat{A}_{ij} = 1|z_i, z_j) = \vartheta(z_i^T z_j) \qquad (6)$$

$$p\left(\widehat{X} = [k_1, \ldots, k_n]^T|Z\right) = \prod_{i=1}^{N} P(\widehat{X}_i = k_i|z_i) = \prod_{i=1}^{N} softmax(WZ + b) \qquad (7)$$

where $\vartheta$ is the logistic sigmoid function and $\widehat{A}_{ij}$ indicates the element of $\widehat{A}$. $k_i$ indicates the $i^{th}$ operation. We optimize the GIN autoencoder by maximizing the lower bound $\mathfrak{L}$ of variational parameters as:

$$\mathfrak{L} = \mathbb{E}_{q(Z|X,A)}\left[\log p\left(\widehat{X}, \widehat{A}|Z\right)\right] - KL[q(Z|X, A)||p(Z)] \qquad (8)$$

where we assume the adjacent matrix $A$ and the operation matrix $X$ are conditionally independent here. The $\mathbb{E}$ term indicates the expectation, and the $KL$ term measures the differences between the posterior distribution $q(\cdot)$ and the prior distribution $p(\cdot)$. Then, the full-batch gradient descent was performed and the parameterization scheme was used to generate random noises during the training process as the regularization [26].

**Search Strategy on Embeddings.** With the obtained RNN-based embeddings cell on the PAE-related fMRI data, we then employed down-stream search methods to evaluate our model on pre-trained embeddings.

During the GR-NAS-RL process, the state is the 16-dimension embedding. The action is the movement on one of the embedding's 16 dimensions. The reward is the reconstruction loss from current state. Pre-trained embedding is passed to the ST-DARTS RNN cell to evaluate current state and then obtain the next action and state based on the $L^2$ distance to minimize the reward.

In order to prove the robustness and stability of our framework, we also took use of Bayesian optimization (BO) as an alternative search strategy for comparison. During the

GR-NAS-BO process, the deep networks for global optimization was used to search for the optimal architecture on the 16-dimensional embeddings. One-layer adaptive basis regression network is employed for modeling the distribution over functions with 128 hidden dimensions.

We use Adam optimizer here and set the learning rate to be 0.01. After the searching process, the derived best ST-DARTS RNN cell architecture will be fed into the GR-NAS model for future spatio-temporal function network learning for PAE.

### 2.5  PAE Spatio-Temporal Functional Network Learning

After GR-NAS searching on the embeddings, optimal ST-DARTS RNN cells are achieved for PAE subtypes' spatio-temporal functional network. With the latent ST-DARTS RNN cells, the temporal network dynamics were achieved, and the spatial networks were derived with the Elastic Net regression further [8, 26]. We used the Pearson's correlation coefficient (PCC) to evaluate how consistent temporal networks are with the true brain states that are stimulated by the task design. For each extracted spatial network, we use the Dice coefficient (DC) [27] to measure the similarity between two networks (the derived FBNs $Net^{(i)}$ and the benchmark networks).

More specifically, $Net^{(0)}$ is the benchmark derived from the true brain state series that stimulated by tasks with Elastic Net regression, and $Net^{(1)}$ to $Net^{(32)}$ are the brain spatial networks that are derived PAE-related FBNs. The Elastic Net regression is an effective way to regress the temporal series to the spatial features that could take advantage of both Lasso and Ridge regressions [8, 26].

In the temporal functional network learning process, the epoch was set as 200 to guarantee convergence, the batch size was set as 128. Each cell has 8 nodes that are set in the same ways as in [19, 20, 22]. The hidden layer size was set as 32, and the initial learning rate was set as 20. We implemented the proposed GR-NAS model with PyTorch 1.4.1 on a single RTX 2080 GPU.

## 3  Results

The framework has been applied to the dataset of three groups of PAE related participants: Control, Non-Dys PAE and PAE. The severity of PAE is in the order of Control < Non-Dys PAE < PAE. The common networks are learned for all three group and the group-wise statistic is applied to each group separately.

### 3.1  Optimized ST-DARTS Cell Architecture

We implemented GR-NAS to embed and then learn the best cell architecture with the RL search strategy. With the learned cell architecture, we can produce the spatio-temporal functional brain networks from PAE fMRI datasets. In order to get the cell architecture, GR-NAS takes approximately one GPU day. Under the same super-parameter setting, our GR-NAS model was compared with ST-DARTS based on the PAE-related task fMRI data. As shown in Fig. 2, for different groups, our GR-NAS model could achieve greater

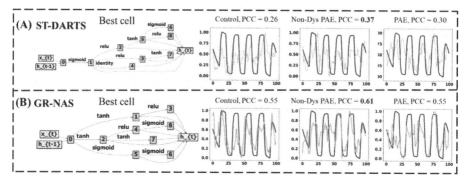

**Fig. 2.** Comparisons across different cell architectures derived from ST-DARTS and GR-NAS. The best cell architectures learned by the search models are shown on the left of each row. And the most task stimuli-correlated temporal networks learned by such cell architectures are shown on the right. The blue curves denote the task-design convolved with HRF and the orange curves denote the learned temporal network dynamics.

performance than ST-DARTS, which means our model can learn the temporal functional networks better.

The temporal results with the searched cell genotypes are shown in Fig. 2. The task design curve convolved with hemodynamic response function (HRF) is visualized as blue curves, which is used for calculating the PCCs with the learned temporal networks from GR-NAS model. For ST-DARTS, the PCC is around 0.3 and the best PCC is 0.37 for the Non-Dys PAE group. The PCCs for all the groups under GR-NAS are all higher than 0.55 and the best PCC is 0.61 for the Non-Dys PAE group. Apparently, the GR-NAS model performs much better than the SOTA original ST-DARTS.

### 3.2   Spatio-Temporal Functional Networks of Subtypes of PAE

In order to illustrate the brain spatial networks, we show the most typically derived group spatial functional brain networks in Fig. 3. To avoid the effect of the noises, we set the $z$-score maps with a threshold $>1.65$. As reported in [6, 24], the activation regions tend to shrink by the increment of severity of PAE effect, which means the number of activated voxels would decrease from Control group to the severe PAE group.

As shown in Fig. 3, based on the activation patterns and the quantitative voxel numbers of the activation region across three groups under three different methods share the same pattern: Control > Non-Dys PAE > PAE, which is consistent with the characteristic of PAE [6]. More specifically, the temporal and parietal brain regions are activated clearly with all the three models, and the area of activation has been decreased from Control to the PAE patients. This is consistent with the previous literature, which has already shown that such temporal and parietal brain regions' activation decrease is related to the mental disease [28]. For the ST-DARTS model, the number of activated voxels does not decrease between the Control group and the Non-Dys PAE group. However, according to the visualization of the brain patterns, the key regions are cut down. With GR-NAS, the pattern is shown more clearly, which proves that GR-NAS could obtain more variable and neuroscientific brain networks.

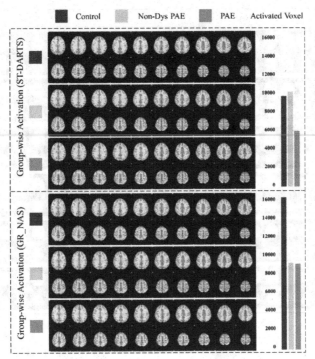

**Fig. 3.** Spatial network of group-wise activation. With GR-NAS, the voxel number (V) of the group networks decreases across three groups, i.e., V(Control) > V(Non-Dys PAE) > V(PAE). The top graph shows the activation of spatial map for each group.

## 3.3 Specific Network Analysis

In order to support the sub-networks identified by the group-wise analysis, we selected the top three temporal correlated and anti-correlated networks of the Control group that exhibited high correlation with the task design HRF and the top three networks that exhibited high anti-correlation with the task design HRF. For each selected top temporal network in the Control group, we selected the most DC-correlated networks in the Non-Dys PAE group and the PAE group based on the DC. As shown in Table 1, most of the DCs between the Non-Dys PAE group and the Control group are greater than those between the PAE group and the Control group.

With GR-NAS, the highest DC between Control and Non-Dys PAE groups is 0.7, and a 0.02 decrease is occurred between the Control and PAE group. This also proves that the disease severity of the Non-Dys PAE group is not as severe as that of the PAE group, which produces a straightforward evidence to reveal the PAE mechanism.

## 3.4 Individual-Wise Brain Spatial Networks Analysis

Furthermore, in order to prove our findings are robust and stable, we also show the individual-wise brain spatial networks with both GR-NAS-RL and GR-NAS-BO in Fig. 4. Similar to the group-wise results, the activated brain voxels decrease sharply

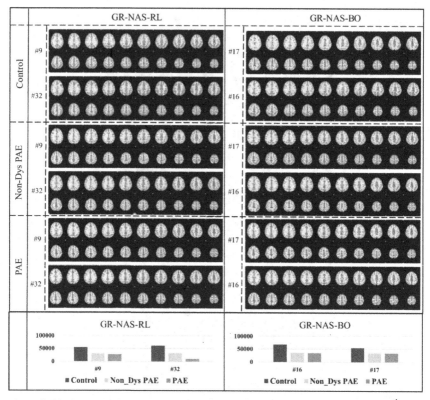

**Fig. 4.** Individual-wise brain spatial networks. The index under the graph means the $i^{th}$ network in that group. The bottom picture shows the activated voxel for each brain network.

**Table 1.** Networks derived with GR-NAS in the Control group, the matched networks in the Non-Dys PAE group and the PAE group selected by DC.

|  | Control | Non-Dys PAE | PAE |
|---|---|---|---|
| The top temporal anti-correlated networks | #25 | #28 (0.63) | #6 (0.63) |
|  | #31 | #3 (0.7) | #19 (0.68) |
|  | #29 | #3 (0.63) | #12 (0.61) |
| The top temporal correlated networks | #4 | #28 (0.61) | #15 (0.61) |
|  | #12 | #4 (0.63) | #12 (0.64) |
|  | #7 | #8 (0.62) | #31 (0.62) |

from the Control group to the Non-Dys PAE group then to the PAE group. Though the activations on the individual-wise brain are affected by the noise and lead to the uneven cluster, the activated brain voxels decreasing tendency is clearly same as the group-wise results. Especially with the GR-NAS-RL method, for both Network #9 and Network

#32, the tendency of the decrease is clearer. The activated regions are clustered around the parietal and temporal areas, which are the key areas for cognitive conception [29]. Quantitatively, for the Network #9 with GR-NAS-RL method, the number of activated voxels decreases from 54101 to 30202 then to 27385, and for the Network #32, the activated voxels goes down from 60667 to 32776 and then to 10682. This is similar with the tendency that the activation regions would shrink with the increment of severity of PAE effects [30]. On the other hand, with the GR-NAS-BO method, the number of activated voxels of Network #17 decreases from 55219 to 34860 and then to 34354, and from 68236 to 35966 and then to 34689 for Network #16. Based on Fig. 4, the difference between Non-Dys PAE and PAE group is consistently less than the difference between the Control group and the Non-Dys PAE group, no matter each kind of search strategy, which may provide evidence for the diagnosis of Non-Dys PAE.

## 4    Discussions and Conclusions

In this paper, we utilized a novel GR-NAS model for optimal brain network decomposition. Unlike previous methods, we embedded the RNN cells into a pre-trained embedding space to preserve the topological information so we can learn optimal architecture in a continuous space. Then, we implemented alternative searching strategies to search for the optimal architectures on the embeddings. Our approaches have been applied to three groups of participants affected by PAE to different degrees, namely, the Control group, the Non-Dys PAE group and the PAE group. The experimental results have suggested that our method can detect the temporal and parietal networks across three groups, while such networks are affected by an increment of PAE severity (i.e., the activated regions shrink according to the PAE degree).

**Acknowledgment.** This work was partially supported by the National Natural Science Foundation of China (Grant No. 62206024).

## References

1. Jones, K.L., Smith, D.W.: Recognition of the fetal alcohol syndrome in early infancy. Lancet **302**, 999–1001 (1973). https://doi.org/10.1016/S0140-6736(73)91092-1
2. Sandler, A.D.: Brain dysmorphology in individuals with severe prenatal alcohol exposure. J. Dev. Behav. Pediatr. **22**, 341 (2001). https://doi.org/10.1097/00004703-200110000-00024
3. Bandoli, G., et al.: Patterns of prenatal alcohol exposure and alcohol-related dysmorphic features. Alcohol. Clin. Exp. Res. **44**, 2045–2052 (2020)
4. Mattson, S.N., Bernes, G.A., Doyle, L.R.: Fetal alcohol spectrum disorders: a review of the neurobehavioral deficits associated with prenatal alcohol exposure. Alcohol. Clin. Exp. Res. **43**, 1046–1062 (2019)
5. Zhao, S., et al.: A multi-stage sparse coding framework to explore the effects of prenatal alcohol exposure. In: Ourselin, S., Joskowicz, L., Sabuncu, M., Unal, G., Wells, W. (eds.) MICCAI 2016. LNCS, vol. 9900, pp. 28–36. Springer, Cham (2016). https://doi.org/10.1007/978-3-319-46720-7_4

6. Lv, J., et al.: Assessing effects of prenatal alcohol exposure using group-wise sparse representation of FMRI data. Psychiatry Res. **233**, 254–268 (2015). https://doi.org/10.1016/j.gde.2016.03.011

7. Huang, H., et al.: Modeling task fMRI data via deep convolutional autoencoder. IEEE Trans. Med. Imaging **37**, 1551–1561 (2018). https://doi.org/10.1109/TMI.2017.2715285

8. Li, Q., Dong, Q., Ge, F., Qiang, N., Wu, X., Liu, T.: Simultaneous spatial-temporal decomposition for connectome-scale brain networks by deep sparse recurrent auto-encoder. Brain Imaging Behav. **15**, 2646–2660 (2021). https://doi.org/10.1007/s11682-021-00469-w

9. Zhao, Y., et al.: 4D Modeling of fMRI data via spatio-temporal convolutional neural networks (ST-CNN). IEEE Trans. Cognit. Dev. Syst. **12**, 451–460 (2020). https://doi.org/10.1109/TCDS.2019.2916916

10. Zhao, L., et al.: Embedding human brain function via transformer. Presented at the (2022). https://doi.org/10.1007/978-3-031-16431-6_35

11. Zhao, L., Dai, H., Jiang, X., Zhang, T., Zhu, D., Liu, T.: Exploring the functional difference of Gyri/Sulci via hierarchical interpretable autoencoder. Presented at the (2021). https://doi.org/10.1007/978-3-030-87234-2_66

12. Yu, X., Zhang, L., Zhao, L., Lyu, Y., Liu, T., Zhu, D.: Disentangling spatial-temporal functional brain networks via twin-transformers (2022)

13. Li, Q., et al.: Simultaneous spatial-temporal decomposition of connectome-scale brain networks by deep sparse recurrent auto-encoders. In: Chung, A., Gee, J., Yushkevich, P., Bao, S. (eds.) IPMI 2019. LNCS, vol. 11492, pp. 579–591. Springer, Cham (2019). https://doi.org/10.1007/978-3-030-20351-1_45

14. Li, Q., Zhang, W., Zhao, L., Wu, X., Liu, T.: Evolutional neural architecture search for optimization of spatiotemporal brain network decomposition. IEEE Trans. Biomed. Eng. **69**, 624–634 (2022). https://doi.org/10.1109/TBME.2021.3102466

15. Zhang, W., et al.: Identify hierarchical structures from task-based fMRI data via hybrid spatiotemporal neural architecture search net. In: Shen, D., et al. (eds.) MICCAI 2019. LNCS, vol. 11766, pp. 745–753. Springer, Cham (2019). https://doi.org/10.1007/978-3-030-32248-9_83

16. Liu, S., Ge, F., Zhao, L., Wang, T., Ni, D., Liu, T.: NAS-optimized topology-preserving transfer learning for differentiating cortical folding patterns. Med. Image Anal. **77**, 102316 (2022). https://doi.org/10.1016/j.media.2021.102316

17. Dai, H., Ge, F., Li, Q., Zhang, W., Liu, T.: Optimize CNN model for FMRI signal classification via Adanet-based neural architecture search. In: 2020 IEEE 17th International Symposium on Biomedical Imaging (ISBI), pp. 1399–1403. IEEE (2020). https://doi.org/10.1109/ISBI45749.2020.9098574

18. Li, Q., Zhang, W., Lv, J., Wu, X., Liu, T.: Neural architecture search for optimization of spatial-temporal brain network decomposition. In: Martel, A.L., et al. (eds.) MICCAI 2020. LNCS, vol. 12267, pp. 377–386. Springer, Cham (2020). https://doi.org/10.1007/978-3-030-59728-3_37

19. Liu, H., Simonyan, K., Yang, Y.: DARTS: differentiable architecture search. In: International Conference on Learning Representations – ICLR, pp. 1–12 (2019)

20. Li, Q., Wu, X., Liu, T.: Differentiable neural architecture search for optimal spatial/temporal brain function network decomposition. Med. Image Anal. **69**, 101974 (2021). https://doi.org/10.1016/j.media.2021.101974

21. Wozniak, J.R., et al.: Functional connectivity abnormalities and associated cognitive deficits in fetal alcohol spectrum disorders (FASD). Brain Imaging Behav. **11**, 1432–1445 (2017)

22. Dai, H., et al.: Graph representation neural architecture search for optimal spatial/temporal functional brain network decomposition. In: Lian, C., Cao, X., Rekik, I., Xu, X., Cui, Z. (eds.) MLMI 2022. LNCS, vol. 13583, pp. 279–287. Springer, Cham (2022). https://doi.org/10.1007/978-3-031-21014-3_29

23. Xu, K., Jegelka, S., Hu, W., Leskovec, J.: How powerful are graph neural networks? In: International Conference on Learning Representations – ICLR, pp. 1–17 (2019)

24. Santhanam, P., Li, Z., Hu, X., Lynch, M., Coles, C.: Effects of prenatal alcohol exposure on brain activation during an arithmetic task: an fMRI study. Alcohol. Clin. Exp. Res. **33**, 1901–1908 (2009)

25. Graves, A.: Generating Sequences with Recurrent Neural Networks. http://arxiv.org/abs/1308.0850

26. Zou, H., Hastie, T.: Regularization and variable selection via the elastic net. J. Roy. Stat. Soc. B **67**, 301–320 (2005). https://doi.org/10.1037/h0100860

27. Dice, L.R.: Measures of the amount of ecologic association between species. Ecology **26**, 297–302 (1945)

28. Calhoun, V.D., Eichele, T., Pearlson, G.: Functional brain networks in schizophrenia: a review. Front. Hum. Neurosci. **3**, 1–12 (2009). https://doi.org/10.3389/neuro.09.017.2009

29. Barch, D.M., et al.: Function in the human connectome: task-fMRI and individual differences in behavior. Neuroimage **80**, 169–189 (2013). https://doi.org/10.1016/j.neuroimage.2013.05.033

30. Santhanam, P., Coles, C.D., Li, Z., Li, L., Lynch, M.E., Hu, X.: Default mode network dysfunction in adults with prenatal alcohol exposure. Psychiatry Res Neuroimaging **194**, 354–362 (2011)

# A Novel Application of a Mutual Information Measure for Analysing Temporal Changes in Healthcare Network Graphs

David Ben-Tovim[1], Mariusz Bajger[2], and Shaowen Qin[2]([⊠]) [iD]

[1] College of Medicine and Public Health, Flinders University, Bedford Park, Australia
[2] College of Science and Engineering, Flinders University, Tonsley, SA 5042, Australia
shaowen.qin@flinders.edu.au

**Abstract.** We have previously demonstrated that network graphs generated using patient administrative data can represent health services functional structures through which hospital inpatient care is delivered. However, hospitals have to respond to changes in the external environment within which they function. These changes affect the network graphs and can be captured with a sequence of graphs of the same system over time. We developed a novel method for measuring temporal changes in healthcare network graphs based on the concept of mutual information, modified to allow for modest variations in node identities. The method is straightforward to apply, and we demonstrate its efficacy in a series of monthly snapshots of the functional structures in two hospitals.

**Keywords:** Mutual Information · Network Graphs · Modularity · Healthcare Services Provision · Patient Flow

## 1 Introduction

Transformative health care transforms health related concerns into valued health care outcomes. It is the product of interactions between teams of health care providers with varying backgrounds, interacting with each other and with the physical resources available in institutions and services in a variety of ways, and in a variety of locations, both intramurally, and increasingly, at a distance from the service. It is difficult to describe complicated systems of this type in a scientifically robust manner [1]. Network Science, a rapidly growing area of scientific and technical advance has been described as a relational science [2]. We have explored the utility of a Network Science approach for mining the voluminous data produced by hospitals and health services and for representing and analyzing the systems involved [3].

We have previously described a set of network graphs that bring together the physical and human components of the functional structures through which hospital inpatient care is delivered, in a strategy that puts patients at the center of the relevant interactions. We have data-mined hospitals' patient-focused administrative data systems to create bipartite network graphs [3] whose nodes are named hospital wards and clinical units, with the patients whose care links wards and units, forming the edges of the graphs.

© The Author(s), under exclusive license to Springer Nature Switzerland AG 2023
X. Yang et al. (Eds.): ADMA 2023, LNAI 14178, pp. 92–102, 2023.
https://doi.org/10.1007/978-3-031-46671-7_7

Clinical Units are basic structures for the provision of medical (that is mainly, but not exclusively, doctor-based) care. Named wards not only refer to a physical resource, but stand for the clinical services provided by nursing and other disciplines based in a hospital ward, or other, geographical location. Clinical units may, by contrast, treat patients in a variety of different locations.

In Fig. 1, we provide a network graph of the interactions through which inpatient care is delivered by one of hospitals we have studied. It shows that the functional structures though which care is delivered are modular in nature, when a module is a community, in this instance of wards and clinical units, within a graph whose community components interact more strongly with each other than they do with nodes in other communities [4]. Nevertheless, links between communities are also identifiable. Modularity as a graph metric [5] can be generated by a comparison between an algorithm derived network graph within which an identifiable modular structure emerges, and a random graph of a similar number of nodes and edges [6], but linked at random. Modularity scores vary between 0–1, with scores between 0.3 to 0.7 being most common in modular systems. The Fig. 1 graph's modularity score was 0.71. A group of domain experts found that the graphs were readily interpretable as identifying important aspects of the real world health system they represented. A variety of those aspects of interest have been identified and studied [3] and further work is ongoing.

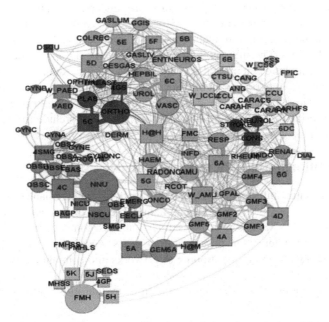

**Fig. 1.** Graph showing modular structures in a hospitals inpatient service with modules distinguished by colours. Vertices corresponding to wards are indicated by rectangles while clinical units by ovals. Acronyms used on each vertex correspond to its original name in the Hospital.

The in-patient services of hospitals care for the most severely ill patients. The functional structures underlying the delivery of that care have evolved over time [7]: whilst technologies for treatment may change, underlying structures persist. However, hospitals have to respond to changes in the external environment within which they function. A pandemic is one such environmental challenge. An important issue in relation to the utility of Network Science analyses of hospital systems is the extent to which network graphs, which are state measures, are able to represent changes within hospitals as they respond to external or internal demands. Network graphs identify structures at the time that the data was produced. Whilst the graph format is not necessarily well-suited to time-series type analyses, a variety of approaches have been developed for studying how network graphs of existing systems change over time [8, 9].

One strategy is to create multiple system-snapshots that can then function as temporal graphs of the same system over time. Conceptually, the multiple snapshots may be thought of as individual 'layers' within a multilayer network whose layers are time-ordered [10]. In network graph theory, multilayer networks in which only the edges change are described as multiplex networks, and networks where nodes, and therefore edges, appear and disappear, are multilayer networks. A number of different ways of analyzing both forms of these complex network graphs have been developed, some of which make use of tensor analysis [10]. However, such approaches are mathematically difficult, and may not pinpoint when an important change in the organization of functional systems actually occurs, a necessary precondition for understanding the relevant events.

We have developed an analytic strategy for looking at temporal changes in network graphs based on the concept of mutual information, modified to allow for modest variations in node identities. The strategy is straightforward to apply, and we demonstrate its application when looking at temporal variations in a series of monthly snapshots of the functional structures in two large public hospitals [3]. The snapshots covered the first period of the Covid-19 pandemic in the communities served by the hospitals that were studied. We describe the concept of mutual information as a prelude to describing the methods and results of our investigations.

## 2 Mutual Information Measure Modified

The concept of mutual information originates in work of Claude Shannon in 1948 [11] and since then it has proved to be beneficial in multiple and diverse applications across many disciplines. It has been found to be useful in statistics, communication theory, complexity analysis, computer vision, physics and network science, where it is predominantly used for comparing clustering/partitioning of networks into communities/modules (see e.g., [10, 12–15]).

In this study we use the following definition of mutual information (MI) [12, 13, 15]:

$$MI(P, Q) = \frac{-2 \sum_{i \in P, j \in Q} N_{ij} \ln\left(\frac{N_{ij}N}{N_i N_j}\right)}{\sum_{i \in P} N_i \ln\left(\frac{N_i}{N}\right) + \sum_{i \in Q} N_j \ln\left(\frac{N_j}{N}\right)}$$

where P, Q are the two partitions (used here to describe whole graphs) of the same network, N is the total number of nodes, $N_i$ is the number of nodes in module $p_i$ in the partition P, $N_j$ is the number of nodes in module $q_j$ in partition Q, and $N_{ij}$ is the number of nodes that are both in $p_i$ and $q_j$. The equation is symmetric and defines a measure which can be used to compare the similarity of partitions. If the two partitions are identical, MI = 1, and if they are completely uncorrelated, MI = 0. Observe that the definition is based on a confusion matrix N, where the rows correspond to modules in P, the columns correspond to the modules in Q, and $N_{ij}$ is the number of nodes in module i that is also found in module j. MI(P, Q) represents the amount of information conveyed by partition P about partition Q.

The definition readily applies to partitions of networks where the number of nodes in a network is constant. This is however not the case with complex networks describing the functionality of working hospitals which by their nature are dynamic with potentially operationally significant changes in number of nodes between monthly snapshots. Where nodes are added or removed the above definition suffers from two issues, firstly, as we observed, the $N_{ij}$ may become zero, for some modules $p_i$ in P, $q_j$ in Q, and as a consequence $\ln(N_{ij})$ has no value, secondly, the symmetry condition may fail which is critical for MI to be a sensible measure of similarity between two partitions. We rectify those issues by adopting a sensible convention and by re-defining MI so that it does preserve a symmetry property in networks with varying number of nodes.

We set by definition $0 * \ln(0) = 0$, and define mutual information for partitions with varying number of nodes (MIV) as follows:

$$MIV(P, Q) = \frac{MI(P, Q) + MI(Q, P)}{2}$$

Observe that MIV(P, Q) = MIV(Q, P) and these values can be sensibly calculated regardless of $N_{ij}$ being potentially equal zero for some modules. It is worth mentioning that by averaging the values we obtain symmetry property for MIV and when the number of nodes does not change MIV(P, Q) = MI(P, Q).

## 3  Method

The current studies make use of hospital data derived from mandated patient level data systems (patient administrative data sets) for two Australian public general hospitals, Hospital 1 and 2. The data sets comply with detailed national guidelines [16]. All the analyses reported here were undertaken on anonymized data sets, and the studies were performed in conformity to an institutional ethics review of the methods involved. The institutions involved do not make their data available for public access.

The data covered the whole of 2019, and up to September in 2020 for both hospitals. In the case of Hospital 2, the data collection was extended to cover the remaining months of 2020. Both Hospitals were major general public hospitals with strong community roots. Hospital 2 was a designated state -wide COVID response hospitals, whilst Hospital 1 retained the care of COVID patients as part of its community remit. For both hospitals, anonymised hospital wide snapshots of inpatient data for all patients present in each hospital at midnight of the 15th day of a month were extracted. Bipartite graphs with clinical units and named wards as nodes, and patients linking the units and wards as edges, were prepared. The underlying adjacency matrices were analyzed using the open-source software Gephi [17] (Version 0.9.2). The Louvain community detection algorithm [18] for detecting modules was applied, and the resultant graphs displayed using the ForceAtlas2 algorithm.

## 4   Results

The monthly snapshot graphs have been described previously [3]. To detect substantial temporal variations in our network graphs we calculated MIV values for the two hospitals for 24 months period - Hospital 2, and for 21 months period for Hospital 1, starting from January 2019. Table 1 shows a subset of MIV values table for Hospital 2.

We then used the 'heatmap' data visualization technique to visually analyze the magnitude of MIV values. Heatmaps by their construction are symmetrical about the diagonal with each cell's numerical value corresponding to a specific shade of colour of the cell – the key is provided on the right-hand site of the map. Figures 2 and 3 show MIV heatmaps for both hospitals. It is observable that a pattern of two darker squares (positioned in the upper left and lower right corners) appears for Hospital 2 data.

**Table 1.** MIV values for Hospital 2 (first 6 months).

| | Date | 2019-01-15 | 2019-02-15 | 2019-03-15 | 2019-04-15 | 2019-05-15 | 2019-06-15 |
|---|---|---|---|---|---|---|---|
| 0 | 2019-01-15 | 1.00 | 0.81 | 0.78 | 0.74 | 0.72 | 0.79 |
| 1 | 2019-02-15 | 0.81 | 1.00 | 0.81 | 0.74 | 0.76 | 0.80 |
| 2 | 2019-03-15 | 0.78 | 0.81 | 1.00 | 0.82 | 0.77 | 0.80 |
| 3 | 2019-04-15 | 0.74 | 0.74 | 0.82 | 1.00 | 0.79 | 0.80 |
| 4 | 2019-05-15 | 0.72 | 0.76 | 0.77 | 0.79 | 1.00 | 0.77 |
| 5 | 2019-06-15 | 0.79 | 0.80 | 0.80 | 0.80 | 0.77 | 1.00 |
| 6 | 2019-07-15 | 0.71 | 0.74 | 0.77 | 0.76 | 0.74 | 0.77 |
| 7 | 2019-08-15 | 0.76 | 0.75 | 0.74 | 0.75 | 0.74 | 0.77 |

To quantify our observation, we plot MIV distributions across the whole observation period against the starting month (January 2019) column. Figure 4 shows the plots for the two hospitals. It is noticeable that for Hospital 2 a sudden fall of about 0.3 happened in observation month 16 (April 2020).

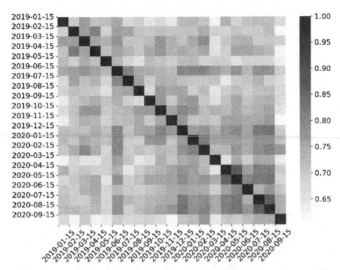

**Fig. 2.** MIV heatmap for Hospital 1.

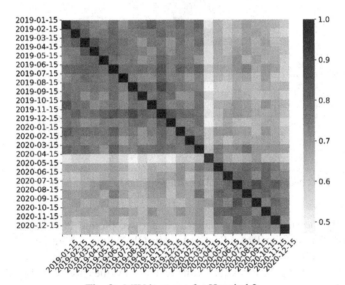

**Fig. 3.** MIV heatmap for Hospital 2.

To better understand the pattern change we calculated the Pearson Correlation Coefficient (PCC) between each pair of columns in the MIV heatmaps. Figures 5 and 6 show the PCC heatmaps for the hospitals. In Hospital 1 there is no clear change of pattern visible while in Hospital 2 we again clearly identify significant change from April 2022.

The correlation changes from positive to negative. Figure 7 shows the scatter plots of data within two months period surrounding April 2020. The change of correlation coefficient sign is very clear.

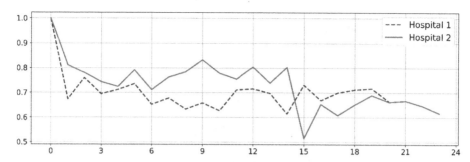

**Fig. 4.** Comparison of MIV distributions for both hospitals against January 2019 column

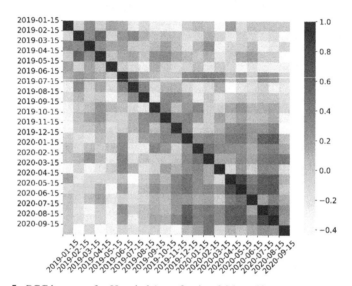

**Fig. 5.** PCC heatmap for Hospital 1 confirming fairly uniform MIV distribution

To quantify the strength of the pattern change we plotted PCC values in January 2019 (the first month of the data collection) against the whole period. Figure 6 shows the two plots. It is visible that for Hospital 1 there is only minor fluctuation of PCC values across the whole period of 21 months, with changes in values not exceeding 0.3. In case of Hospital 2 we observe a substantial change in April 2020 where PCC value drops from positive value of 0.7 to negative 0.7, that is, about 1.4 absolute value change. The negative value stays for the remaining 8 months of the data collection period (Fig. 8).

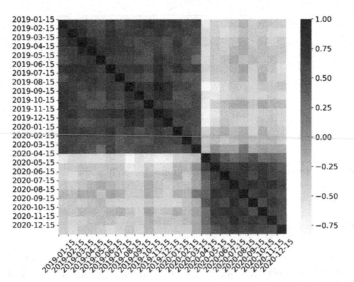

**Fig. 6.** PCC heatmap for Hospital 2 showing a change of pattern in MIV distribution from April 2020

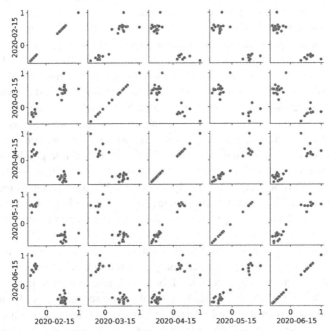

**Fig. 7.** Scatter plots of January 2019 column against Feb – June columns in 2020. The change of PCC sign after April 2020 is clearly visible.

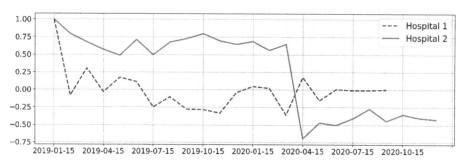

**Fig. 8.** Qualitative comparison of PCC distributions for both hospitals against January 2019 column.

## 5 Discussion

Network graphs of healthcare systems such as hospital inpatient services bring together nodes with identifiable physical locations (and the staffs associated with those locations) and nodes of person-based clinical units, all linked by patients who are their joint responsibility. In so doing, they represent the multi-functional structures through which the care needed by the most seriously ill members of our communities is delivered. The real world modular systems identified by the healthcare network graphs have evolved over a considerable period. They provide a resilient platform for clinical care. They have to. Many aspects of inpatient care are time urgent. There is not time to renegotiate delegations and the distribution of resources for each patient. Systems need to be predictable and reliable. But they also need to be flexible enough to respond to challenges, such as pandemics, that will materially influence the organization and delivery of care. The question that we sought to answer is whether a Network Science approach could be structured to represent the resilience and adaptability of healthcare systems.

MIV is introduced here as a modification of the Mutual Information measure commonly used in Network Science to compare alternative algorithmic graphs of the same network. In the current program of work, we have used MIV to link a series of network graphs as a de facto time series of monthly snapshots. We have assessed the extent to which comparisons of MIV values can provide insights into changes within the functional structures involved. In our previous work [3], we demonstrated that the hospitals we studied retained a modular structure over periods of observation of up to 24 months, with only modest changes in the value of the modularity metric. We tested the extent to which MIV provides a more finer grained analysis of the functional structures involved.

Two hospitals were studied. Hospital 1 is a substantial community hospital. The MIV measure shows that the hospitals functional structures showed only modest and unsystematic changes over the whole period of observation. In the language of quality control statistics, what was being observed was a period of common cause variation; the hospital responding to predicable variations in its existing work load before and during the initial period of the COVID-19 pandemic. Whilst Hospital 2 also continued to have an underlying resilient modular structure, with modest changes in modularity over a 24 month period, the MIV analysis identified a moment of substantial internal reorganization that might otherwise might have been overlooked. The heatmap presentations of the MIV values showed that a sudden change in the distribution of nodes and edges occurred in April 2020, changes that showed that a network structure had emerged that differed from previous functional structures. Again, in process control terminology, this was special cause variation. The accompanying PCC change was substantial and served as a valid indicator of significant network change. April 2020 coincided with the introduction of substantial restrictions on Elective admissions in Hospital 2, in anticipation of changes in demands for care as a consequence of the COVID pandemic. The PCC plot shows that the reorganized internal structures were maintained for the remaining period of study, as the MIV values were internally consistent over that period, and differed from the periods before then. The specifics of the reorganization of the internal structures of the hospital are best investigated by more detailed interactions with the staffs involved. But the potential for a straightforward and computationally accessible, approach that integrates Network Science and Information theory to play a role in identifying moments of institutional adaptation in healthcare provision is clear.

There are a number of limitations to the present study. It is based on a small sample of hospitals, and the MIV metric applies to the changes of a specific system over time whose physical structure has not been extensively redeveloped. The study is essentially methodological in nature and needs replication in different healthcare settings. It may however have a broader range of applications.

## 6   Conclusion

Network Science provides an extensive suite of measures with which to represent and analyze systems of many kinds. Network graphs are state graphs. They represent a data source at the point the data was obtained. By integrating and modifying Mutual Information, an Information Theory based measure of variations in network graphs structures, we have developed a novel method for examining structural aspects of healthcare network graphs in temporal sequences. MIV is straightforward to compute and provides an accessible methodology for looking at dynamic aspects of network graph evolution. The MIV metric substantially expands the utility of Network Science approaches for understanding how healthcare is delivered in the real world, and may have wider uses.

## References

1. Barabási, A.-L., Albert, R.: Emergence of scaling in random networks. Science **286**(5439), 509–512 (1999)
2. Brandes, U., Robins, G., McCranie, A., Wasserman, S.: What is network science? Netw. Sci. **1**(1), 1–15 (2013)
3. Ben-Tovim, D., Bajger, M., Bui, V.D., Qin, S.: Network graph analysis of hospital and health services functional structures. In: Li, B., et al. (eds.)  ADMA 2022. LNCS, vol. 13087, pp. 33–44. Springer, Cham (2022). https://doi.org/10.1007/978-3-030-95405-5_3
4. Baldwin, C.Y., Clark, K.B., Clark, K.B.: Design Rules: The Power of Modularity, vol. 1. MIT Press, Cambridge (2000)
5. Newman, M.E.: Modularity and community structure in networks. Proc. Natl. Acad. Sci. **103**(23), 8577–8582 (2006)

6. Newman, M.E., Barabási, A.-L.E., Watts, D.J.: The Structure and Dynamics of Networks. Princeton University Press (2006)
7. Badash, I., Kleinman, N.P., Barr, S., Jang, J., Rahman, S., Wu, B.W.: Redefining health: the evolution of health ideas from antiquity to the era of value-based care. Cureus **9**(2) (2017)
8. Boccaletti, S., Latora, V., Moreno, Y., Chavez, M., Hwang, D.-U.: Complex networks: structure and dynamics. Phys. Rep. **424**(4–5), 175–308 (2006)
9. Boccaletti, S., et al.: The structure and dynamics of multilayer networks. Phys. Rep. **544**(1), 1–122 (2014)
10. Yang, S.: Networks: An Introduction by MEJ Newman: Oxford, UK: Oxford University Press, 720 pp. $85.00. Taylor & Francis (2013)
11. Shannon, C.E.: A mathematical theory of communication. Bell Syst. Tech. J. **27**, 379–423 (1948)
12. Sawardecker, E., Amundsen, C., Sales-Pardo, M., Amaral, L.A.: Comparison of methods for the detectionof node group membership in bipartite networks. Eur. Phys. J. B **72**(4), 671–677 (2009)
13. Sawardecker, E.N., Sales-Pardo, M., Amaral, L.A.N.: Detection of node group membership in networks with group overlap. Eur. Phys. J. B **67**(3), 277–284 (2009)
14. Newman, M.E., Cantwell, G.T., Young, J.-G.: Improved mutual information measure for clustering, classification, and community detection. Phys. Rev. E **101**(4), 042304 (2020)
15. Danon, L., Diaz-Guilera, A., Duch, J., Arenas, A.: Comparing community structure identification. J. Stat. Mech. Theory Exp. **2005**(09), P09008 (2005)
16. Hospital data collection. https://www1.health.gov.au/internet/main/publishing.nsf/Content/health-casemix-data-collections-about
17. Bastian, M., Heymann, S., Jacomy, M.: Gephi: an open source software for exploring and manipulating networks. In: Proceedings of the International AAAI Conference on Web and Social Media, pp. 361–362 (2009)
18. Blondel, V.D., Guillaume, J.-L., Lambiotte, R., Lefebvre, E.: Fast unfolding of communities in large networks. J. Stat. Mech. Theory Exp. **2008**(10), P10008 (2008)

# Drugs Resistance Analysis from Scarce Health Records via Multi-task Graph Representation

Honglin Shu[1], Pei Gao[2,3], Lingwei Zhu[4], Zheng Chen[5(✉)], Yasuko Matsubara[5], and Yasushi Sakurai[5]

[1] Hong Kong Polytechnic University, Hong Kong SAR, China
[2] Sony AI, Tokyo, Japan
[3] Nara Institute of Science and Technology, Ikoma, Japan
[4] University of Alberta, Edmonton, Canada
[5] Osaka University, ISIR, Osaka, Japan
chen.zheng.bn1@gmail.com

**Abstract.** Clinicians prescribe antibiotics by looking at the patient's health record with an experienced eye. However, the therapy might be rendered futile if the patient has drug resistance. Determining drug resistance requires time-consuming laboratory-level testing while applying clinicians' heuristics in an automated way is difficult due to the categorical or binary medical events that constitute health records. In this paper, we propose a novel framework for rapid clinical intervention by viewing health records as graphs whose nodes are mapped from medical events and edges as correspondence between events in given a time window. A novel graph-based model is then proposed to extract informative features and yield automated drug resistance analysis from those high-dimensional and scarce graphs. The proposed method integrates multi-task learning into a common feature extracting graph encoder for simultaneous analyses of multiple drugs as well as stabilizing learning. On a massive dataset comprising over 110,000 patients with urinary tract infections, we verify the proposed method is capable of attaining superior performance on the drug resistance prediction problem. Furthermore, automated drug recommendations resemblant to laboratory-level testing can also be made based on the model resistance analysis.

**Keywords:** Drug Resistance Analysis · Health Records · Graph Neural Networks · Multi-task Learning

## 1 Introduction

Multi-drug resistance, i.e., the insusceptibility to multiple antimicrobial agents in pathogenic bacteria, is currently endangering the efficacy of antibiotic treatment and has become a staple global concern in public health [30]. Antibiotics

---

H. Shu and P. Gao—Joint first authors.

© The Author(s), under exclusive license to Springer Nature Switzerland AG 2023
X. Yang et al. (Eds.): ADMA 2023, LNAI 14178, pp. 103–117, 2023.
https://doi.org/10.1007/978-3-031-46671-7_8

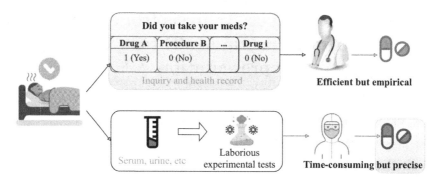

**Fig. 1.** Clinicians prescribe antibiotics by checking the patient's health record consisting of binary medical events. Drug resistance is determined in microbiology labs after time-consuming testing. Hence, the purpose of this work is to investigates *"how to provide a precise drug recommend by binary health record data?"*

are usually chosen empirically and their resistance needs to be determined via laboratory-level drug resistance testing using specimens such as serum, urine, cerebrospinal fluid, etc. as shown in Fig. 1. [23]. While laboratory testing has shown its indisputable efficacy in clinical practice, rapid clinical intervention such as drug recommendation becomes impossible due to the long hours in microbiology laboratories that can last up to 72 h. Further, it is often difficult to apply in large scale the heuristics of human clinicians [6].

The community has seen a large number of research attempts on developing data-driven clinical assistants to accurately identify candidates for the therapy. Among these works, there is a trend on combining machine learning (ML) methods with Health Record (HR) data for more accurate and adaptive prediction of drug resistance [34]. Although there is promise on the reported performance, those traditional statistical ML models have not yet been widely adopted in real clinic scenes. In general, statistical ML models generate probabilistic predictions with a focus on minimizing the number of misclassifications. But what is more relevant for causal inference in clinical practice lies in understanding the data-representation process. A good representation makes it easier to extract crucial information when building classifiers or any kind of predictors, as well as to contribute to better model performance, interpretability and credibility.

The HR data can be represented as a series of patient status information (medical events) with a time dimension (time windows). These medical events are usually recorded as a row in a table with each element being "binary" to denote whether a medical event *happens or not*. Such $\{0,1\}$ encoded tabular data, though intuitive, incurs intractable computational burden e.g., when using dictionaries for mapping categorical features to vector indices [2]. In general, high dimensional binary features can cause computability concerns for conventional machine learning models [1]. Although there are some attempts on solving this issue like feature hashing [24], they run at the cost of losing latent correlation between medical event features due to irreversibility and unlinkability [27].

The troublesome inexpressive binary data turn out to have surprisingly suitable representations as graphs: since each row of the HR is a series of binary elements showing a patient's historical medical events, we can consider it as a graph with events being nodes and edges reflecting binary relationship between two nodes, i.e., an edge exists when two medical events happened at the same time. Graph Neural Networks (GNNs) [32] have recently made significant progress in processing graph data and finding hidden information. The majority of GNNs [11,15,29,31,33] use a message passing mechanism [8]. The Graph Isomorphism Network (GIN) [33], for example, uses an aggregation function to iteratively update the node representations before generating the vector representation of the graph via the readout function [20,35]. GNN can effectively learn from graph structure information to more expressive latent vector representations using this form to improve the performance of downstream tasks. The medical event graph can help to learn a more telling latent representation by providing additional structure information.

However, applying GNNs to health records is far from being straightforward. Another majoy difficulty arises from the fact that the emergence of drug resistance is a low probability event, hence supervisory signals one can exploit to train the model is highly imbalanced, with far more negative samples than positive ones. As a result, noise in the supervised signals has a significant impact on supervised learning, causing biased learning and overfitting to the partial label distribution. To solve this problem, we take inspiration from the recently popular multi-task learning (MTL) literature. MTL with neural networks is accomplished by learning shared representations from various supervisory signals of many tasks to reduce the impact of noise on any single task [28].

In this paper we propose a novel framework for handling massive health records (more than 110,000 patients) from which we extract useful information such as logical correspondence between highly sparse medical events. By viewing each patient as a giant graph with binary medical events (and their connections) being nodes (and edges), our framework is capable of leveraging graph properties to extract informative features which conventional methods struggle to do. We further propose to alleviate the impact of imbalanced supervisory signals resulted from the sparsity of the dataset by multi-task learning. Our novel architecture consists of a shared input phase and a multi-head output that yields prediction for each of the *laboratory-recommended antibiotics*. In summary, the contributions of this paper are:

- We propose a novel graph learning framework that extracts informative features from high-dimensional sparse health records by transforming patient status to graphs. The multi-head architecture based on MTL outputs predictions for all labels exploiting one model and alleviates great impact imposed by imbalanced supervisory signals.
- On a massive dataset consisting of more than 110,000 patients we verify the proposed framework is capable of learning informative correspondence between medical events and outputting antibiotics recommendations highly similar to laboratory labels.

## 2    Graph Neural Networks and Multi-Tasks Learning

**Graph Neural Networks (GNNs)** leverage the graph structure information and node attributes to generate the latent representation of each node $\mathbf{h}_v$ or the entire graph $\mathbf{h}_{\mathcal{G}}$. A message passing mechanism is used by the majority of modern GNNs to iteratively aggregate the representation of each node's neighbors and update the latent representation of each node [8]. The latent representation $\mathbf{h}_v^k$ of node $v$ after $k$ iterations of aggregation is summarized as:

$$\mathbf{a}_v^k = \texttt{AGGREGATE}^{(k)} \left( \mathbf{h}_u^{k-1} \mid u \in \mathcal{N}(v) \right), \tag{1}$$

$$\mathbf{h}_v^k = \texttt{UPDATE}^{(k)} \left( \mathbf{h}_v^{k-1}, \mathbf{a}_v^k \right), \tag{2}$$

where $\mathcal{N}(v)$ is a set of neighbors of node $v$ and $\mathbf{h}_v^0$ is initialized as node attributes or one-one encoding when node attributes is not available. Latent representation $\mathbf{h}_v^k$ is used to do node-level tasks such as node classification or link prediction. For graph classification, we need to combine all node representations in graph $\mathcal{G}$ to generate a representation $h_{\mathcal{G}}$ of the entire graph via the readout function:

$$\mathbf{h}_{\mathcal{G}} = \texttt{READOUT} \left( \mathbf{h}_v^k \mid v \in \mathcal{G} \right). \tag{3}$$

The expressivity of latent representations depends on the choice of the aggregation, update, and readout functions. Many different aggregation function have been suggested. In Graph Convolutional Networks (GCN) [15], the average neighbor messages aggregation is employed to learn node representation:

$$\mathbf{h}_v^k = \sigma \left( \mathbf{W}_k \sum_{u \in \mathcal{N}(v)} \frac{1}{|\mathcal{N}(v)|} \mathbf{h}_u^{k-1} \right), \tag{4}$$

where $\mathbf{W}_k$ is $k$-th layer trainable weight matrix and $\sigma(\cdot)$ is the activation function e.g., ReLU. Simplifying Graph Convolution (SGC) [31] eliminates superfluous complexity by progressively reducing nonlinearities and weight matrices between succeeding layers. Element-wise max-pooling is used in GraphSAGE [11] as opposed to mean-pooling Eq. (4). Many variants [3,8,29,32,33] can be summarized similarly as Eq. (1) and Eq. (2). To obtain a global representation of the entire graph, readout function can be sum, mean, and max operation for all nodes [33]. Besides, readout function can also be the graph-level pooling layer with graph coarsening mechanism [20,35], which is a hierarchical approach to read out graph representation by iteratively dividing the entire graph into multiple subgraphs as cluster nodes to construct coarsened graphs. Many other readout functions can generate the graph representation [5,17].

**Multi-Task Learning (MTL)** refers to more than one loss function being optimized during model training. MTL can improve the model's generalizability on all tasks by allowing the model to share the domain-specific representations it has learned during training on a series related tasks [28]. UberNet [16] is a typical hard parameter sharing strategy in which different CNNs of tasks are jointly trained to handle low-, mid-, and high-level vision tasks such as boundary detection, surface normals estimation, saliency estimation, semantic segmentation, and so forth. Other parameter sharing approaches include a shared encoder and numerous task-specific decoders [4,10,14,25]. In soft parameter sharing, each task has its own independent parameters in the representation extraction layers and the decision layers, they express similarity by restricting the discrepancies between the parameters of various tasks. As a way of soft feature fusion, cross-stitch networks [22] employ a linear combination of activations in each layer of the task-specific networks. Other soft parameter sharing approaches are developed with dimensionality reduction methods with feature fusion layers [7], or attention mechanism to share feature [19]. The issue of soft parameter sharing approaches is scalability, as the size of network tends grow linearly with the number of tasks.

# 3    Problem Formulation

Our objective is to investigate how to accurately evaluate whether a patient has resistance to a variety of medications based on their medical history. Let matrix $\mathbf{X} \in \mathbb{R}^{M \times N}$ be a dataset and $\mathbf{y} = \{\mathbf{y}_n \in \mathbb{R}^M : n \in \{1, 2, ..., T\}\}$ denotes the set of labels for the patients' drug resistance, where $M$ is the number of patients and $N$ medical events. The elements of matrix $\mathbf{X}$ are binary, i.e., if a specific medical event $i$ occurred in certain samples $j$, the corresponding element $\mathbf{X}_{i,j}$ is 1, otherwise it is 0. The $i$-th element $\mathbf{y}_n^{(i)}$ of $\mathbf{y}_n$ is 1 if patient $i$ has resistance to drug $n$, otherwise it is 0. $T$ represents the number of distinct drug resistances that must be predicted (i.e., number of tasks).

The matrix $\mathbf{X}$ is significantly sparse and high-dimensional. Conventional methods often struggle to effectively estimate the conditional probability distribution $P(\mathbf{y}_n \mid \mathbf{X}, \theta)$, where $\theta$ denotes the parametrization of the model. To alleviate the issue of sparsity and high-dimensionality, we convert each row of $\mathbf{X}$ into a medical event graph $\mathcal{G}_i$ with $i \in \{1, 2, ..., M\}$, where the nodes in $\mathcal{G}_i$ denote the historical medical events and the connectivity of two nodes depends on whether these two medical events happened at the same time. We use a graph-based encoder $f_{\mathcal{G},\theta}(\cdot)$ to learn an expressive graph vector representation since one can mine more information from its topological structure than digging from raw data. Therefore, the problem is modeled as a graph classification task equivalent to a graph isomorphism problem:

$$\theta_n = \arg\max_\theta \sum_{i=1}^M P\left(\mathbf{y}_n^{(i)} \mid f_{\mathcal{G},\theta}(\mathcal{G}_i)\right), \ n \in \{1, 2, ..., T\}, \tag{5}$$

i.e., the solution to $n$-th prediction problem is obtained as the maximizer of the conditional probability over all training instances.

More importantly, since the events of drug resistance occur infrequently, the resistance labels of patients to all drugs $\mathbf{y}$ is unbalanced: *the number of positive samples is substantially smaller than the number of negative samples*. This type

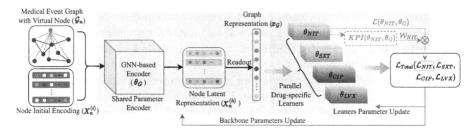

**Fig. 2.** System overview of the proposed GNN-based dynamic weighted multi-task learning framework. (Color figure online)

of supervision signals can mislead the model into biased latent representations and subsequently overfitting to the imbalanced label distribution. In real-world scenarios, such high precision low recall models is not informative for drug resistance analyses.

Naturally, the medical decision for assessment of drug resistance is to find a trade-off between multiple drugs. Hence, to alleviate the negative effect from the biased supervisory signals of any single task, we can formulate the problem as a multi-objective optimization problem with graph-based shared parameter encoder, which can help the model filter out noises in single-task labels and force the model to extract more generalizable graph latent representations. Our proposed objective formulation is to minimize a weighted summation of losses on the prediction of different drugs resistance:

$$
\arg\min_{\theta_{\mathcal{G}},\theta_1,\theta_2,...,\theta_T} \mathcal{L}_{Total}\left(\theta_{\mathcal{G}},\theta_1,\theta_2,...,\theta_T\right)
$$
$$
= \arg\min_{\theta_{\mathcal{G}},\theta_1,\theta_2,...,\theta_T} \sum_{n=1}^{T} w_n \mathcal{L}_n\left(\theta_{\mathcal{G}},\theta_n\right), \tag{6}
$$

where $\theta_{\mathcal{G}}$ denote the shared parameters of graph-based encoder and $\theta_n$ the independent parameters of task-specific learners. $w_n$ is a weight coefficient of each loss to control the difficulty and speed in model learning.

## 4   Proposed Framework

As illustrated in Fig. 2, we propose a novel and effective multi-task graph learning framework leveraging a GNN-based encoder as the bottom parameter-sharing module, parallel task-specific learners as drug-specific resistance predictors, and dynamically weighted adaption as a mechanism to balance learning speed and problem difficulty.

**GNN-Based Encoder.** Given a vector $x_i \in \mathbb{R}^N$ denoting $N$ historical medical events of the $i$-th patient, we convert it to a medical event graph $\mathcal{G}_i$. Let us recall that there are $N$ nodes in graph $\mathcal{G}_i$ to represent the $N$ medical events. An edge between two nodes exists only when the two corresponding medical events occur at the same time, where occurring at the same time indicate two medical events are both 1 value in the original tabular data. For example, a patient has accepted N medical events. In the medical event graph of the patient, if a edge exists between i-th medical event and j-th medical event, the element of i-th and j-th medical events should be 1 in original tabular data.

Since the vector $x_i$ may be very sparse, the corresponding graph $\mathcal{G}_i$ is far from fully connected, and can lead to very poor graph representations if conventional massage-passing-based GNNs are naively used.

To takcle this issue, we add to the graph virtual nodes connected to all other nodes. The virtual nodes allow isolated connected components to indirectly update the node representation through it. We adopt the GIN convolution [33] as the GNN-based encoder. The nodes in graph $\mathcal{G}_i$ are using one-hot encoding only, which implies only the graph structure information is exploited to classify a graph. GIN imitates the Weisfeiler-Lehman test [26] that upper-bounds the graph classification problem with only structural information. Hence, the $k$-th layers of GNN-based encoder can be expressed as:

$$h_v^{(k)} = \text{MLP}^{(k)} \left( h_v^{(k-1)} + \sum_{u \in \mathcal{N}(v)} h_u^{(k-1)} \right), \tag{7}$$

where $h_v^{(k)}$ is the vector representation of node $v \in \mathcal{G}_i$ and $\mathcal{N}(v)$ is the set of nodes adjacent to node $v$. $\text{MLP}^{(k)}$ is an injection function consisting of $n$ fully connected layers. After $K$ iterations of updating, we can separately readout the node latent representation of each layer via mean pooling to finally generate the a set of graph representations:

$$h_{\mathcal{G}_i} = \{h_{\mathcal{G}_i}^{(1)}, h_{\mathcal{G}_i}^{(2)}, ..., h_{\mathcal{G}_i}^{(K)}\} = \{\text{MEAN}(\{h_v^k \,|\, v \in \mathcal{G}_i\}) \,|\, \forall k\}. \tag{8}$$

**Task-specific Learners.** Our model consists of independent tasks-specific learners. After the encoder, we simply append linear layers as task-specific learners for predicting drug resistance. Given a set of graph representations $h_{\mathcal{G}_i}$, the probability of resistance of drug $t$ is calculated as:

$$P\left(y_t^{(i)} = 1 \,\Big|\, f_{\theta_{\mathcal{G}}}(\mathcal{G}_i)\right) = \frac{1}{1 + e^{-\sum_{k=1}^{K} \mathbf{X}_t h_{\mathcal{G}_i}^{(k)}}}, \tag{9}$$

where $\mathbf{X}_t$ are the task-specific weighted matrix for drug $t$.

**Dynamically Weighted Adaptation.** In multi-task learning, the weights in loss summation will affect task learning difficulty. We find that the difficulty of assessing resistance to different drugs also varies. Taking inspiration from the focal loss function [18] for sample-level imbalance problem, we introduce Dynamic Task Prioritization (DTP) [9] into the weight adjustment while training our parallel task architecture. DTP adjusts the weight of each task iteratively based on the key performance indicators (KPIs) $\hat{K}_t^\tau$. The KPIs $\hat{K}_t^\tau$ is computed via moving average:

$$\hat{K}_t^\tau = \alpha K_t^\tau + (1 - \alpha)\hat{K}_t^{\tau-1}, \tag{10}$$

where $K_t^\tau \in [0, 1]$ is the performance indicator of task $t$ on step $\tau$, e.g., recall. $\alpha$ is a decaying factor. Since the number of positive samples is significantly smaller than the amount of negative samples, *recall* has emerged as a key indicator of the learned model's quality as well as the difficulty of the task to be learnt. More specifically, after calculating $\hat{K}_t^\tau$, the weight $w_t^\tau$ for task $t$ is updated as:

$$w_t^\tau = -(1 - \hat{K}_t^\tau)^\gamma \log \hat{K}_t^\tau, \tag{11}$$

where $\gamma$ is a hyper-parameter to adjust the weight. Higher weights $w_t^\tau$ reflect higher learning difficulty on task $t$. The total loss in our proposed MTL framework can hence be summarized as:

$$\mathcal{L}_{Total} = \sum_{t=1}^{T} -(1 - \hat{K}_t^\tau)^\gamma \log \hat{K}_t^\tau \, \mathcal{L}_t \, (\theta_\mathcal{G}, \theta_t) \,. \tag{12}$$

**Optimization.** The weighted sum of loss functions for different tasks has certain favorable properties for multi-task optimization. First, it is differentiable, which allows for direct optimization using stochastic gradient descent. Second, the chain rule ensures updates of the task-specific learners do not interfere with each other during the optimization process (See Eq. (13)), and the loss of tasks can update the GNN-based encoders (Eq. (14)).

$$\frac{\partial}{\partial \theta_t} \mathcal{L}_{Total} = -(1 - \hat{K}_t^\tau)^\gamma \log \hat{K}_t^\tau \frac{\partial}{\partial \theta_t} \mathcal{L}_t \, (\theta_\mathcal{G}, \theta_t) \tag{13}$$

$$\frac{\partial}{\partial \theta_\mathcal{G}} \mathcal{L}_{Total} = \sum_{t=1}^{T} -(1 - \hat{K}_t^\tau)^\gamma \log \hat{K}_t^\tau \frac{\partial}{\partial \theta_\mathcal{G}} \mathcal{L}_t \, (\theta_\mathcal{G}, \theta_t) \tag{14}$$

Here, the encoders (green part in Fig. 2) share the parameters at the same time.

## 5    Experiments and Discussion

### 5.1    Dataset Description

We validate the proposed method on a large-scale AMR-UTI dataset [21]. AMR-UTI dataset contains electronic health record (EHR) information from over 110,000 patients with urinary tract infections (UTI) treated at Massachusetts General Hospital and Brigham & Women's Hospital in Boston, MA, USA between 2007 and 2016. Each patient in the dataset provided urine cultures for antibiotic drug resistance testing.

We include only the observations with empiric antibiotic prescriptions from the AMR-URI dataset. Exactly one of the first-line antibiotics, nitrofurantoin (NIT) or TMP-SMX (SXT), or one of the second-line antibiotics, ciprofloxacin (CIP) or levofloxacin (LVX) was prescribed. Here, 11136 observations composed our dataset. We remove observations that do not have any health event. For each observation, a feature is constructed from its EHR as a binary indicator

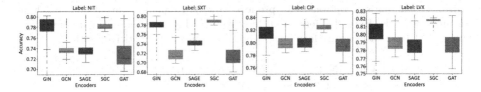

**Fig. 3.** Comparison between the proposal (GIN) and four other graph encoders.

for whether the patient was undergoing a particular medical event in a specified time window. The first part of the medical event is the past clinical history associated with antibiotic resistance, including recurrent UTIs, hospitalizations and resistance of previous infections. Besides, the risk of an infection being resistant to different antibiotics is associated with patient demographics and comorbidities. Known comorbidities associated with resistance include the presence of a urinary catheter, immunodeficiency and diabetes [12]. Surgery, placement of a central venous catheter (CVC), mechanical ventilation, hemodialysis, and parenteral nutrition were included in the prior procedure description.

## 5.2   Experiments Setting

We split the dataset into train/validation/test subsets with 0.7/0.1/0.2 proportion. When showing experiment results, we compare the proposed method against several conventional multi-label learning approaches in drug resistance prediction [36] such as Binary Relevance with Naive Bayes (BRNB), Label Powerset with Naive Bayes (LPNB), Classifier Chain with Naive Bayes (CCNB), adaptive KNN (MLKNN), Multi-Layer Perceptron (MLP), and a baseline [13] that combine with Logistic Regression, Random Forest, and Decision Tree. AUROC is used as a metric for performance comparing. The threshold is set as 95% due to we wonder if the methods can effectively predict positive samples.

We also conduct extensive experiments to evaluate the impact of different graph neural network encoders (i.e., GIN, GCN, GraphSAGE, SGC, GAT), multi-task optimization strategies (i.e., Dynamic Weight Adaptioin and Fixed Uniform Weight), and virtual node (yes/no). We set the number of convolution layers of GIN to 7, the neighbor pooling and readout function to *sum pooling* and *mean readout*, respectively. We use the configurations from their respective publications and set the number of layers of GCN, GAT, and SGC to 2 and GraphSAGE to 3. Batch size is set to 32 and learning rate to 0.01 after trial-and-error tuning. The strategy of learning rate decaying is step decaying.

## 5.3   Results and Analyses

We firstly compare against other conventional ML approaches in Table 1. As expected, the conventional approaches performed poorly in obtaining a sensible trade-off between precision and recall due to the high dimensionality and sparsity of the data, which finally lead to the their terrible performance on AUROC (95%

**Table 1.** Comparison between the proposed method against conventional machine learning methods. Here, ↑ refers to an improvement between the optimal convention and ours, ↓ refers a lower performance, while ↕ denotes a competitive result.

| Methods/Labels | NIT | SXT | CIP | LVX |
|---|---|---|---|---|
| | AUROC (95% CI) | | | |
| MLP | 0.52 | 0.58 | 0.64 | 0.67 |
| BRNB | 0.60 | 0.50 | 0.66 | 0.65 |
| LPNB | 0.54 | 0.51 | 0.52 | 0.59 |
| CCNB | **0.60** | 0.50 | 0.66 | 0.65 |
| MLKNN | 0.53 | 0.52 | 0.60 | 0.60 |
| Baseline | 0.56 | **0.59** | **0.64** | **0.64** |
| Proposed Method | 0.59 (0.01)↕ | **0.61** (0.02)↑ | **0.70** (0.06)↑ | **0.72** (0.08) ↑ |

CI). Our proposed method significantly outperformed the conventional methods on three label systems (SXT, CIP, and LVX), which is 0.02, 0.06, and 0.08 within AUROC (95% CI) superior to the experimental conventional methods. In contrary, on the NIT system the proposed method performed only slightly under CCNB: 0.01 lower and BRNB: 0.01 lower in AUROC (95% CI) score.

The characteristics of the data pose a challenge to graph encoders. To thoroughly investigate which encoder performs the best, we evaluate 5 popular GNN encoders on the validation set in Fig. 3. In our experiment, since we found that NIT is more difficult to be learned than other label systems. The result shows that GIN outperformed all other encoders on NIT label system and was competitive on the other label systems. Therefore, we might safely put that GIN is one of the most suitable graph encoders for the chosen scenario.

Figure 4 displays the result of our comprehensive ablation study. We showed *precision* and *recall* of the four tasks on the testing set leveraging all possible combinations of graph-based encoders, multi-objective optimization strategies, and virtual node. As shown in the first row of Fig. 4, our proposed method significantly improved the Recall on NIT, CIP, and LVX. At the same time, it retained as much precision as possible at the cost of only slight loss on SXT Recall, which implies our proposed method can effectively alleviate the problem of label imbalance to exert the few positive samples. Compared with the ablation choices, it is visible that their performance suffered from significant drop. This observation suggests the importance of DTP and VN without which the performance suffered. Hence it can be concluded that DTP and VN improve the proposed method from both the optimization-level (i.e., DTP dynamic control the learning process) and the model-level (i.e., virtual node play as a global agent to connect the isolate component). Combining these experimental results, we can easily find that NIT is more difficult to learn effectively than other drugs. The results in Fig. 5 and Fig. 6 also verify this point of view, and further prove the effectiveness of our proposed DTP and VN. That is, in the case of using DTP and/or VN, the model can more effectively learn the key features of NIT and improve the final prediction results without hurting the model performance on the other three drugs.

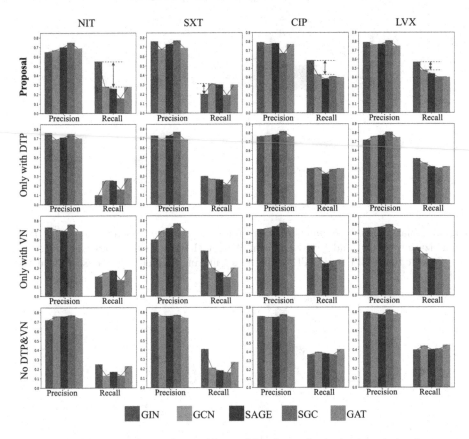

**Fig. 4.** Comparison between all possible combinations of encoders, optimization strategy and the virtual nodes.

## 5.4   Drug Resistance Phenotype Classification

We visualize the drug resistance phenotypes in Fig. 7 by thresholding probabilities output by the model. For different observation we set a series of distinct thresholds for probabilities. Observations with model probabilities above the threshold is classified as *non-susceptible* and *susceptible* for those below the threshold. As the threshold increases, more observations are classified into the *non-susceptible* phenotype. The two-color background of the radar graphs represents the proportion of *non-susceptible* phenotype to the susceptible one, decided by antibiotic drug resistance testing in microbiology laboratory (NIT: 88.17% SXT: 79.57%, CIP: 93.88%, LVX: 94.03%). A critical threshold allows the model phenotype distribution to approximate the laboratory phenotype distribution. Such a threshold holds the promise to translate model probabilities into direct drug treatment recommendations.

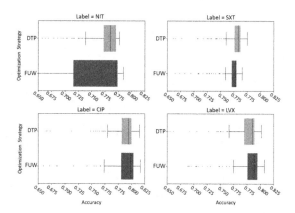

**Fig. 5.** Ablation study of multi-task optimization strategy between Dynamic Task Prioritization (DTP) and Fixed Uniform Weight (FUW). DTP is more favorable on the difficult tasks NIT and SXT, while no significant difference was observed on the rest.

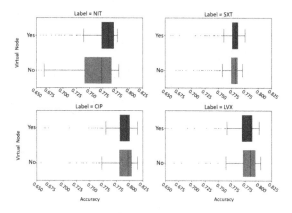

**Fig. 6.** Ablation study of the virtual nodes. Virtual nodes connect isolated components indirectly and allow them to update their representations. On the difficult NIT and SXT tasks adding virtual nodes improved the final performance, while no significant difference was observed on the rest.

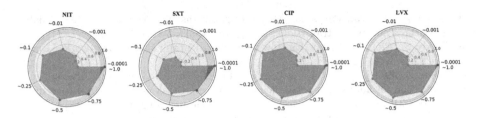

**Fig. 7.** Cumulative probabilities of drug resistance output by the model. Numbers around the circles indicate thresholds. The two-color background represents the proportion of non-susceptible phenotype decided by antibiotic drug resistance testing in microbiology laboratory.

# 6  Discussion

We proposed a graph-based multi-tasks learning framework and showed that it can alleviate the imbalance label distribution in drug resistance prediction problem. There a potential problem should be further explore. In our experiment, when the model perform well on NIT label system, the model performance on other label systems has decreased to varying degrees. Hence, we consider the proposed method can effectively alleviate the imbalance distribution problem but it is not the optimal solution. A multi-task learning approach that can simultaneously improve the performance of all drug resistance predictions without incurring performance losses on other label systems needs to be further explored.

# 7  Conclusion

In this paper we proposed to view health records consisting of binary or categorical medical events as graphs: nodes are mapped from the events and edges reflect whether two events happened at the same time. A novel graph-based deep learning architecture was then proposed to extract informative features such as correspondence between medical events from those high dimensional and sparse graphs. On a massive dataset comprising over 110,000 patients with urinary tract infections we verified that the proposed method was capable of attaining superior performance on the drug resistance prediction problem. We further showed that automated drug recommendations could be made on top of the resistance analyses output by the model. Extensive ablation studies within the graph neural networks literature as well as against conventional baselines demonstrated the effectiveness of the proposed binary health record data as graphs framework.

**Acknowledgement.** This work was supported by JSPS KAKENHI Grant-in-Aid for Scientific Research Number JP21H03446, NICT 03501, JST-AIP JPMJCR21U4.

# References

1. Attenberg, J., Weinberger, K., Dasgupta, A., Smola, A., Zinkevich, M.: Collaborative email-spam filtering with the hashing trick. In: CEAS (2009)

2. Bertsimas, D., Van Parys, B.: Sparse high-dimensional regression: exact scalable algorithms and phase transitions. Ann. Stat. **48**(1), 300–323 (2020)
3. Brody, S., Alon, U., Yahav, E.: How attentive are graph attention networks? CoRR abs/2105.14491 (2021)
4. Chen, Z., Badrinarayanan, V., Lee, C., Rabinovich, A.: GradNorm: gradient normalization for adaptive loss balancing in deep multitask networks. In: Proceedings of the 35th International Conference on Machine Learning, ICML (2018)
5. Diehl, F.: Edge contraction pooling for graph neural networks. CoRR abs/1905.10990 (2019)
6. Ferrer, R., et al.: Empiric antibiotic treatment reduces mortality in severe sepsis and septic shock from the first hour: results from a guideline-based performance improvement program. Crit. Care Med. **42**(8), 1749–1755 (2014)
7. Gao, Y., Ma, J., Zhao, M., Liu, W., Yuille, A.L.: NDDR-CNN: layerwise feature fusing in multi-task CNNs by neural discriminative dimensionality reduction. In: CVPR, pp. 3205–3214 (2019)
8. Gilmer, J., Schoenholz, S.S., Riley, P.F., Vinyals, O., Dahl, G.E.: Neural message passing for quantum chemistry. In: ICML, pp. 1263–1272 (2017)
9. Guo, M., Haque, A., Huang, D.-A., Yeung, S., Fei-Fei, L.: Dynamic task prioritization for multitask learning. In: Ferrari, V., Hebert, M., Sminchisescu, C., Weiss, Y. (eds.) ECCV 2018. LNCS, vol. 11220, pp. 282–299. Springer, Cham (2018). https://doi.org/10.1007/978-3-030-01270-0_17
10. Guo, P., Lee, C., Ulbricht, D.: Learning to branch for multi-task learning. In: Proceedings of the 37th International Conference on Machine Learning, ICML 2020, 13–18 July 2020, Virtual Event. Proceedings of Machine Learning Research, vol. 119, pp. 3854–3863. PMLR (2020)
11. Hamilton, W.L., Ying, Z., Leskovec, J.: Inductive representation learning on large graphs. In: NeurIPS, pp. 1024–1034 (2017)
12. Ikram, R., Psutka, R., Carter, A., Priest, P.: An outbreak of multi-drug resistant escherichia coli urinary tract infection in an elderly population: a case-control study of risk factors. BMC Infect. Dis. **15**(1), 1–7 (2015)
13. Kanjilal, S., Oberst, M., Boominathan, S., Zhou, H., Hooper, D.C., Sontag, D.A.: A decision algorithm to promote outpatient antimicrobial stewardship for uncomplicated urinary tract infection. Science Transl. Med. **12** (2020)
14. Kendall, A., Gal, Y., Cipolla, R.: Multi-task learning using uncertainty to weigh losses for scene geometry and semantics. In: 2018 IEEE Conference on Computer Vision and Pattern Recognition, CVPR (2018)
15. Kipf, T.N., Welling, M.: Semi-supervised classification with graph convolutional networks. In: ICLR (2017)
16. Kokkinos, I.: UberNet: training a universal convolutional neural network for low-, mid-, and high-level vision using diverse datasets and limited memory. In: 2017 IEEE Conference on Computer Vision and Pattern Recognition, CVPR (2017)
17. Lee, J., Lee, I., Kang, J.: Self-attention graph pooling. In: Proceedings of the 36th International Conference on Machine Learning, ICML 2019, Long Beach, California, USA, 9–15 June 2019. Proceedings of Machine Learning Research, vol. 97, pp. 3734–3743. PMLR (2019)
18. Lin, T., Goyal, P., Girshick, R.B., He, K., Dollár, P.: Focal loss for dense object detection. In: ICCV, pp. 2999–3007 (2017)
19. Liu, S., Johns, E., Davison, A.J.: End-to-end multi-task learning with attention. In: CVPR, pp. 1871–1880 (2019)
20. Ma, Y., Wang, S., Aggarwal, C.C., Tang, J.: Graph convolutional networks with eigenpooling. In: ACM SIGKDD, pp. 723–731 (2019)

21. Michael, O., Soorajnath, B., Helen, Z., Sanjat, K., Sontag, D.: AMR-UTI: antimicrobial resistance in urinary tract infections (version 1.0.0). PhysioNet (2020)
22. Misra, I., Shrivastava, A., Gupta, A., Hebert, M.: Cross-stitch networks for multi-task learning. In: 2016 IEEE Conference on Computer Vision and Pattern Recognition, CVPR 2016, Las Vegas, NV, USA, 27–30 June 2016, pp. 3994–4003. IEEE Computer Society (2016)
23. Reller, L.B., Weinstein, M., Jorgensen, J.H., Ferraro, M.J.: Antimicrobial susceptibility testing: a review of general principles and contemporary practices. Clin. Infect. Dis. **49**(11), 1749–1755 (2009)
24. Seger, C.: An investigation of categorical variable encoding techniques in machine learning: binary versus one-hot and feature hashing (2018)
25. Sener, O., Koltun, V.: Multi-task learning as multi-objective optimization. In: NeurIPS, pp. 525–536 (2018)
26. Shervashidze, N., Schweitzer, P., van Leeuwen, E.J., Mehlhorn, K., Borgwardt, K.M.: Weisfeiler-Lehman graph kernels. J. Mach. Learn. Res. **12**, 2539–2561 (2011)
27. Tulyakov, S., Farooq, F., Mansukhani, P., Govindaraju, V.: Symmetric hash functions for secure fingerprint biometric systems. Pattern Recogn. Lett. **28**(16), 2427–2436 (2007)
28. Vandenhende, S., Georgoulis, S., Gansbeke, W.V., Proesmans, M., Dai, D., Gool, L.V.: Multi-task learning for dense prediction tasks: a survey. IEEE Trans. Pattern Anal. Mach. Intell. **44**(7), 3614–3633 (2022)
29. Velickovic, P., Cucurull, G., Casanova, A., Romero, A., Liò, P., Bengio, Y.: Graph attention networks. In: ICLR (2018)
30. Ventola, C.L.: The antibiotic resistance crisis: part 1: causes and threats. Pharm. Ther. **40**(4), 277 (2015)
31. Wu, F., Souza, A., Zhang, T., Fifty, C., Yu, T., Weinberger, K.Q.: Simplifying graph convolutional networks. In: ICML, pp. 6861–6871 (2019)
32. Wu, Z., Pan, S., Chen, F., Long, G., Zhang, C., Yu, P.S.: A comprehensive survey on graph neural networks. IEEE Trans. Neural Netw. Learn. Syst. **32**, 4–24 (2019)
33. Xu, K., Hu, W., Leskovec, J., Jegelka, S.: How powerful are graph neural networks? In: ICLR (2019)
34. Yelin, I., et al.: Personal clinical history predicts antibiotic resistance of urinary tract infections. Nat. Med. **25**(7), 1143–1152 (2019)
35. Ying, Z., You, J., Morris, C., Ren, X., Hamilton, W.L., Leskovec, J.: Hierarchical graph representation learning with differentiable pooling. In: NeurIPS, pp. 4805–4815 (2018)
36. Zhang, M.L., Zhou, Z.H.: A review on multi-label learning algorithms. IEEE Trans. Knowl. Data Eng. **26**, 1819–1837 (2014)

# Text Classification

# ParaNet:Parallel Networks with Pre-trained Models for Text Classification

Yujia Wu, Xin Guo$^{(\boxtimes)}$, Yi Wei, and Xingli Chen

School of Information Science and Technology, Sanda University, Shanghai 201209, China
wuyujia@whu.edu.cn, guoxin@sandau.edu.cn

**Abstract.** The application of linguistic knowledge derived from pre-trained language models has demonstrated considerable potential in text classification tasks. Despite this, effectively learning the distance between samples and different labels for supervised learning tasks remains a practical challenge. In this study, we propose a novel approach, termed Parallel Networks with Pre-trained Models (ParaNet), which learns distance information between input samples and different labels within the same space. Specifically, ParaNet utilizes a Parallel Networks network architecture comprising two distinct Transformer Encoders to extract sample features and label features separately. By fine-tuning the network parameters, ParaNet can achieve the closest possible distance between the sample and its corresponding label, while simultaneously achieving the farthest possible distance between the sample and a label that does not belong to it. To fully exploit label information, the model leverages the semantic knowledge of the pre-trained model by adding templates to the labels. Our experimental analysis of eight benchmark text classification datasets demonstrates that ParaNet significantly improves classification accuracy, with an average accuracy rate increase from 89.1% to 89.64%.

**Keywords:** Text Classification · Transformer Encoders · Parallel Networks · Pre-Trained Language Models

## 1 Introduction

In recent years, the field of text classification [1–3] has witnessed significant advancements due to the rapid development of deep learning technology [4,5]. Pre-trained language models (PLMs) have emerged as a powerful tool in this regard, as they are capable of acquiring a vast amount of knowledge through self-supervised learning on large-scale corpora. By leveraging the knowledge gained by PLMs and fine-tuning them for downstream task models, we can enhance the overall performance of these models. BERT, a widely used PLM, is based on Transformer Encoder and has been trained on the English Wikipedia and BooksCorpus English corpus data to acquire a substantial amount of semantic knowledge [6].

© The Author(s), under exclusive license to Springer Nature Switzerland AG 2023
X. Yang et al. (Eds.): ADMA 2023, LNAI 14178, pp. 121–135, 2023.
https://doi.org/10.1007/978-3-031-46671-7_9

In the context of text classification tasks, utilizing pre-trained language models (PLMs), to obtain feature vector representations from input text sequences, is a transfer learning method that enables the completion of text classification tasks with only a small number of samples [7].Therefore, building text classification models using PLMs is currently the mainstream approach. Croce et al. [8] proposed a semi-supervised generative adversarial network text classification method that fine-tunes the BERT architecture using unlabeled data in a generative adversarial setting. Qin et al. [9] proposed a BERT-based feature projection method that projects neutral features to achieve high accuracy classification, and then improve the performance of BERT-based models for text classification.

Despite the excellent feature representation capability of PLMs, some useful information is encoded in the labels. Therefore, Mekala et al. [10] used BERT to create a contextualized corpus for generating pseudo-labels to implement a context-based weakly supervised text classification method. Chen et al. [11] proposed a label-aware data augmentation method based on dual contrast learning for text classification tasks. This method uses labels as augmented samples and employs contrast learning methods to learn the association between input and augmented samples. Giovanni et al. [12] implicitly exploit the semantic information of labels for text classification tasks by generating labels at the time of prediction. Hu et al. [13] extend the label word space using an external knowledge base and refine it using PLM before using it for prediction to improve the performance of zero-shot and few-shot. Mueller et al. [14] proposed a label semantic-aware pre-training model that utilizes labels to improve the generalization ability and computational efficiency of few-shot text classification. While these approaches have demonstrated good results by utilizing label semantic information, further improvements are still possible.

However, the approaches mentioned above were unable to learn the distance information between samples and labels, which is a critical aspect of supervised learning tasks. While using label information as a data enhancement method has been shown to be feasible [15], effectively learning the distance between samples and different labels remains a practical challenge. To address this issue, this paper proposes a novel approach, namely Parallel Networks with Pre-trained Models (ParaNet), which learns the distance information between input samples and different labels simultaneously within the same space. By leveraging the sample and label features extracted by PLM and learning the distance information between them, ParaNet effectively utilizes the distance information to classify input samples. Moreover, to fully exploit the label information, ParaNet incorporates prompt templates for data supplementation of labels to obtain richer semantic information of text.

Our experimental results on eight publicly available text classification datasets demonstrate that ParaNet outperforms fine-tuning models based on PLM. We attribute the strong performance of ParaNet to the diversity of acquired features. In summary, our three main contributions are as followed:

(1) To the best of our knowledge, we are the first to propose using PLM to obtain distance information between samples and different labels for text classification tasks;

(2) We introduce prompt templates for data supplementation of labels to fully exploit the semantic knowledge of PLM;

(3) We conduct experiments on eight publicly available datasets, demonstrating that ParaNet generally outperforms the baseline algorithm, with an average accuracy rate improvement from 89.1% to 89.64%.

## 2   Related Work

### 2.1   Text Classification

The current mainstream approach for building text classification models involves using PLMs [8–11]. Zhang et al. [16] propose a meta-learning framework that simulates learning scenarios in zero-sample by using pre-existing classes and virtual non-existing classes. Gera et al. [17] propose a plug-and-play approach that uses a self-training approach to complete learning of the target task. Zhang et al. [18] proposed a random text technique to generate high-quality comparison samples to improve the accuracy of zero-shot text classification. Few-shot text classification refers to using a pre-trained model with only a small number of samples to complete the model training and improving the performance of classification by a prompt fine-tuning approach. Min et al. [19] introduced a noise channel approach for language model prompting to achieve few-shot text classification with limited updates of language model parameters through contextual presentation or prompt tuning. Zha et al. [20] proposed a self-supervised hierarchical task clustering method that unravels potential relationships between tasks by dynamically learning knowledge from different clusters to improve interpretability. Zhao et al. [21] proposed an explicit and implicit consistent regularization to enhance the language model. Zhang et al. [22] proposed a meta-learning model by assigning label word learning to the base learner and template learning to a meta-learner to achieve robust few-shot text classification.

Wang et al. [23] proposed a unified prompt tuning framework to improve few-shot text classification in BERT-style models by explicitly capturing prompt semantics from non-target NLP datasets. Shnarch et al. [24] proposed improving the performance of such models by adding intermediate unsupervised classification tasks between pre-training and fine-tuning phases. Zhao et al. [25] proposed a memory imitation meta-learning approach to enhance the model's dependence on task adaptation support sets. However, in low-resource situations, a small validation set may not be sufficiently representative, and for this reason, Choi et al. [26] proposed an early stopping method using unlabeled samples to better estimate the class distribution of unlabeled samples. Approaches have also been proposed to improve text classification performance by incorporating visual feature information into sentence modeling through an image retrieval mechanism [27]. The use of PLMs for prompt tuning in text classification has shown promise, allowing for zero-shot or few-shot classification tasks without the need for extensive downstream data adjustments. We propose a fusion model that leverages PLMs by learning the distance between samples and their corresponding labels. This approach involves retraining the PLM-based model to construct the fusion model.

## 2.2    Pre-trained Language Model

With the rapid development of PLM, PLM-based models have become the dominant approach for handling downstream natural language processing tasks. Models for PLM are also getting larger, and many works are exploring the upper limit of parameter size of the models, such as Megatron-turing NLG [28] with 530 billion parameters. BERT [6] is one of the most influential PLM models, which uses MLM (Masked Language Model) and NSP (Next Sentence Prediction) for bi-directional joint training, and by self-supervised training enables the PLM to learn a large amount of semantic knowledge. In order to reduce the number of parameters, AlBERT [29] reduces the size of the model by sharing parameters across layers and uses SOP (Sentence Order Prediction) task instead of NSP task in BERT to improve the performance of the model. ELECTRA [30] replaces MLM in BERT with replacement token detection to solve the problem of BERT in MASK's pre-training phase and fine-tuning phase inconsistently while fully considering the computational efficiency and absolute performance of the model. XLNet [31] uses three major mechanisms of aligned language model, dual-stream attention and recurrent, which also make the autoregressive model also available with bidirectional contextual information.

A new PLM based on Transformer Decoder has been developed that is very effective at handling language generation tasks. One such model is GPT-3 [32], a large-scale language model that has become a hot topic pursued by academia and industry. Using a method that uses human feedback for fine-tuning to align the language model with user intent on various tasks by using a set of prompts written by annotators and fine-tuned using supervised learning, and then further using reinforcement learning to fine-tune the model from human feedback, researchers have developed InstructGPT [33]. Based on this, a language model called ChatGPT has been implemented which is capable of reasoning and conversational tasks very well. However, due to its inability to be fine-tuned based on text classification datasets, only ChatGPT-based text classification tasks with few-shot can be implemented at present, and the performance it achieves hardly exceeds that of fine-tuned models based on other open-source PLMs [34]. Despite the effectiveness of ChatGPT in various natural language processing tasks, it still has limitations in text classification. Although PLMs have shown promising results, they do not consider the distance information between samples and labels in supervised learning. we propose a novel method that focuses on extracting distance information between samples and labels to enhance the performance of text classification models.

## 3    Method

This section outlines the proposed ParaNet model framework. Given a text sequence, the sequence is initially passed to a Transformer Encoder for encoding. The corresponding label, which increases the sequence length using a prompt template, is also passed to another Transformer Encoder for encoding. Subsequently, a multi-headed attention mechanism is employed to learn the distance information between the two features.

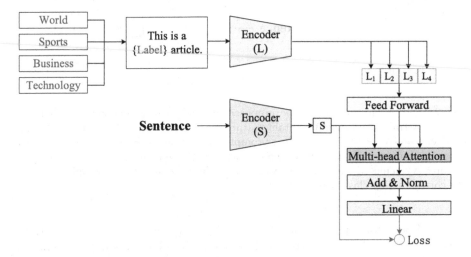

**Fig. 1.** The proposed method's general framework involves inputting sentences to an encoder to obtain sentence features $S$. For all labels, prompt templates are utilized to increase the length and input to another encoder to produce $L_j$. During training, an end-to-end training method is employed to achieve the closest distance between $S$ and the features produced by the corresponding labels.

## 3.1 Formalizations

The proposed framework for text classification involves a given original text sequence, denoted by $D = \{i_1, i_2, \cdots, i_L\}$, where $k$ represents the class label. The text sequence $D$ comprises $L$ tokens, representing a sequence of length $L$. Each token $i^L \in R^N$ is represented by an $N-$dimensional word embedding. The entire training process of the model defines a function $f_k : D \rightarrow \hat{K}$, with the output of the model being the final class $\hat{K}$. Therefore, for any text sequence $D$, we can obtain the final classification result $\hat{k}$ through the function $f_k : d \rightarrow \hat{k}$.

Figure 1 illustrates the general framework of ParaNet proposed in this paper. The model's input is a text sequence $D$, which generates word token $i^L$. Sentence features $S$ and label features $L_j (j = 1, 2, ...k)$ in the text sequence are extracted by two different Transformer Encoders, respectively. Each text sequence feature is represented by $[CLS]$, which represents the overall semantic information of the sentence. For each label, the label's length is increased using a prompt template, and then the label feature $L_j (j = 1, 2, ...k)$ is obtained after the encoder. Finally, a multi-head attention layer is employed to fuse the features $S$ and $L_j (j = 1, 2, ...k)$ to learn the distance information between samples and labels. During the training process, the model is trained to minimize the distance between

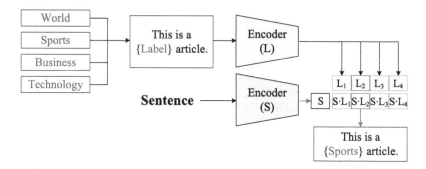

**Fig. 2.** The testing process of the proposed method involves passing each sentence through a trained encoder to obtain feature $S$. Different labels are passed through another trained encoder after adding prompt information to obtain different label features $L_j$. Subsequently, the distance $S \cdot L_j$ between the sentence and the different label features is calculated, and the label with the smallest distance to the sample is used as the corresponding classification result.

feature $S$ and corresponding label $L_j(j = 1, 2, ...k)$ and maximize the distance between it and other non-corresponding labels. For model testing, test samples are input to one encoder to obtain sentence features $S$. Meanwhile, the sequences obtained after adding data to all labels using the prompt template are input to another encoder to obtain label features $L_j(j = 1, 2, ...k)$. The distances between $S$ and $L_j(j = 1, 2, ...k)$ are then calculated separately, and the label with the smallest distance is used as the final classification probability, as depicted in Fig. 2.

### 3.2   Extraction of Text Features

The text sequence encoder inputs a text sequence $D$ and obtains the word embedding $E$ of the text sequence $D$. Subsequently, $E$ is input to a text encoder to obtain the feature representation of the text sequence. The feature extraction of the text sequence can be accomplished using different pre-trained models, such as BERT [6] or ELECTRA [30], among others. $S$ represents the text features extracted using Encoder, which accepts the input of the text sequence $D$ and obtains the text feature representation through as follows:

$$S = Encoder_S \left( D * \mathbf{w}^{\mathbf{T}} \right) \tag{1}$$

We use the hidden state of the last layer in Encoder as the feature representation. The input sequence $D$ is initially mapped to the word embedding layer to obtain the word embedding $E$ which is represented as follows:

$$E = D * \mathbf{w}^{\mathbf{T}} \tag{2}$$

To obtain the word embedding $E$, a matrix $\mathbf{w}$ is multiplied with the input sequence $D$. Subsequently, the word embedding $E$ is passed to the Encoder to obtain the features $S$ with relational information of the context. For a different label $C_j (j = 1, 2, ...k)$, a set of sentences $H_j (j = 1, 2, ...k)$ is obtained using a prompt template, represented as follows:

$$H_j = M(C_j) \tag{3}$$

Each sentence $H_j$ representing a label class is fed into a text encoder to obtain a feature representation representing the label, as follows:

$$L_j = Encoder_L(H_j) \tag{4}$$

$L_j$ passes through a feedforward layer and is then fed to a multi-head attention layer with sentence features $S$. The attention score is calculated to obtain a feature $Z_j$ that represents the distance information between the sentence and the different labels, as shown as follows:

$$Z_j = MLP(MultiHead(K, Q, V)) \tag{5}$$

where $Q$ represents sentence feature $S$ obtained by multiplying it with a trainable parameter sharing matrix $W^Q$, $K$ and $V$ are obtained by multiplying $L_j$ with two trainable parameter sharing matrices $W^K$ and $W^V$, respectively. $MultiHead$ denotes multi-head attention, and MLP represents a linear layer.

$Z_j$ is a feature vector that includes distance information between sentences and labels. During the training phase, $Z_j$ compares the distance with the feature $S$ that represents the sentence, and the calculation method is described in Sect. 3.3.

### 3.3    Loss Function

The training objective of ParaNet aims to minimize the distance between the sentence features $S$ and the corresponding feature vectors $Z_j$ that fuse the labels, while keeping the feature vectors of non-matching labels as far away as possible. This objective is achieved by constructing a loss function.

We utilize cosine similarity to measure the distance between two sets of vectors, and the distance $\rho(S, Z_j)$ between the sentence feature $S$ and the label feature $Z_j$ is defined as follows:

$$\rho(S, Z_j) = \frac{S \cdot Z_j}{\|S\| \times \|Z_j\|} \tag{6}$$

where $\|\cdot\|$ denotes the vector distance.

When $\rho(S, L_j) = 1$ indicates that the two sets of vectors are closest to each other, with the best similarity. Conversely, when $\rho(S, L_j) = -1$ indicates that the two sets of vectors are farthest from each other, with the worst similarity. The training objective of ParaNet is to minimize the distance between samples and labels of the same classes as close to 1 as possible and maximize the distance between labels of different classes close to $-1$. Therefore, the loss function is defined as follows:

$$Loss = \sum_{i=1}^{k} L\left(S, Z_j, y\right)^i \tag{7}$$

where $L\left(S, Z_j, y\right)$ is calculated as shown as follows:

$$L\left(S, Z_j, y\right) = \frac{(1-y)}{2}\left(\rho\left(S, Z_i\right) + 1\right)^2 + \frac{y}{2}\left(\rho\left(S, Z_i\right) - 1\right)^2 \tag{8}$$

Through end-to-end training, ParaNet minimizes the distance between samples and feature vectors generated by corresponding labels and maximizes the distance from feature vectors generated by non-corresponding labels.

## 4    Experiment

To evaluate the proposed method in this paper, experiments were conducted on widely used text classification datasets, including eight benchmark datasets.

**Table 1.** Statistics for the eight text classification datasets.

| Dataset | class | Avglen | size | template |
|---------|-------|--------|------|----------|
| AGNews | 4 | 45 | 127600 | This is a {label} article |
| MPQA | 2 | 3 | 10604 | The answer to the question is {label} |
| SUBJ | 2 | 23 | 9999 | The review is {label} |
| TREC | 6 | 10 | 5891 | This question is about the type of {label} |
| MR | 2 | 20 | 10662 | This movie review expresses a {label} sentiment |
| SST1 | 5 | 18 | 11855 | This movie review expresses a {label} sentiment |
| SST2 | 2 | 19 | 9613 | This movie review expresses a {label} sentiment |
| CR | 2 | 18 | 3770 | This customer review expresses a {label} sentiment |

### 4.1    Datasets

The proposed method was validated on the following eight datasets:

**AG News:** AG News is a news text dataset from academia that includes four categories. It uses the headline and description text of each news item and contains 127,600 samples [35].

**MPQA:** Multi-Perspective Question Answering (MPQA) is a viewpoint tendency dataset that uses the viewpoint tendency in MPQA as a classification task and contains two categories, positive and negative, including 10,604 samples from news articles from various news sources. Of these, 3,311 are positive texts and 7,293 are negative texts [36].

**SUBJ:** The Subjectivity dataset is an opinion dataset that classifies user comments according to whether they are subjective or objective. A sample is usually a sentence and includes 9,999 samples. These sentences are subjective and objective and are divided into two categories [37].

**TREC:** The TREC question dataset classifies sentences according to the question, which is divided into six categories, and it has 5,891 samples [38].

**MR:** The Movie Review (MR) dataset includes reviews of various released movies on the web by users. These reviews are used to identify different sentiment levels and are divided into two categories, positive and negative. It contains 5,331 positive and 5,331 negative data [39].

**SST1:** The Stanford Sentiment Treebank (SST) is also a movie review dataset that includes online reviews posted by movie users about movies. This dataset is divided into five categories according to the sentiment level of the sentences: very positive, positive, neutral, negative, and very negative. It contains 11,855 texts [40].

**SST2:** SST-2 is the same as SST-1, but it removes the neutral comments and then divides the comments into two categories, positive and negative. It contains 9,613 dichotomous texts [40].

**CR:** The Customer Review (CR) dataset includes customer reviews where each sample is labeled as positive or negative. It contains 3,770 samples [41].

This paper presents a summary of the statistics for each dataset in Table 1. The table includes the number of target classes, average sentence length, dataset size, and prompt template used by the proposed method.

## 4.2   Baselines

To verify the performance of the proposed algorithm, we conducted experiments comparing it to several recent benchmark algorithms, which are as follows:

**BERT:** Bert-base consists of 12 layers, 768 hidden layers, 12-attention heads, and 110M parameters. We use Bert-based Uncased, where Uncased indicates that the text is lowercase before WordPiece tokenization [6].

**ELECTRA:** ELECTRA-base does not mask the input but samples plausible tokens from the generator and then replaces the tokens on the original input. It has 110M parameters [30].

**RoBERTa:** RoBERTa has the same architecture as Bert-base, with minor adjustments to the embedding and settings for the Roberta pre-training model [42].

**Muppet:** Muppet-RoBERTa-base is a pre-trained model that is re-trained on top of RoBERTa, which has significant gains on small datasets [43].

### 4.3 Experimental Settings and Results

The experiments were conducted on a GPU server equipped with 2 Xeon 4210 CPUs, 128GB RAM, and 4 NVIDIA GV100 GPUs with 32GB*4=128G video memory. The proposed algorithm was implemented in Ubuntu 18.04 OS using the Python language and the PyTorch framework.

We fine-tuned the pre-trained models Bert-base, ELECTRA-base, RoBERTa-base, and Muppet-RoBERTa-base using the AdamW optimizer with a weight decay factor of 0.01. The model training was set to 30 epochs, and a learning rate of 1e-05 was used. The value of b was set to 1, and the dropout rate of all layers was set to 0.4. The batch size for all datasets was set to 64. In our experiments, the training set accounted for 80%, while the test and validation sets each accounted for 10%. We used accuracy as the evaluation metric.

The experimental results are summarized in Table 2 and Table 3. The black bold in the table indicates the best result. Table 2 focuses on the news classification, question classification, and subjectivity classification datasets, while Table 3 shows the four sentiment datasets. The notation $X + Y$ indicates the use of $X$ as the sentence feature extractor, with only $[CLS]$ used as the feature representation. $Y$ is used as the label feature extractor after adding the template, and in this paper, we use Bert-base as the label feature extractor. In all tables

**Table 2.** This paper presents the outcomes of a method proposed for text classification, which were compared against a baseline method on four datasets. The sentence feature extractor denoted as $X$ and the label feature extractor denoted as $Y$ were utilized in each block after incorporating a template. The label feature extractor exclusively used in this study is Bert-base.

| Method | AG News | MPQA | SUBJ | TREC | Avg. |
|---|---|---|---|---|---|
| BERT | 94.42± 0.20 | 91.06±0.18 | 96.48±0.26 | 94.63±0.64 | 94.15±0.32 |
| BERT+Y | 94.48±0.12 | 90.72±0.23 | 96.88±0.19 | 94.57±0.43 | 94.16±0.24 |
| RoBERTa | 94.92±0.30 | 91.36±0.20 | 97.30±0.42 | 95.59±0.27 | 94.79±0.30 |
| RoBERTa+Y | 94.95±0.49 | 91.40±0.53 | **97.36±0.18** | 95.72±0.39 | 94.86±0.40 |
| ELECTRA | 95.32±0.20 | 91.23±0.15 | 97.12±0.30 | 95.82±0.50 | 94.87±0.29 |
| ELECTRA+Y | **95.59±0.16** | 91.62±0.34 | 97.14±0.21 | 96.20±0.28 | 95.14±0.25 |
| Muppet | 95.35±0.22 | 92.83±0.15 | 97.16±0.18 | 96.26±1.05 | 95.40±0.40 |
| Muppet+Y | 95.52±0.11 | **92.96±0.11** | **97.36±0.23** | **97.15±0.84** | **95.75±0.32** |

of our experiments, accuracy(%) was utilized as the evaluation metric. The presented results in the tables denote the mean accuracy of five experiments, with their corresponding standard deviation values indicated alongside each outcome.

Table 2 provides several observations. Our proposed method achieves the optimal accuracy on the MPQA, SUBJ, and TREC datasets by using Muppet-RoBERTa-base as the word feature extractor. On the AG News dataset, using ELECTRA-base as the sentence feature extractor achieves the optimal accuracy. By using various pre-trained models, our method can improve the results of the baseline feature extractor. On these four datasets, the average accuracy improved from 94.80% to 94.98%, while the standard deviation decreased from 0.33 to 0.30. This validates the effectiveness of the proposed model and demonstrates that learning the distance of the sample and the label, and the distance information can improve the accuracy of text classification.

**Table 3.** The results of our proposed method compared to the baseline method on the four sentiment classification datasets. The experimental configuration remained consistent with Table 2.

| Method | MR | SST1 | SST2 | CR | Avg. |
|---|---|---|---|---|---|
| BERT | 87.45±0.18 | 52.49±0.13 | 93.55±0.17 | 92.31±0.42 | 81.45±0.23 |
| BERT+Y | 88.14±0.23 | 53.47±0.17 | 95.34±0.20 | 92.41±0.30 | 82.34±0.23 |
| RoBERTa | 90.00±0.41 | 54.26±0.13 | 94.96±0.28 | 92.68±0.61 | 82.98±0.36 |
| RoBERTa+Y | 90.15±0.39 | 56.25±0.30 | 95.84±0.16 | 92.73±0.30 | 83.74±0.29 |
| ELECTRA | 90.24±0.15 | 55.19±0.13 | 96.77±0.16 | 92.31±0.42 | 83.63±0.22 |
| ELECTRA+Y | 91.56±0.30 | 56.88±0.30 | 96.88±0.37 | 93.05±0.44 | 84.59±0.35 |
| Muppet | 94.41±0.22 | 59.51±0.19 | 95.38±0.12 | 92.94±0.61 | 85.56±0.29 |
| Muppet+Y | **94.43±0.14** | **59.56±0.30** | **97.29±0.16** | **94.80±0.72** | **86.52±0.33** |

Table 3 shows that on the four sentiment classification datasets, including MR, SST1, SST2, and CR, our method achieves the optimal accuracy when using Muppet-RoBERTa-base as the word feature extractor. The average accuracy improved by 0.90% from 83.40% to 84.30%. This indicates that the proposed method also showed better performance on the sentiment classification dataset.

Overall, on all eight datasets, our proposed method achieves significant performance improvement compared to the baseline method. This indicates that fusing the distance information between samples and different labels is very effective in extracting more complete semantic information.

### 4.4  Ablation Experiments and Analysis

The label paired with the sample is typically only one word, while the input text for PLM is a complete sentence. To bridge this distribution gap, we found that using a prompt template customized for each task can significantly improve classification performance. For example, using the prompt template "This is a {label} article." for the AG News dataset improved performance over just using the baseline label text. We conducted experiments on eight datasets, and the results showed that increasing the text length by customizing the prompt template for each task can improve classification performance, as shown in Table 4. The black bold in the table indicates the best result. The prompt templates utilized in our experiments are enumerated in the final column of Table 1.

Table 4 shows that adding prompt templates and increasing the length of the labels improves classification accuracy relative to the baseline method on all eight datasets. In each block, +*template* indicates the label after the template is used, and Bert-base is used as the label feature extractor in this paper. Accuracy (%) is the evaluation metric, and each result in the table is the average accuracy of five experiments with the standard deviation next to it.

**Table 4.** The results of our proposed method with and without added templates relative to the baseline approach. In each block, +*template* indicates the label after the template is used, and Bert-base is used as the label feature extractor in this paper.

| Dataset | BERT | RoBERTa | ELECTRA | Muppet |
|---|---|---|---|---|
| **AG News** | 94.38±0.21 | 94.62±0.36 | 95.41±0.19 | 95.40±0.13 |
| +*template* | **94.48±0.12** | **94.95±0.49** | **95.59±0.16** | **95.52±0.11** |
| **MPQA** | 90.62±0.34 | 91.21±0.39 | 91.60±0.60 | 92.02±0.22 |
| +*template* | **90.72±0.23** | **91.40±0.53** | **91.62±0.34** | **92.96±0.11** |
| **SUBJ** | 96.84±0.11 | 97.10±0.16 | 97.06±0.20 | 97.28±0.19 |
| +*template* | **96.88±0.19** | **97.36±0.18** | **97.14±0.21** | **97.36±0.23** |
| **TREC** | 94.53±0.25 | 95.69±0.09 | 95.96±0.33 | 97.11±0.27 |
| +*template* | **94.57±0.43** | **95.72±0.39** | **96.20±0.28** | **97.15±0.84** |
| **MR** | 88.09±0.48 | 89.96±0.15 | 90.81±0.39 | 94.24±0.32 |
| +*template* | **88.14±0.23** | **90.15±0.39** | **91.56±0.30** | **94.43±0.1** |
| **SST1** | 53.42±0.13 | 56.12±0.30 | 56.79±0.13 | 59.46±0.29 |
| +*template* | **53.47±0.17** | **56.25±0.30** | **56.88±0.30** | **59.56±0.30** |
| **SST2** | 95.23±0.32 | 95.73±0.37 | 96.63±0.38 | 97.19±0.16 |
| +*template* | **95.34±0.20** | **95.84±0.16** | **96.88±0.37** | **97.29±0.16** |
| **CR** | 92.15±0.30 | 92.36±0.22 | 92.84±0.42 | 94.43±0.42 |
| +*template* | **92.41±0.30** | **92.73±0.30** | **93.05±0.44** | **94.80±0.72** |

Our experimental results show that adding prompt templates and increasing the length of the labels improves classification accuracy regardless of which pretrained model is used. This illustrates that a complete sentence extracts richer semantic information from the text relative to a word for pre-training models. Therefore, adding prompt templates to the labels enables the model to make full use of the semantic knowledge of the pre-trained model, thereby improving text classification accuracy.

## 5   Conclusion

In this paper, we propose a novel Parallel Networks with Pre-trained Models that uses two different Transformer Encoders to extract sample features and label features separately to learn the distance information between samples and labels and improve text classification representation. Adding prompt templates to the labels enables the model to make full use of the semantic knowledge of the pre-trained model. Through extensive empirical experiments on eight benchmark datasets, we demonstrate the effectiveness of our proposed method.

**Acknowledgement.** This work was Sponsored by Natural Science Foundation of Shanghai(No.22ZR1445000) and Research Foundation of Shanghai Sanda University(No.2020BSZX005,No.2021BSZX006).

## References

1. Wu, Y., Li, J., Wu, J., Chang, J.: Siamese capsule networks with global and local features for text classification. Neurocomputing **390**, 88–98 (2020)
2. Wu, Y., Li, J., Song, C., Chang, J.: Words in pairs neural networks for text classification. Chin. J. Electron. **29**(3), 491–500 (2020)
3. Wu, Y., Li, J., Chen, V., Chang, J., Ding, Z., Wang, Z.: Text classification using triplet capsule networks. In: 2020 International Joint Conference on Neural Networks (IJCNN), pp. 1–7 (2020)
4. Wan, J., Lai, Z., Liu, J., Zhou, J., Gao, C.: Robust face alignment by multi-order high-precision hourglass network. IEEE Trans. Image Process. **30**, 121–133 (2020)
5. Wan, J., Xi, H., Zhou, J., Lai, Z., Pedrycz, W., Wang, X., Sun, H.: Robust and precise facial landmark detection by self-calibrated pose attention network. IEEE Trans. Cybern. (2021)
6. Devlin, J., Chang, M.W., Lee, K., Toutanova, K.: Bert: pre-training of deep bidirectional transformers for language understanding. In: Proceedings of the 2019 Conference of the North American Chapter of the Association for Computational Linguistics: Human Language Technologies, pp. 4171–4186 (2019)
7. Hong, S., Jang, T.Y.: Lea: Meta knowledge-driven self-attentive document embedding for few-shot text classification. In: Proceedings of the 2022 Conference of the North American Chapter of the Association for Computational Linguistics: Human Language Technologies, pp. 99–106 (2022)
8. Croce, D., Castellucci, G., Basili, R.: Gan-Bert: generative adversarial learning for robust text classification with a bunch of labeled examples. In: Proceedings of the 58th Annual Meeting of the Association for Computational Linguistics, pp. 2114–2119 (2020)
9. Qin, Q., Hu, W., Liu, B.: Feature projection for improved text classification. In: Proceedings of the 58th Annual Meeting of the Association for Computational Linguistics, pp. 8161–8171 (2020)

10. Mekala, D., Shang, J.: Contextualized weak supervision for text classification. In: Proceedings of the 58th Annual Meeting of the Association for Computational Linguistics, pp. 323–333 (2020)

11. Chen, Q., Zhang, R., Zheng, Y., Mao, Y.: Dual contrastive learning: text classification via label-aware data augmentation. arXiv preprint arXiv:2201.08702 (2022)

12. Paolini, G., et al.: Structured prediction as translation between augmented natural languages. In: International Conference on Learning Representations (2021)

13. Hu, S., et al.: Knowledgeable prompt-tuning: Incorporating knowledge into prompt verbalizer for text classification. In: Proceedings of the 60th Annual Meeting of the Association for Computational Linguistics, pp. 2225–2240 (2022)

14. Mueller, A., et al.: Label semantic aware pre-training for few-shot text classification. In: Proceedings of the 60th Annual Meeting of the Association for Computational Linguistics, pp. 8318–8334 (2022)

15. Chalkidis, I., Fergadiotis, M., Kotitsas, S., Malakasiotis, P., Aletras, N., Androutsopoulos, I.: An empirical study on large-scale multi-label text classification including few and zero-shot labels. In: Proceedings of the 2020 Conference on Empirical Methods in Natural Language Processing (EMNLP), pp. 7503–7515 (2020)

16. Zhang, Y., Yuan, C., Wang, X., Bai, Z., Liu, Y.: Learn to adapt for generalized zero-shot text classification. In: Proceedings of the 60th Annual Meeting of the Association for Computational Linguistics, pp. 517–527 (2022)

17. Gera, A., Halfon, A., Shnarch, E., Perlitz, Y., Ein-Dor, L., Slonim, N.: Zero-shot text classification with self-training. In: Conference on Empirical Methods in Natural Language Processing, pp. 1107–1119 (2022)

18. Zhang, T., Xu, Z., Medini, T., Shrivastava, A.: Structural contrastive representation learning for zero-shot multi-label text classification. In: Findings of the Association for Computational Linguistics: EMNLP 2022, pp. 4937–4947 (2022)

19. Min, S., Lewis, M., Hajishirzi, H., Zettlemoyer, L.: Noisy channel language model prompting for few-shot text classification. In: Proceedings of the 60th Annual Meeting of the Association for Computational Linguistics, pp. 5316–5330 (2022)

20. Zha, J., Li, Z., Wei, Y., Zhang, Y.: Disentangling task relations for few-shot text classification via self-supervised hierarchical task clustering. arXiv preprint arXiv:2211.08588 (2022)

21. Zhao, L., Yao, C.: EICO: improving few-shot text classification via explicit and implicit consistency regularization. In: Findings of the Association for Computational Linguistics: ACL 2022, pp. 3582–3587 (2022)

22. Zhang, H., Zhang, X., Huang, H., Yu, L.: Prompt-based meta-learning for few-shot text classification. In: Proceedings of the 2022 Conference on Empirical Methods in Natural Language Processing, pp. 1342–1357 (2022)

23. Wang, J., et al.: Towards unified prompt tuning for few-shot text classification. arXiv preprint arXiv:2205.05313 (2022)

24. Shnarch, E., et al.: Cluster & tune: boost cold start performance in text classification. In: Proceedings of the 60th Annual Meeting of the Association for Computational Linguistics, pp. 7639–7653 (2022)

25. Zhao, Y., et al.: Improving meta-learning for low-resource text classification and generation via memory imitation. In: Proceedings of the 60th Annual Meeting of the Association for Computational Linguistics, pp. 583–595 (2022)

26. Choi, H., Choi, D., Lee, H.: Early stopping based on unlabeled samples in text classification. In: Proceedings of the 60th Annual Meeting of the Association for Computational Linguistics, pp. 708–718 (2022)

27. Zhang, Z., et al.: Universal multimodal representation for language understanding. IEEE Trans. Pattern Anal. Mach. Intell. **01**, 1–18 (2023)

28. Smith, S., et al.: Using deepspeed and megatron to train megatron-turing NLG 530B, a large-scale generative language model. arXiv preprint arXiv:2201.11990 (2022)
29. Lan, Z., Chen, M., Goodman, S., Gimpel, K., Sharma, P., Soricut, R.: Albert: a lite BERT for self-supervised learning of language representations. In: International Conference on Learning Representations (2020)
30. Clark, K., Luong, M.T., Le, Q.V., Manning, C.D.: Electra: pre-training text encoders as discriminators rather than generators. arXiv preprint arXiv:2003.10555 (2020)
31. Yang, Z., Dai, Z., Yang, Y., Carbonell, J., Salakhutdinov, R., Le, Q.V.: XLNET: generalized autoregressive pretraining for language understanding. In: Proceedings of the 33rd International Conference on Neural Information Processing Systems, pp. 5753–5763 (2019)
32. Brown, T.B., et al.: Language models are few-shot learners. In: Proceedings of the 34th International Conference on Neural Information Processing Systems, pp. 1877–1901 (2020)
33. Ouyang, L., et al.: Training language models to follow instructions with human feedback. Adv. Neural. Inf. Process. Syst. **35**, 27730–27744 (2022)
34. Qin, C., Zhang, A., Zhang, Z., Chen, J., Yasunaga, M., Yang, D.: Is chatGPT a general-purpose natural language processing task solver? arXiv preprint arXiv:2302.06476 (2023)
35. Zhang, X., Zhao, J., LeCun, Y.: Character-level convolutional networks for text classification. In: Proceedings of the 28th International Conference on Neural Information Processing Systems, pp. 649–657 (2015)
36. Wiebe, J., Wilson, T., Cardie, C.: Annotating expressions of opinions and emotions in language. Lang. Resour. Eval. **39**, 165–210 (2005)
37. Pang, B., Lee, L.: A sentimental education: Sentiment analysis using subjectivity summarization based on minimum cuts. In: Proceedings of the 42nd Annual Meeting of the Association for Computational Linguistics, pp. 271–278 (2004)
38. Li, X., Roth, D.: Learning question classifiers. In: Proceedings of the 19th international conference on Computational Linguistics, pp. 1–7 (2002)
39. Pang, B., Lee, L.: Seeing stars: exploiting class relationships for sentiment categorization with respect to rating scales. In: Proceedings of the 43rd Annual Meeting on Association for Computational Linguistics, pp. 115–124 (2005)
40. Socher, et al.: Recursive deep models for semantic compositionality over a sentiment treebank. In: Proceedings of the 2013 Conference on Empirical Methods in Natural Language Processing, pp. 1631–1642 (2013)
41. Hu, M., Liu, B.: Mining and summarizing customer reviews. In: Proceedings of the Tenth ACM SIGKDD International Conference on Knowledge Discovery and Data Mining, pp. 168–177 (2004)
42. Liu, Y., et al.: Roberta: s robustly optimized BERT pretraining approach. arXiv preprint arXiv:1907.11692 (2019)
43. Aghajanyan, A., Gupta, A., Shrivastava, A., Chen, X., Zettlemoyer, L., Gupta, S.: Muppet: massive multi-task representations with pre-finetuning. In: Proceedings of the 2021 Conference on Empirical Methods in Natural Language Processing, pp. 5799–5811 (2021)

# Open Text Classification Based on Dynamic Boundary Balance

Ganlin Xu[1], Jianzhou Feng[1(✉)], and Qikai Wei[2]

[1] School of Information Science and Engineering, Yanshan University,
Qinhuangdao 066000, China
`fjzwxh@ysu.edu.cn`
[2] School of Computer and Communication Engineering, University of Science
and Technology Beijing, Beijing 100000, China

**Abstract.** Open classification is the problem where there exist some unseen/unknown classes in the test set, i.e., these unknown/unseen classes don't appear when the model is trained. Existing work often maps samples to high-dimensional space to make decisions, which leads to unobservable and inexplicable results. To address the issue, we shift perspectives to two-dimensional space and put forward a two-stage learning method built on the dynamic decision boundaries balance. We refer it to open classification with dynamic boundary balance (OCD2B). First, we construct a vanilla classifier via known classes with BERT model. Then, we use the prior knowledge of known classes to dynamically determine the decision boundaries between known classes and unknown classes in low-dimensional space. We propose a novel boundary loss function as a boundary balance strategy to reduce open space risk and empirical risk. Experimental results on two standard datasets show that our method achieves performance gain over existing methods, providing easily observable results. In particular, the larger the ratio of unseen classes is, the more obvious the performance advantage the model achieves.

**Keywords:** open text classification · two-stage learning method · decision boundary

## 1 Introduction

Traditional text classification task requires the test set to contain the same classes (known/seen classes) as the training set. [8] referred to it as a closed-world assumption. This opinion has been widely adopted in many fields of natural language processing, such as emotion analysis [21], spam recognition [15] and news classification [7]. However, the dynamic open environment frequently exits some scenes that the model never study before. It is critical to distinguish these unknown scenes as much as possible from the known scenes. For example, in a dialogue system, user intents are often complex and diverse, so the system cannot learn all the user intents. It is a challenging task because it is difficult

X. Yang et al. (Eds.): ADMA 2023, LNAI 14178, pp. 136–150, 2023.
https://doi.org/10.1007/978-3-031-46671-7_10

to obtain prior knowledge of unknown classes for lack of unknown class samples [12]. As a result, the classifier cannot identify unknown classes during testing. A better classifier is required not only to classify known classes but also to obtain a novel mechanism to discover unknown classes. [8] referred to it as open (world) classification.

According to [20], open text classification is a $m+1$ classification task, where $m$ is the number of known classes. $l_i$ is the $i^{th}$ known class label given a label set $L = \{l_1, l_2, ..., l_i, ..., l_m\}$ of a training set. The labels of all unknown classes are defined as $l_{m+1}$; thus, all classes that do not belong to $L$ are labeled as unknown classes. Our task is to classify the $m$-class known classes into their corresponding classes correctly while identifying the $(m+1)^{th}$ class as suggested in [17, 20, 22], where the $(m+1)^{th}$ class represents the unknown class.

In recent years, some progress has been made in the research on open text classification, in which the primary difficulty is how to discover the unknown class. To this end, some methods in the literature utilize fixed or adaptive thresholds for discovering unknown classes [1, 2, 6, 9, 16, 19, 25, 26]. Besides, some models treat unknown classes as outliers and use the outlier detection algorithm, such as LOF, to detect these outlier samples [4, 12, 27]. Although the above work achieved excellent performance, they involve making decisions on samples in high-dimensional space, leading to final results unobservable and inexplicable. Because the plane decision boundary in the two-dimensional space is more visible than the curved decision boundary in high-dimensional representation space, we shift perspectives to a two-dimensional plane and propose a two-stage method based on the dynamic decision boundary balance. To perform open text classification, the model utilizes the prior knowledge of the known classes to dynamically adjust the decision boundary on the sigmoid function.

The main contributions are as follows:

- We propose OCD2B, a simple yet effective approach for open text classification based on dynamic balance of decision boundary, which maps samples to a two-dimensiona space and provides a better observable and explicable results.
- We propose a novel boundary loss function to dynamically adjusts the decision boundaries by using the prior knowledge of known classes.
- The experimental results on two standard datasets show that our method exhibits better performance than the previous methods.

## 2   Method

The structure of OCD2B is presented in Fig. 1 and the whole model includes two parts. The classification net classifies known classes, and the open classifier can identify unknown classes using a dynamic boundary balance strategy. These components work together to complete the open text classification.

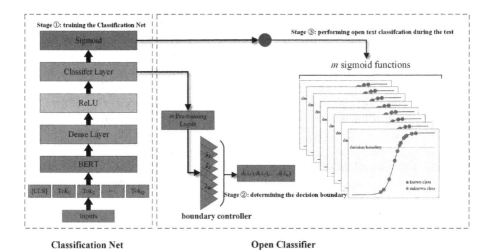

**Fig. 1.** The model architecture of OCD2B. The classification net is a vanilla classifier, which is used to classify known classes and provide prior knowledge of known classes for decision boundaries. The open classifier utilizes the prior knowledge of known classes and dynamically adjusts the decision boundary, which aimed to distinguish the unknown classes. The classification net and open classifier work together to perform open text classification.

## 2.1 Classifying Known Classes by Classification Net

The classification net is constructed based on the BERT model, and it is used to classify known classes and provide the prior knowledge of known classes for decision boundaries. The architecture of the classifier is shown on the left of Fig. 1, we train the classification net in the stage ①. The token $[CLS, Tok_1, Tok_2, ..., Tok_Q]$ is input into the classification net, and then output token embedding $[C, T_1, Tok_2, ..., Tok_Q] \in R^{(Q+1) \times H}$. As suggested in [25], we perform mean-pooling on these token embeddings to obtain averaged vector $\boldsymbol{h}$:

$$\boldsymbol{h} = mean - pooling([C, T_1, T_2, .., T_Q]) \tag{1}$$

where $Q$ is the length of the sequence, $H$ is the size of the hidden layer.

Then, we lower $\boldsymbol{h}$ to a $m$-dimension dense vector $p_{1:m}$ via two fully linear module, one intermediate ReLU activation function and one outer sigmoid activation function $\sigma$.

$$p_{1:m} = \sigma(W_2(ReLU(W_1\boldsymbol{h} + b_1)) + b_2) \tag{2}$$

where $p_{1:m}$ is the probability sequence of one sample corresponding to each known class. $m$ is the number of seen/training categories. $W_1$, $b_1$, $W_2$ and $b_2$ are trainable weights, respectively. We select the binary cross-entropy loss function $\mathcal{L}_c$, as follows:

$$\mathcal{L}_c = -\sum_{i=1}^{n}\sum_{j=1}^{m}\mathbb{I}(t_i = l_i)log(p(t_i = l_i)) + (1 - \mathbb{I}(t_i \neq l_i))log(1 - p(t_i = l_i)) \quad (3)$$

where $n$ is batch size, $t_i$ is the real sample label, $l_i$ is the expected sample label, and $\mathbb{I}(\cdot)$ is defined as follows:

$$\mathbb{I}(\cdot) = \begin{cases} 1, t_i = l_i \\ 0, t_i \neq l_i \end{cases} \quad (4)$$

As shown in Fig. 1, after training the classification net, we extract the $m$ dimension pre-training logits from the classifier layer as the prior knowledge of known classes to determine the decision boundaries.

## 2.2   Open Classifier with Dynamic Boundary Balance

After training the classification net, the open classifier can execute the boundary balance strategy to obtain the decision boundaries, which distinguish between known classes and unknown classes. [20] points out that the probability distributions of known classes are often in the "higher" part of the sigmoid function, but output probability distributions of unknown classes without participating in the training of classification net are in the "lower" part. Based on this assumption and premise, open text classification can be implemented by introducing a dividing line as the decision boundary on the sigmoid function as shown in Fig. 2.

**Boundary Loss Function.** Pre-training logits that classification net output are input into the *boundary controller* to adaptively determine decision boundaries $d(\lambda) = <d(\lambda_1), d(\lambda_2),..., d(\lambda_m)>$ for every known class. Specifically, for *boundary controller*, we first randomly initialized decision threshold $\lambda = <\lambda_1, \lambda_2, ..., \lambda_m>$. Next, we propose a boundary loss function $\mathcal{L}_\lambda$ to dynamically adjust the $\lambda$ in the stage ②. Then, through the adaptive training, we obtain the decision boundaries $d(\lambda)$ to distinguish between known classes and unknown classes. The boundary loss function $\mathcal{L}_\lambda$ is as follows:

$$\mathcal{L}_\lambda = \sum_{i=1}^{n} k\delta(y_i - \lambda_{y_i}) + (1 - \delta)(\lambda_{y_i} - y_i) \quad (5)$$

where $y_i$ is the output of train set at the classifier layer, $\lambda_{y_i}$ denotes the decision threshold of the known class corresponding to $y_i$, and $n$ is batch size. During the initialization of $\lambda_{y_i}$, its range is within $(-\infty, +\infty)$, which conforming to the Gaussian distribution. $\delta$ is defined as follows:

$$\delta = \begin{cases} 1, y_i \geq \lambda_{y_i} \\ 0, y_i < \lambda_{y_i} \end{cases} \quad (6)$$

During the training of the open classifier, $\lambda$ is updated by the following:

$$\lambda = \lambda - \eta\frac{\partial \mathcal{L}_\lambda}{\lambda} \quad (7)$$

where $\eta$ is the learning rate. we dynamically select $\lambda$ by Algorithm 1.

---

**Algorithm 1.** Decision Threshold Selection Algorithm

---

**Require:** initial decision threshold $\lambda_{1:m}$ and token $[CLS, Tok_1, Tok_2, ..., Tok_n]$ of sentences.

**Ensure:** final decision threshold $\grave{\lambda}_{1:m}$.

1: **while** minimize $\mathcal{L}_\lambda$ **do**
2:     $[CLS, Tok_1, Tok_2, ..., Tok_n]$ is input to pretrained classification net to obtain the $m$ Pre-training logits.
3:     $m$ Pre-training logits are input to boundary controller.
4:     compute boundary loss $\mathcal{L}_\lambda$ by Eq 5.
5:     update $\grave{\lambda}_{1:m} \leftarrow \lambda_{1:m}$ by Eq 7 during back propagation.
6: **return** $\grave{\lambda}_{1:m}$

---

**The Role of Decision Boundary.** The adaptive decision boundary can balance open space risk and empirical risk to perform open classification. We assume that if $(y_i-\lambda_i) \geq 0$, the known samples will be below the corresponding decision boundaries and identified as unknown classes, which will cause the empirical risk. However, if $(\lambda_i-y_i) < 0$, more unknown samples will be above the corresponding decision boundaries and identified as known classes. So we need "move up" the decision boundaries to reduce the open space risk.

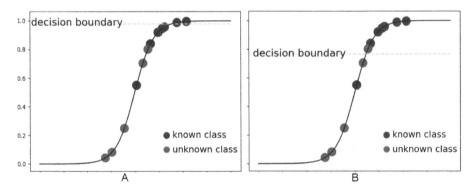

**Fig. 2.** (A) shows $k = 1$, and (B) shows $k = 4$. Blue points represent unknown classes, and purple points represent known classes.

In Eq. 5, we define $k$ as a balance factor to make decision boundaries adaptive to known class space. The left loss $(y_i-\lambda_i)$ in Eq. 5 will increase with the increase in $k$ and reduce the open space risk. For example, as shown in Fig. 2, when $k = 1$, a large number of known class samples are under the decision boundary, which affects the classification performance of the model. When $k = 4$, the decision boundary will "move down", and it can better distinguish known classes and unknown classes. However, If the $k$ is too large, more unknown classes will be identified as known classes, which leads to empirical risk. Thus, compared to other models, we shift perspectives from high-dimensional space to two-dimensional sigmoid plane, which provides a better observable and explicable results.

The "boundary loss" is calculated repeatedly through Eq. 5 to minimize $\mathcal{L}_\lambda$. Finally, the decision boundary can achieve balance between known and unknown classes. It can reduce the open space risk and empirical risk at the same time.

**Open Classification.** After determining the decision boundaries, each known class corresponds to a decision boundary $d(\lambda_i)$. In stage ③, the output probability of the unknown classes becomes below the corresponding decision boundary, and the known classes become above the decision boundary. If the predicted probability value of a test sample is less than the decision boundary of the corresponding known class, then it belongs to the unknown class. Otherwise, it belongs to the known class with the highest probability value. The formula for implementing open classification is as follows:

$$\hat{y} = \begin{cases} unknown\ class, & if\ d(y_i) < d(\lambda_{y_i}) \\ argmax_{l_i \in L} d(\lambda_{y_i}), & otherwise \end{cases} \tag{8}$$

where $L$ denotes the set of labels for known classes, $m$ denotes the number of training classes, and $d(\cdot)$ is the sigmoid function.

The sigmoid activation function has three advantages. First, it can make the range of the decision boundary within $(0, 1)$. Second, it is totally differentiable with different $\lambda$. Finally, it can introduce nonlinear factors and be convenient for derivation.

## 3  Expriment

In this section, two datasets are firstly introduced. Then we introduce evaluation metrics and experimental settings. Next, we introduce the six baselines. In Sect. 4.4, we analyzed the experimental results on two datasets.

### 3.1  Datasets

Two datasets are used to better compare with other models to complete the experiments. Both datasets belong to the short text dataset.

- **OOS**   It is a dataset of semantic intention classification, covering common intent in daily life [11]. It contains 150 classes. Each class consists of 150 labeled sentences. We select 100 sentences as the training set, 20 sentences as the validation set and 30 sentences as the test set for each class. The maximum token length in the dataset is 28, and the average token length is 8.31.
- **BANKING**   It is a fine-grained problem consultation dataset in the banking field [5]. It contains 77 classes, and the number of queries in each class is different, with a total of 13083 queries. We select 9003 queries as the training set, 1000 queries as the validation set, and 3080 queries as the test set. The maximum token length of the queries is 79, and the average token length is 11.91.

## 3.2 Evaluation Metrics and Experimental Settings

We hold that open text classification is an extension of multi-text classification. Previous works conducted by [20] and [12] take macro-F1 score as the only evaluation metric. In this paper, we use macro-F1 score and accuracy as evaluation metrics.

In the experiment, This study varies the number of training classes and uses 25%, 50%, and 75% classes for training and all classes for testing. Taking OOS as an example, for 25% classes, we use 38 classes for training and all 150 classes for testing. We count 10 times and average the results in every setting. The experimental results are shown in Tables 1 and 2, and the best results are highlighted in bold.

We use the BERT model (BERT-uncased, with a 12-layer transformer, 768 hidden sizes, and 12 self-attention heads) implemented by Pytorch. The learning rate of the classification net is 2e-5, and the learning rate of an open classifier is 0.05. The batch size during the training is 64, and the batch size during evaluation and testing is 32. We use Adam as an optimizer in the classification net and open classifier. In addition, We utilize a grid search algorithm for selecting $k$ and our method achieves the best results when $k = 14$.

## 3.3 Baselines

Our OCD2B is compared with the following baseline model and the experimental results are shown in Tables 1 and 2, respectively.

- **MSP**    The model utilizes the probability score by softmax as the classification basis after the last linear layer and then selects the class with the highest probability for comparison at 0.5 [9]. If it is lower than 0.5, then it is judged as an unknown class; otherwise, it belongs to one of the known classes.
- **DOC**    The algorithm replaces softmax with sigmoid and determines the threshold for each known class based on the statistical method as the decision boundary of each class to find unknown classes [20].
- **OpenMax**    OpenMax is an open set recognition by CNNs with a softmax output layer in computer vision, we adapt it for open text classification [3]. Firstly, it uses logits as the feature space and fits a Weibull distribution. Then, it recalibrates the confidence scores with the OpenMax Layer to perform open text classification.
- **LOF**    The model checks the low-density outlier samples as the unknown samples in a density-based manner [4].
- **DeepUnk**    It uses margin loss to increase inter-class variance and reduce intra-class variance and then uses the outlier detection algorithm LOF to find new classes [12].
- **($K$+1)-Way**    This is the method of data augmentation, mainly using a combination of inliers and a sampling of other data as outliers for data augmentation [24].
- **ARPL**    The model learns reciprocal point representations using maximizing the difference between unknown and known instances.

- **ADB**      The BERT model is used to extract text features, and the mean value of the vector is used as the class center [25]. A post-processing method of the defined loss function is proposed to determine the decision boundary. Finally, it carries out open classification by calculating the distance between data points and each class.
- **OCD2B** ($k = 1$)      It is the basic OCD2B ($k = 1$). We compare it with conventional OCD2B ($k = 14$) to show the effect of balance factor $k = 1$ for the selection of decision boundaries.

**Table 1.** Accuracy and macro-F1 score of open classification with different known class proportions on OOS.

| Methods | 25% | | 50% | | 75% | |
|---|---|---|---|---|---|---|
| | Accuracy | F1 | Accuracy | F1 | Accuracy | F1 |
| MSP | 47.02 | 47.62 | 62.96 | 70.41 | 74.07 | 82.38 |
| DOC | 74.97 | 66.37 | 77.16 | 78.26 | 78.73 | 83.59 |
| OpenMax | 68.50 | 61.99 | 80.11 | 80.56 | 76.80 | 73.16 |
| LOF | 87.77 | 78.13 | 85.22 | 83.86 | 85.07 | 87.20 |
| DeepUnk | 81.43 | 71.16 | 83.35 | 82.16 | 83.71 | 86.23 |
| ($K$+1)-way | 86.98 | 76.58 | 83.71 | 82.85 | 85.31 | 87.90 |
| ARPL | 82.25 | 71.62 | 78.33 | 79.59 | 82.66 | 86.80 |
| ADB | 87.59 | 77.19 | 86.54 | 85.05 | **86.32** | **88.53** |
| OCD2B ($k = 1$) | 90.27 | 72.49 | 84.10 | 74.53 | 73.33 | 69.29 |
| OCD2B | **91.97** | **81.63** | **88.86** | **85.42** | 85.94 | 86.76 |

**Table 2.** Accuracy and macro-F1 score of open classification with different known class proportions on BANKING.

| Methods | 25% | | 50% | | 75% | |
|---|---|---|---|---|---|---|
| | Accuracy | F1 | Accuracy | F1 | Accuracy | F1 |
| MSP | 43.67 | 50.09 | 59.73 | 71.18 | 75.89 | 83.60 |
| DOC | 56.99 | 58.03 | 64.81 | 73.12 | 76.77 | 83.34 |
| OpenMax | 49.94 | 54.14 | 65.31 | 74.24 | 77.45 | 84.07 |
| LOF | 66.73 | 63.38 | 71.13 | 76.26 | 77.21 | 83.64 |
| DeepUnk | 64.21 | 61.36 | 65.31 | 74.24 | 78.53 | 84.31 |
| ($K$+1)-way | 75.43 | 68.31 | 74.66 | 78.13 | 79.90 | 85.22 |
| ARPL | 76.50 | 63.77 | 75.29 | 78.24 | 79.26 | 85.18 |
| ADB | 78.85 | 71.62 | 78.86 | 80.90 | 81.08 | 85.96 |
| OCD2B ($k = 1$) | **85.89** | 65.98 | 79.20 | 76.96 | 69.38 | 73.59 |
| OCD2B | 80.81 | **73.11** | **80.31** | **81.65** | **81.52** | **86.20** |

### 3.4   Experimental Results and Analysis

The results of BANKING and OOS are shown in Tables 1 and 2, respectively. The tables show the following observations:

1. Under all settings, OCD2B outperforms other baselines except for ADB by a large margin, which proves the effectiveness of our method.
2. For 25% setting, our model achieved significant performance advantages compared to other baselines. The accuracy and f1-score are ahead of other models on OOS. On BANKING, the f1-score is higher than other baselines but accuracy (80.81) is less than variant OCD2B ($k = 1$) (85.89). We analyze that for $k = 1$, the decision boundaries are in a "higher" position, leading to some known classes being incorrectly classified as unknown classes but a large number of unknown classes is correctly recognized. As a consequence, the accuracy score is very high but the f1-score is very low.
3. For 50% settings, our model is significantly ahead of others. Compared with the second-best results, our method improves accuracy on OOS by 2.32%, on BANKING by 1.45% and improves f1-score on OOS by 0.37%, on BANKING by 0.75%. We hold that the selection of decision boundaries plays an important role in the setting, which considerably reduces the inaccurate classification where one known class is incorrectly identified as other known classes or unknown classes are incorrectly identified as known classes.
4. For 75% settings, our method still outperforms most baselines but is slightly worse than some models. For example, OCD2B is slightly worse than those ADB on OOS but acquires advantages on BANKING. We analyze that the reason is that the traditional classification performance of the classification net declines with the increase of known classes. That is, some known classes will be classified as other known classes. As displayed in Fig. 3, with the number of known categories increasing, the model's classification performance drops (Table 3).

**Table 3.** Accuracy of classification net perfroming traditonal classfication on validation set and test set, respectively. Traditional classification refer to test set and training set have the same class space.

| Methods | 25% | | 50% | | 75% | |
|---|---|---|---|---|---|---|
| | OOS | BANKING | OOS | BANKING | OOS | BANKING |
| Validation set | 98.55 | 98.88 | 97.87 | 96.09 | 95.94 | 94.30 |
| Test set | 97.69 | 95.49 | 96.75 | 93.63 | 95.78 | 92.04 |

## 4   Discussion and Analysis

### 4.1   Fine-Grained Expriments

Further, we also perform fine-grained experiments. Tables 4 and 5 show the macro F1-score on open intent and known intents respectively. We can observe that our method still achieves better performance in most settings compared with other models. In the 75% setting on OOS, our model (86.68) is slightly worse than but close to ADB on known classes (88.58). In addition, the basic OCD2B ($k = 1$) achieves the best score in the 25% and 50% settings on the open class of BANKING. That is because the decision boundaries are "higher" position the same as the 25% setting of open classification.

**Table 4.** Results of open classification with different known class proportions on OOS. "Known" and "Open" denote the macro F1-score over known classes and open class, respectively.

| Methods | 25% | | 50% | | 75% | |
|---|---|---|---|---|---|---|
| | Known | Open | Known | Open | Known | Open |
| MSP | 47.53 | 50.88 | 70.58 | 57.62 | 82.59 | 59.08 |
| DOC | 65.96 | 81.98 | 78.25 | 79.00 | 83.69 | 72.87 |
| OpenMax | 61.62 | 75.76 | 80.54 | 81.89 | 73.13 | 76.35 |
| LOF | 77.77 | 91.96 | 83.81 | 87.57 | 87.24 | 82.81 |
| DeepUnk | 70.73 | 87.33 | 82.11 | 85.85 | 86.27 | 81.15 |
| ($K$+1)-way | 76.19 | 91.44 | 82.82 | 85.84 | 87.95 | 82.39 |
| ARPL | 71.19 | 87.83 | 79.59 | 79.48 | 86.89 | 77.23 |
| ADB | 76.80 | 91.84 | 85.00 | 88.65 | **88.58** | 83.92 |
| OCD2B ($k = 1$) | 71.92 | 94.07 | 74.35 | 88.14 | 69.24 | 75.24 |
| OCD2B | **81.44** | **94.93** | **85.35** | **91.07** | 86.68 | **84.61** |

**Table 5.** Results of open classification with different known class proportions on BANKING. "Known" and "Open" denote the macro F1-score over known classes and open class, respectively.

| Methods | 25% | | 50% | | 75% | |
|---|---|---|---|---|---|---|
| | Known | Open | Known | Open | Known | Open |
| MSP | 50.55 | 41.43 | 71.97 | 41.19 | 84.36 | 39.23 |
| DOC | 57.85 | 61.42 | 73.59 | 55.14 | 83.91 | 50.60 |
| OpenMax | 54.28 | 51.32 | 74.76 | 54.33 | 84.64 | 50.85 |
| LOF | 62.89 | 72.64 | 76.51 | 66.81 | 84.15 | 54.19 |
| DeepUnk | 60.88 | 70.44 | 77.74 | 69.53 | 84.75 | 58.54 |
| ($K$+1)-way | 67.61 | 81.52 | 78.29 | 72.38 | 85.62 | 62.13 |
| ARPL | 62.75 | 83.17 | 78.36 | 73.55 | 85.63 | 59.34 |
| ADB | 70.94 | 84.56 | 80.96 | 78.44 | **86.29** | 66.47 |
| OCD2B ($k = 1$) | 64.66 | **91.18** | 76.85 | **81.04** | 73.83 | 60.01 |
| OCD2B | **72.20** | 85.90 | **81.76** | 80.26 | 86.16 | **68.62** |

## 4.2   Analysis on Balance Factor $k$

Our OCD2B model has a balance factor $k$ to control how model select a proper decision boundaries (Sect. 2.2). We analyze the effect of $k$ in Fig. 3. As seen, too small or large $k$ will cause performance degradation and a moderate $k$ around (13–15) yields the highest performance. From the plots, we also see that balance factor $k$ is insensitive to the datasets, justifying our design of Eq. 5.

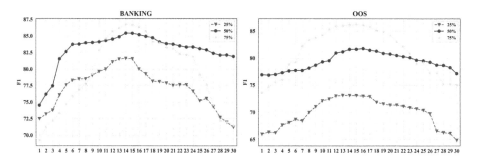

**Fig. 3.** Performances on the test set of different balance factor $k$.

## 4.3   Convergence Rate of Decision Boundary

As shown in Fig. 4, we record the trianing boundary loss $\mathcal{L}_\lambda$, accuracy and f1-score of open classifier during training for 50% known classes on BANKING. Decision boundaries tend to converge when f1-score on validation set reaches the peak. We can observe that our method make decision boundaries converge quickly by approximately 10 epochs.

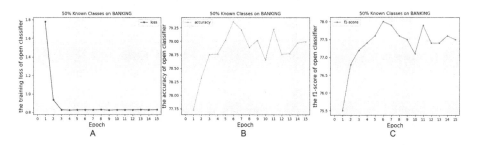

**Fig. 4.** (A), (B) and (C) shows the training loss, accuracy and f1-score of open classifier during training, respectively.

# 5   Related Work

There are some previous methods for open classification. [18] initially propose the concept of open space risk in computer vision to evaluate open classification. They recognize unknown images using the SVM's hyperplane of binary classification. The concept of open space risk was subsequently applied to the field of natural language processing. [17] fit the probability distributions for each class based on statistical Extreme Value Theory (EVT) and use a Weibull-calibrated multiclass SVM for an open classification task. [10] build a Weibull-calibrated SVM classifier using EVT that further improves the performance. However, these methods determine decision boundaries by using unknown class samples. [8] reduce the open space risk by setting the fixed boundary of each known class in the sphere. However, traditional machine learning [8,18] only focused on the n-gram information among sentences instead of high dimensional semantic features. Results show that the performance of these methods is poorer than the deep learning method [14].

Because machine learning is limited in capturing high-level semantic features, many researchers employ deep neural networks to solve open classification tasks. [3] propose a new method called OpenMax in computer version for open set recognition and one weak assumption is that it still needs prior knowledge of unknown classes. [14] extract text features by Convolutional Neural Network (CNN), and then the distance between the sentence and the class center was mapped into a probability value through the Weibull distribution. [23] focus on the unsupervised out-of-domain detection. They propose a supervised contrastive learning objective to minimize intra-class variance and maximize inter-class variance. To effectively find new classes, [8] and [9] utilize fixed thresholds or boundaries to distinguish between unknown and known classes, which does not explore the distribution of different classes and leading to poor performance. [1] proposes a variational auto-encoder to find inputs with new contexts independent of distribution, which solves some challenges for OOD intent detection. [2] uses negative and adaptive instances to produce effective comprehensive open classes without prior knowledge and additional datasets.

Further, some methods begin by attempting adaptive thresholds for detecting unknown classes. [20] replaced softmax with sigmoid in the last layer of CNN, and a threshold is determined for each known class as the decision boundary based on statistics. [16] transforms the original problem into a pair-wise matching problem and outputs how probable two samples belong to the same class. [19] takes advantage of the sequence-to-sequence language model BART to create distributionally shifted examples from the training examples, which are aimed at learning an open representation. [25] utilize the BERT model to extract text features, and then propose a post-processing method to reduce open space and empirical risk. [6] treat the open classification task with respect to multi-class aggregation and model the extra class with related points. [26] models textual distance-aware representations and variable decision boundaries to solve open text classification.

Some methods regard unknown classes as outliers and adopt outlier detection techniques, such as LOF [4], to find these samples. [12] utilize margin loss to increase the inter-class variance and reduce the intra-class variance for unknown intent detection. Then density-based detection algorithm called local outlier factor LOF is used to detect unknown classes. However, it does not take open space into consideration for distinguishing the open intent. [13] propose a post-processing method to extract text features using a neural network to discover the unknown intention of a dialogue system. [27] start from the nature of out-of-domain intent classification and further utilize K-Nearest Neighbors of in-domain intents to obtain discriminative semantic features for out-of-domain detection.

## 6   Conclusion

In this paper, we propose a simple yet effective method called OCD2B for open text classification, which makes decisions in low-dimensional space and provided better explanation results. First, we utilize the BERT to construct a traditional classifier classification net that can effectively separate known classes from unknown classes. Then, on this basis, we utilize the trained threshold to obtain the decision boundary to perform open text classification. Finally, we showed that OCD2B performs better than the state-of-the-art methods from both text classification domains.

Subsequently, we plan to continue to improve the accuracy of traditional classification when maintaining the ability to "separate" known classes from unknown classes. We also consider the meta-learning mechanism. The ability of lifelong learning is added based on our model. When sufficient unknown classes of a certain class are learned, the model adds it to the known class set to have the learning ability of "people". More importantly, we will continue to explore the classification of unknown samples that belongs to unknown classes.

**Acknowledgements.** This work is supported by the Zhongyuanyingcai program-funded to central plains science and technology innovation leading talent program (No. 204200510002), the General program of Hebei Natural Science Foundation (No. F2022203028), Program for Top 100 Innovative Talents in Colleges and Universities of Hebei Province (CXZZSS2023038), the General program of National Natural Science Foundation of China (No. 62172352) and the Central leading local science and Technology Development Fund Project (No. 226Z0305G).

## References

1. Akbari, M., Mohades, A., Shirali-Shahreza, M.H.: A hybrid architecture for out of domain intent detection and intent discovery. arXiv preprint arXiv:2303.04134 (2023)
2. Bai, K., et al.: Open world classification with adaptive negative samples. arXiv preprint arXiv:2303.05581 (2023)
3. Bendale, A., Boult, T.E.: Towards open set deep networks. In: Proceedings of the IEEE Conference on Computer Vision and Pattern Recognition, pp. 1563–1572 (2016)

4. Breunig, M.M., Kriegel, H.P., Ng, R.T., Sander, J.: Lof: identifying density-based local outliers. In: Proceedings of the 2000 ACM SIGMOD International Conference on Management of Data, pp. 93–104 (2000)
5. Casanueva, I., Temčinas, T., Gerz, D., Henderson, M., Vulić, I.: Efficient intent detection with dual sentence encoders. arXiv preprint arXiv:2003.04807 (2020)
6. Chen, G., Peng, P., Wang, X., Tian, Y.: Adversarial reciprocal points learning for open set recognition. IEEE Trans. Pattern Anal. Mach. Intell. 44(11), 8065–8081 (2021)
7. Dilrukshi, I., De Zoysa, K., Caldera, A.: Twitter news classification using SVM. In: 2013 8th International Conference on Computer Science & Education, pp. 287–291. IEEE (2013)
8. Fei, G., Liu, B.: Breaking the closed world assumption in text classification. In: Proceedings of the 2016 Conference of the North American Chapter of the Association for Computational Linguistics: Human Language Technologies, pp. 506–514 (2016)
9. Hendrycks, D., Gimpel, K.: A baseline for detecting misclassified and out-of-distribution examples in neural networks. arXiv preprint arXiv:1610.02136 (2016)
10. Jain, L.P., Scheirer, W.J., Boult, T.E.: Multi-class open set recognition using probability of inclusion. In: Fleet, D., Pajdla, T., Schiele, B., Tuytelaars, T. (eds.) ECCV 2014. LNCS, vol. 8691, pp. 393–409. Springer, Cham (2014). https://doi.org/10.1007/978-3-319-10578-9_26
11. Larson, S., et al.: An evaluation dataset for intent classification and out-of-scope prediction. arXiv preprint arXiv:1909.02027 (2019)
12. Lin, T.E., Xu, H.: Deep unknown intent detection with margin loss. arXiv preprint arXiv:1906.00434 (2019)
13. Lin, T.E., Xu, H.: A post-processing method for detecting unknown intent of dialogue system via pre-trained deep neural network classifier. Knowl.-Based Syst. 186, 104979 (2019)
14. Neal, L., Olson, M., Fern, X., Wong, W.K., Li, F.: Open set learning with counterfactual images. In: Proceedings of the European Conference on Computer Vision (ECCV), pp. 613–628 (2018)
15. Puniškis, D., Laurutis, R., Dirmeikis, R.: An artificial neural nets for spam e-mail recognition. Elektronika ir Elektrotechnika 69(5), 73–76 (2006)
16. Qin, Q., Hu, W., Liu, B.: Text classification with novelty detection. arXiv preprint arXiv:2009.11119 (2020)
17. Scheirer, W.J., Jain, L.P., Boult, T.E.: Probability models for open set recognition. IEEE Trans. Pattern Anal. Mach. Intell. 36(11), 2317–2324 (2014)
18. Scheirer, W.J., de Rezende Rocha, A., Sapkota, A., Boult, T.E.: Toward open set recognition. IEEE Trans. Pattern Anal. Mach. Intell. 35(7), 1757–1772 (2012)
19. Shu, L., Benajiba, Y., Mansour, S., Zhang, Y.: Odist: open world classification via distributionally shifted instances. In: Findings of the Association for Computational Linguistics: EMNLP 2021, pp. 3751–3756 (2021)
20. Shu, L., Xu, H., Liu, B.: Doc: deep open classification of text documents. arXiv preprint arXiv:1709.08716 (2017)
21. Vo, B.K.H., Collier, N.: Twitter emotion analysis in earthquake situations. Int. J. Comput. Linguist. Appl. 4(1), 159–173 (2013)
22. Xu, H., Liu, B., Shu, L., Yu, P.: Open-world learning and application to product classification. In: The World Wide Web Conference, pp. 3413–3419 (2019)
23. Zeng, Z., et al.: Modeling discriminative representations for out-of-domain detection with supervised contrastive learning. arXiv preprint arXiv:2105.14289 (2021)

24. Zhan, L.M., Liang, H., Liu, B., Fan, L., Wu, X.M., Lam, A.: Out-of-scope intent detection with self-supervision and discriminative training. arXiv preprint arXiv:2106.08616 (2021)
25. Zhang, H., Xu, H., Lin, T.E.: Deep open intent classification with adaptive decision boundary. In: Proceedings of the AAAI Conference on Artificial Intelligence, vol. 35, pp. 14374–14382 (2021)
26. Zhang, H., Xu, H., Zhao, S., Zhou, Q.: Learning discriminative representations and decision boundaries for open intent detection (2022)
27. Zhou, Y., Liu, P., Qiu, X.: KNN-contrastive learning for out-of-domain intent classification. In: Proceedings of the 60th Annual Meeting of the Association for Computational Linguistics, vol. 1: Long Papers, pp. 5129–5141 (2022)

# A Prompt Tuning Method for Chinese Medical Text Classification

Wenhao Li [iD], Junfeng Zhao[(⊠)], and Hanqing Gao

College of Computer Science, Inner Mongolia University, Hohhot, China
{32109095,32109150}@mail.imu.edu.cn, cszjf@imu.edu.cn

**Abstract.** The field of clinical medicine involves complex data analysis and mining, and text classification task in natural language processing can assist in case screening, etiology analysis, disease prediction, and other aspects in the medical field, thereby improving research efficiency and accuracy. However, obtaining supervised data is difficult due to the fact that medical texts such as cases often contain sensitive patient information. This difficulty partially explains why existing text classification methods do not perform well in medical tasks. Although the prompt tuning performs better than traditional neural networks and fine tuning methods in few-shot learning and interpretability. Chinese requires at least two characters to express complex semantics, which makes it challenging to apply the prompt tuning method that is primarily designed for English in Chinese text classification tasks. To address this issue, we propose **K**nowledge Enhanced **M**ulti-Token **P**rompt **T**uning (KMPT). KMPT first uses multiple tokens as label words to have complete Chinese semantics, and then uses external knowledge to expand the label words set, improving coverage and reducing bias. The experimental results on the Chinese medical dataset CHIP-CTC show that KMPT outperforms baseline methods in Chinese medical text classification tasks and has better interpretability and convergence speed than fine tuning.

**Keywords:** Prompt Tuning · Medical Text · Text Classification

## 1 Introduction

The medical field is an information-dense, highly specialized, and complex field that involves a variety of textual information, including but not limited to medical records, case reports, examination reports, and electronic medical records. This textual information plays an important role in medical practice and research, but how to efficiently use this data poses great challenges for physicians and researchers. Artificial intelligence and machine learning have garnered widespread attention in the field of clinical medicine. These technologies have been applied in various fields such as medical image analysis and medical data mining. Natural Language Processing (NLP) can help physicians and researchers process massive amounts of medical textual data, thereby improving the accuracy and efficiency of clinical decision-making. Text classification is an important task in NLP, which aims to categorize text data into different categories. In medical field, it can be applied to assist medical researchers in case selection, etiology analysis, and disease prediction.

Text classification methods can be divided into two categories: traditional feature-based methods and deep learning methods. Traditional methods extract text information based on rules or manually designed features. Deep learning methods offer several advantages over traditional methods. These methods can automatically learn features for classification, which results in better representation and higher accuracy. Additionally, the multi-layer structure of deep learning networks can improve the representation ability and the optimization of the model under large-scale data improves generalization performance. Deep learning methods are also capable of handling noise and uncertainty that can enhance classification performance. However, medical texts usually contain patient private information, and it is difficult to obtain sufficient supervised data. Since neural networks require a large amount of data to learn and extract high-order features, the performance of deep learning is usually limited in the case of less available supervised data or sample imbalance.

Recently, pre-trained language models (PLMs), such as GPT [1] and BERT [2], have shown impressive results in NLP tasks, thanks to their exceptional language understanding and generation capabilities. PLMs are trained through self-supervised learning from large-scale unlabeled corpora during pre-training and further fine tuned with additional labeled data in downstream tasks. This approach has produced state-of-the-art results on many datasets and tasks. However, general domain PLMs often underperform when applied to specific domains such as medicine. To address this, domain-specific PLMs have emerged. These models are pre-trained on a domain-specific corpus, allowing them to better handle the nuances of that particular field. Domain-Specific PLMs have shown better performance in medical applications compared to general domain PLMs [3, 4].

Although fine tuning has achieved tremendous success, there are two key problems. First, there is a significant gap between the objective forms of pre-training and fine tuning. The pre-training tasks of PLMs are usually either sequential language models, which predict the next word based on the context, or masked language models, which predict the masked position of a word based on the context. When fine tuning these PLMs for downstream tasks, additional structures (such as fully connected layers and multi-layer perception) are often added. This gap may lead to negative transfer, overfitting, and low training efficiency. Moreover, the method of adding additional structures to PLMs does not have sufficient interpretability. Second, the additional structures added to fine tune PLMs for downstream tasks require sufficient labeled data for tuning, and their performance is not satisfactory in the presence of sample imbalance and insufficient sample sizes [5].

Prompt tuning is a way to eliminate the gap between pre-training and fine tuning by adding additional language prompts to transform downstream tasks into cloze tasks, which is consistent with the pre-training objective of common PLMs, such as BERT. This reduces the negative impact of additional structures on performance and increases data efficiency, parameter efficiency, and interpretability. In short, prompt tuning transforms input $x$ into a cloze task using a template that includes a [MASK] token. Then PLMs output a prediction for the word in the [MASK] position based on context, and we generally refer to the word filled in the [MASK] position as a label word. Finally, a mapping function called the verbalizer [6] maps the label word to a specific category. Most studies on prompt tuning typically only include one [MASK] token in the prompt templates, as these templates are designed for downstream tasks based on English natural language understanding, such as GLUE.

However, there is a significant difference between English and Chinese, where in English, a complete word can usually be expressed with a single token, whereas the same semantics in Chinese require two or more tokens. For simple classification tasks such as sentiment analysis, we can design a label words set using the emotional polarity that can be expressed by a single token in Chinese, and then map the label words to categories using a verbalizer, such as: {" 好"(nice), " 棒"(great)} → positive. However, when we try to use prompt tuning to solve complex classification tasks, such as medical texts in Chinese, the complex semantics cannot be expressed by a single token, and the semantics of label words are essential for prediction results.

To address the issues mentioned above, in this paper, we propose **K**nowledge Enhanced **M**ulti-Token **P**rompt **T**uning (KMPT) to solve the challenges of prompt tuning on Chinese tasks. Specifically, KMPT supports using label words containing multiple tokens to obtain the complete Chinese semantics. Since PLMs predict the word in the [MASK] position based on context semantics, increasing the number of [MASK] tokens in the template alone may increase the difficulty of PLMs to predict the true label word. Therefore, we integrate external knowledge into the verbalizer, expand the label words set using external knowledge to improve its coverage and reduce bias. As far as we are aware, our work is the first to apply prompt tuning to Chinese medical text classification. Specifically, the main contributions of this paper include:

1. We solve the Chinese medical text classification problem with prompt tuning, which utilizes special prompts to transform downstream tasks into Masked Language Model tasks. This method offers better interpretability, lower computational costs, and reduces the need for a large amount of labeled data. It is especially useful in complex multi-classification tasks and few-shot scenarios.
2. We propose KMPT, which allows using multiple tokens as label words and expands the set of label words using external knowledge bases (KBs). This approach more comprehensively and accurately represents Chinese category information and semantics, improving model generalization performance and accuracy.
3. The success of our approach highlights the potential of using prompt tuning and KMPT in other Chinese NLP tasks.

## 2 Related Work

Before the widespread adoption of deep learning, research on text classification methods for general domains mainly focused on feature-based methods [7]. In deep learning, traditional methods were represented by several classic neural networks. Mikolov et al. [8] proposed the neural network-based Skip-gram, CBOW model, and the concept of word vectors. FastText [9] used the CBOW model to represent text as a vector and used a multi-layer perception (MLP) for text classification, demonstrating good performance on multi-classification tasks. TextCNN [10] and TextRNN [11] respectively applied convolutional neural networks (CNN) and recurrent neural networks (RNN) to text classification tasks, avoiding tedious feature engineering and achieving better results than traditional machine learning methods. TextRCNN [12] combined the characteristics of CNN and RNN, and added a pooling layer on the basis of feature extraction by RNN. Hierarchical attention networks (HAN) [13] and attention-based LSTM [14] enhanced model expression ability by introducing the attention mechanism. PLMs represented by BERT and GPT learned rich knowledge from massive unlabeled data in the pre-training phase, and then transferred the knowledge to downstream tasks through fine tuning on specific datasets, achieving excellent performance on numerous text classification tasks [15].

In the field of medical text classification research, Zhang et al. [16] proposed a Capsule network model that combined LSTM and GRU to extract complex Chinese medical text features with unique routing structures. Yao et al. [17] proposed a text representation method for traditional Chinese medicine clinical records, which combined deep learning with knowledge in the field of traditional Chinese medicine. Experimental results showed that this method performed well in traditional Chinese medicine classification compared to other general text representation methods. Hughes et al. [18] proposed a method for automatically classifying clinical text at the sentence level, using a deep CNN to represent complex features. Qing et al. [19] used convolutional layers and bi-directional gated recurrent units (BIGRU) to extract features within and between sentences, and finally classified medical text by a classifier. PLMs pre-trained on domain-specific corpora had better performance than general domain models like BERT. Chinese PLMs pre-trained on biomedical corpora, represented by MedBERT [4] and MC-BERT [20], were typical examples of such models.

Although prompt tuning was also based on PLMs, they could eliminate the gap between fine tuning and pre-training and ultimately achieve higher data efficiency, parameter efficiency, and interpretability. Prompt tuning used natural language prompts as contexts and transformed downstream tasks into cloze tasks. According to the label words predicted by the PLMs in the [MASK] position, a mapping function called verbalizer was used to map the label words token to a specific category. In recent prompt tuning research, PET [6] first proposed using manually-defined templates and verbalizers. Schick et al. [21] studied the automatic exploration of label words based on manually designed templates. Gao et al. [22] proposed to generate all prompt candidates and applied development sets to find the most effective ones. Shin et al. [23] explored gradient-based prompt exploration to automatically generate templates and label words. PTR [24] designed a prompt tuning with rules to address relation classification problems, designing multiple [MASK] tokens at different positions in the template and determining entity relations based on logical rules applied to the model's output. Recently, some works proposed using a series of learnable continuous embeddings as templates, eliminating the requirement of manually designing prompt [25, 26].

# 3   Method

## 3.1   Prompt Tuning

In text classification tasks, the input sequence $x = (x_0, x_1, \ldots, x_n)$ is classified into a class $y \in Y$. The fine tuning methods adding an additional CLS head, such as a MLP, on top of the PLMs to achieve classification:

$$P(y|x) = \frac{\exp\left(f_{MLP}(h(x)_y\right)}{\exp\left(\sum_{i=1}^{n} f_{MLP}(h(x))_i\right)}. \tag{1}$$

The idea behind prompt tuning is to transform the downstream task into a cloze task that is consistent with the pre-training objectives of the PLMs. Specifically, let $M$ be a PLMs obtained from large-scale corpus pre-training. We use a template $T(\cdot)$ to add extra context as prompts to the input text $x$, resulting in $x' = T(x)$. Typically, the template $T(\cdot)$ contains several additional token embeddings and a masked token, represented by <[MASK]>. Then, we input $x'$ into the PLMs to predict the masked token, which aligns with the pre-training objectives (Masked Language Model, MLM) of most BERT-based models. Subsequently, the PLMs predicts the probability distribution over the vocabulary. Finally, a mapping function is applied to convert the output words of the PLMs to class labels. This mapping function is generally referred to as the Verbalizer, while the words output by the PLMs called label words. In Fig. 1, we provide an example. Suppose we need to classify a sentence $x$ into Therapy or Surgery (labeled as 1) or Disease (labeled as 2). We wrap the sentence with $T(\cdot)$ as follows:

$x'$=[CLS]以下这段描述与[MASK]有关: $xx$.

In English, this translates to:

$x'$=[CLS] The following description is related to [MASK]: $x$.

Afterward, the PLM $M$ provides the probability $P_M([MASK] = v|x')$ for each word in the vocabulary $V$ to be filled in the [MASK] position. The Verbalizer then maps the label words $v$ to the label space $Y$, i.e., $g : V \rightarrow Y$, $V_y$ represents the subset of V that maps to a specific label y. The union of these subsets is equal to the whole label words set: $\bigcup_{y \in Y} V_y = V$.

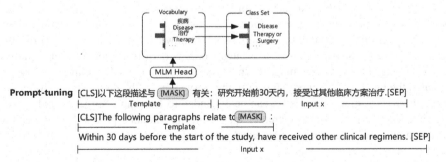

**Fig. 1.** An example of prompt tuning.

The probability of classifying the sentence x into the label y, i.e., $P(y|x')$, is calculated as:

$$P(y|x') = g\left(P_M\left([MASK] = v|x'\right)|v \in V_y\right), \tag{2}$$

The function $g$ converts the probability of predicted label words $v$ into the probability of labels. Suppose $V_1 = \{\text{"Therapy"}\}$ and $V_2 = \{\text{"Disease"}\}$. If the probability of "Therapy" is greater than that of "Disease", we would classify the sentence as Therapy or Surgery.

### 3.2  Knowledge Enhanced Multi-Token Prompt Tuning (KMPT)

As mentioned earlier, prompt tuning has achieved excellent results on common English text classification tasks. Considering the challenges of applying prompt tuning to Chinese text and the complexity of multi-class medical text classification, we designed KMPT (**K**nowledge Enhanced **M**ulti-Token **P**rompt **T**uning).

Inspired by PTR [24], KMPT utilizes multiple [MASK] tokens in the template to address complex text classification problems. For relation classification problems, PTR's template employs three [MASK] tokens—two to determine entity types, and one to establish the relationship between the two entities. By combining the output results of the model with first-order predicate logic, relation classification is achieved.

KMPT follows a similar logical approach. It uses multiple consecutive [MASK] tokens in various positions, which are connected through logical relationships to form complete Chinese words as label words. This enables KMPT to effectively handle the challenges associated with Chinese language text classification tasks and the complexities of multi-class medical text classification.

Based on this foundation, we use external KBs to expand the label words set. There are two primary reasons for doing so. First, since PLMs predict the tokens at the [MASK] positions based on context, simply increasing the number of masked tokens would make predicting the correct label words more challenging for the model. Second, predicting label words is not a single-choice process. Label words generally do not have standard answers, and there might be numerous words fitting the context. Because of these factors, we chose Wantwords[1] and Relatedwords[2] as external KBs to expand the label words. This allows the Verbalizer to have a broader coverage and reduced subjective bias. Notably, this is different from using only class names or similar words when manually setting the Verbalizer. When introducing external knowledge to expand label words, the expanded set includes synonyms and related words with different levels of granularity.

Assuming the Verbalizer maps a label words set to a class {Therapy}→Therapy, initially, the label words set contains only the class name as the unique label words. By using the KBs, the label words set can be expanded to {Therapy, Surgery, Medication, Physiotherapy, …}→Therapy, greatly increasing the model's probability of making correct predictions. For the extended label words sets, we assume that each label word's

---

[1] https://wantwords.net/.

[2] https://relatedwords.org/.

contribution to predicting a label is equal. Thus, we use the average of the predicted scores on $V_y$ as the predicted score for label$y$:

$$\hat{y} = \text{argmax}_{y \in Y} \frac{\sum_{v \in V_y} P_M \left( [MASK] = v | \mathbf{x}' \right)}{|V_y|}. \tag{3}$$

In more specific terms, for a text classification task $T = \{X, Y\}$, KMPT uses a set of conditional functions $F$, where each conditional function $f \in F$ determines whether the input meets certain conditions. For example, the conditional function $f = Position(x, [MASK]_i)$ ascertains whether $x$ is at the position of the $i_{th}$ [MASK] token. The conditional function $f = Vocab(x)$ confirms whether $x$ is a token in the PLMs vocabulary.

Suppose a Chinese label word requires two tokens to represent it in a particular text classification task. We design a template $T(x)$ and a set of label words subsets $V_Y$, where $T(x)$ contains two [MASK] tokens, and $V_y \in V_Y$ denotes the label words subset for each class. The transformed input $\mathbf{x}'$ is obtained from the template $T(x)$, which is then fed into the PLMs to obtain the probability distribution for the two [MASK] positions over the vocabulary. Finally, the conditional functions are used to combine them into a complete label word, determining which class $\mathbf{x}$ belongs to.

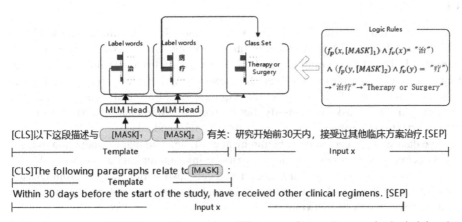

**Fig. 2.** An example of KMPT classifies text into "Therapy or Surgery" categories by judging the complete Chinese semantics composed of multiple masked tokens.

In the Chinese medical text classification task, as shown in Fig. 2, suppose the model outputs two tokens $x$ and $y$, with the correct Chinese label word being "治疗" (therapy). To determine whether the model has output the correct label word, we can formalize this as:

$$(f_p(x, [MASK]_1) \wedge f_v(x) = \text{"治"}) \wedge (f_p(y, [MASK]_2) \wedge f_v(y) = \text{"疗"}) \rightarrow \text{"治疗"},$$

where $f_p$ is the conditional function to judge if a token is at a specific [MASK] token position, and $f_v$ is the conditional function to determine if a token is in the PLMs vocabulary. $f_v(x) = $ "治" implies that "治" is found in the vocabulary. Considering that the template contains multiple [MASK] tokens, we must take all masked positions into account for prediction:

$$p(y|x) = \prod_{j=1}^{n} p([MASK]_j = g_j(y)|T(x)), \tag{4}$$

Here, $n$ represents the number of masked tokens in $T(x)$, and $g_j(y)$ maps the class $y$ to the label words set $V_{[MASK]_j}$ of the $j_{th}$[MASK] position $[MASK]_j$. Ultimately, the learning objective is to maximize:

$$\frac{1}{|X|} \sum_{x \in X} log \prod_{j=1}^{n} p([MASK]_j = g_j(y)|T(x)). \tag{5}$$

## 4  Experiments

We conduct experiments on a Chinese medical text classification dataset to validate the effectiveness of KMPT in Chinese medical text classification tasks.

### 4.1  Dataset

Our experiment utilizes the Clinical Trial Criteria Short Text Classification task (CHIP-CTC) dataset from CBLUE a Chinese Biomedical Language Understanding Evaluation Benchmark. This dataset comprises real screening criteria from the China Clinical Trial Registration Center website, which are free-text descriptions developed by trial investigators to determine if patients meet the requirements of a specific clinical trial. The primary goal of the task is to classify these screening criteria and compare them with case information to ensure patients meet clinical trial standards.

The dataset was generated via hierarchical clustering and manual induction to identify 44 semantic categories, defining descriptive information and annotation rules for each category. This dataset features uneven category distribution, ranging from Disease with 22% to Ethnicity with 0.06% representation. In Table 1, we list the top 3 and bottom 3 categories in terms of data volume.

**Table 1.** The top 3 and bottom 3 categories in terms of data volume

| Category | Count (Ratio) | Min Length | Max Length | Average Length |
|---|---|---|---|---|
| Disease | 6820 (22.25%) | 3 | 213 | 23.90 |
| Multiple | 6092 (19.87) | 7 | 342 | 42.09 |
| Therapy or Surgery | 1991 (6.49%) | 5 | 159 | 21.67 |
| Alcohol Consumer | 23 (0.07%) | 17 | 104 | 56.65 |
| Ethical Audit | 23 (0.07%) | 10 | 21 | 14.5 |
| Ethnicity | 18 (0.06%) | 5 | 15 | 8.70 |

## 4.2 Experiment Settings

Our experiment is based on OpenPrompt [27], an open-source toolkit for prompt tuning. We use the MedBERT [4] directly loaded from the Transformers library, which is a PLM pre-trained on biomedical domain corpora. We optimize MedBERT with the Adam optimizer at a learning rate of 3e-5 and set weight decay to 1e−2. The model is trained for 10 epochs with a batch size of 64.

Macro average F1 is used as the key evaluation metric, which considers both precision and recall. Assuming there are n categories: $c_1,...,c_i,...c_n$, the precision for each category is $P_i$, and the recall is $R_i$. The calculations for macro average are as follows:

$$macro\ average\ precision = \frac{1}{n} \sum_{i=1}^{n} P_i, \tag{6}$$

$$macro\ average\ recall = \frac{1}{n} \sum_{i=1}^{n} R_i, \tag{7}$$

$$macro\ average\ F1 = \frac{1}{n} \sum_{i=1}^{n} \frac{2 \times P_i \times R_i}{P_i + R_i}. \tag{8}$$

For our method and other prompt tuning method, we use manually defined templates, as prior research has shown these yield better performance than optimization-based templates [6]. To introduce randomness, we combine the training set and dev set of the dataset and use different random seeds to split the training, validation, and testing sets by 0.6, 0.2, and 0.2 ratios, repeating the experiment five times. For few-shot learning, we conduct experiments with 1, 8, 16, and 32 shots. For k-shot experiments, we sample k instances from each category in the divided training and validation sets to form new training and validation sets. For categories with fewer than k instances, we use all available instances. Finally, we save model checkpoints based on validation set performance and choose the best checkpoint for testing according to validation performance.

## 4.3 Baselines

In this section, we will introduce the baselines compared in the experiment. These include traditional deep learning methods, fine tuning-based methods, and prompt tuning based methods.

**Traditional Deep Learning Methods.** For text classification, traditional deep learning models typically learn from scratch. We choose TextCNN [10], TextRCNN [11], Fast-Text [9], and Capsule + GRU [16] as baselines. The first three models are based on convolutional neural networks, recurrent neural networks, and the Continuous Bag-of-Words (CBOW) model, respectively. These models are considered classic approaches for text classification.

Additionally, Zhang et al. proposed the Capsule + LSTM model which is specifically designed to address Chinese medical text classification. Furthermore, they conducted experiments using the same dataset, CHIP-CTC, as ours.

**Fine Tuning Pretrained Models.** Traditional fine tuning methods involve feeding the hidden embedding of the [CLS] token from PLMs into a classifier for prediction. For fine tuning the original pretrained models, we directly choose MedBERT [4] as the baseline.

**Prompt-Tuning Pretrained Models.** Conventional prompt tuning methods, represented by PET [6], use class names as the unique label words for each category. However, these methods only allow label words to be a single token, which can be challenging for Chinese tasks. PET proposes to solve this problem by transforming the multi-class problem into a two-category problem. Nonetheless, it is impractical for complex medical text datasets with 44 categories. Instead, we choose WARP [28] as our baseline for the experiment. WARP introduces a soft verbalizer which uses continuous vectors for each category, and employs the dot product between MLM outputs and the category vectors to generate probabilities for each class.

### 4.4 Main Results

In this section, we present the concrete results.

**Full Dataset.** In Table 2, we can see the performance of KMPT and all baseline methods on the complete dataset. KMPT outperforms all baseline methods, achieving an average F1 score of 81.07%, which demonstrates the effectiveness of our approach. It is noteworthy that the prompt tuning based method and the fine tuning use exactly the same PLM, and in the final results, KMPT improves the F1 score by 1.43 percentage points compared to fine tuning.

Comparing the baselines, traditional neural network methods lag behind PLM-based methods due to their randomly initialized parameters, while PLM-based methods benefit from pretraining that allows the models' parameters to start from a comparatively better state. In the case of traditional neural network methods, the gated unit outputs of the RCNN do not reach their best results since the text is relatively short. Also, FastText, which uses a bag-of-words approach, provides limited semantic information; as a result, the performance of RCNN and FastText is inferior to that of CNN. The performance of Capsule + GRU surpasses that of RCNN and FastText, but it still lags behind CNN

**Table 2.** Comparison of KMPT with Different Baseline Methods.

| Model | Precision | Recall | F1 |
|---|---|---|---|
| TextCNN | 81.25 | 69.53 | 73.02 |
| TextRCNN | 76.51 | 64.4 | 67.66 |
| FastText | 77.31 | 60.38 | 65.35 |
| Capsule + GRU | 79.23 | 68.14 | 71.77 |
| Fine Tuning | 81.22 | 79.47 | 79.64 |
| WARP | 82.4 | 77.68 | 78.06 |
| KMPT(ours) | **82.66** | **80.42** | **81.07** |

in terms of effectiveness. WARP uses continuous vectors for each class to perform dot product calculations with MLM output, but its extra prompts are less data-efficient than fine tuning, which leads to lower performance than fine tuning.

**Few-shot.** Table 3 compares the results of baseline methods in k-shot experiments. Traditional neural network methods with randomly initialized parameters fail to converge when k = 1 or k = 8, resulting in poor performance. However, TextCNN achieves performance close to that of fine tuning when the k value reaches 16 or higher. PLM-based methods outperform traditional neural network methods in all cases, with WARP outperforming fine tuning. Nonetheless, the performance gap between the two narrows as the number of shots increases.

When comparing KMPT with the baseline methods, we find that KMPT consistently outperforms all baseline methods. In the experiments with k = 1, 8, and 16, the best baseline F1 scores are improved by 10.64%, 9.6%, and 7.53% on average. KMPT has a significant advantage over WARP in situations with fewer samples, as WARP is an optimization-based method that requires sufficient data for optimization. Thus, when labeled data is limited, its performance is not as strong as that of KMPT.

**Table 3.** F1 scores (%) of 1/8/16/32-shot text classification.

| Shot | TextCNN | TextRCNN | FastTsxt | Fine Tuning | WARP | KMPT(ours) |
|------|---------|----------|----------|-------------|-------|------------|
| 1    | –       | –        | –        | 8.56        | 20.46 | **31.1**   |
| 8    | –       | –        | –        | 31.02       | 43.56 | **53.16**  |
| 16   | 44.34   | 43.07    | 34.55    | 46.75       | 54.55 | **63.98**  |
| 32   | 58.09   | 53.36    | 46.64    | 60.57       | 61.26 | **68.79**  |

In Fig. 3, we observe that KMPT not only has better performance but also converges faster than fine tuning. For low-resource scenarios that may exist in fields such as medicine, KMPT's faster convergence rate has potential positive implications. It can help lower the barrier for applying deep learning-based text classification methods in the medical field.

## 4.5 Supplementary Studies

**Effect of Template Design.** Defining different templates can affect the performance of prompt tuning. We follow the PET [6] approach to designing various templates and selected the best-performing one for KMPT in our experiments. Figure 4 shows the four templates we designed for CHIP-CTC.

To compare the performance of different templates, we conducted experiments on a full dataset split using the same random seed. In Table 4, we found that $T_2$ achieved the highest F1 score of 81.83%, which is 1.3 percentage points higher than the lowest-scoring $T_3$. We used "prefix-style" and "suffix-style" to distinguish between the two template design styles, where "suffix-style" refers to placing the additional prompt after

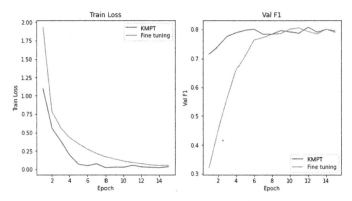

**Fig. 3.** The loss curve on the training set and the F1 score curve on the validation set.

the input $x$, and vice versa for "prefix-style". Notably, "suffix-style" templates appear to be more competitive, with both $T_2$ and $T_4$ outperforming the "prefix-style" templates $T_1$ and $T_3$. Although $T_4$ has a lower F1 score than $T_2$, its precision reaches 84%, which is the highest of the four templates. Our analysis suggests that the "suffix-style" templates, with the input x placed at the beginning, make it easier for the PLMs to summarize the semantics based on the prompt in the template.

$T_1(x)$ = 以下是一条有关[MASK][MASK]的描述：x。
　　　　　Here is a description related to [MASK][MASK]: x.
$T_2(x)$ = x, 以上描述与 [MASK][MASK] 有关。
　　　　　x, the description above is related to [MASK][MASK].
$T_3(x)$ = [ 分类：[MASK][MASK] ] x。
　　　　　[ Category: [MASK][MASK] ] x.
$T_4(x)$ = x 与患者的[MASK][MASK]有关。
　　　　　x is related to the patient's [MASK][MASK].

**Fig. 4.** Templates for CHIP-CTC.

**Table 4.** Comparison of F1 scores (%) between different templates

| Template | Precision | Recall | F1 |
|----------|-----------|--------|-------|
| $T_1$ | 83.53 | 80.50 | 80.85 |
| $T_2$ | 83.75 | **80.80** | **81.83** |
| $T_3$ | 83.12 | 79.46 | 80.53 |
| $T_4$ | **84** | 80.38 | 80.99 |

**Effect of Label Words Length.** In Chinese, two characters typically express one word's meaning, but more characters can be used for a more precise expression. In common Chinese expressions, we usually use two or four characters to convey the same meaning.

Expressions with more characters can complement and enrich the semantics expressed by shorter words. This is reflected in prompt tuning when the template contains a certain number of [MASK] tokens. To study the impact of label words length on performance, we conducted experiments on the full dataset. In Table 5, we used examples from two classes in our dataset to illustrate the label words sets expanded by the external KBs for lengths of 2 and 4, along with the English meaning of each label word in the set. The choice of setting the label words length to 2 and 4 is based on normal Chinese semantics. For example, for the label word "Radiotherapy," we can use the Chinese expressions "放疗" or "放射治疗", which have lengths of 2 and 4 respectively. However, we cannot express the same semantics using words of length 3.

In prompt tuning, we use human-designed templates. The number of [MASK] tokens included in the template during the design process is fixed. If we choose more character-labeled words, we cannot ensure that the same number of characters can be used for label words of each class. At the same time, more [MASK] tokens increase the difficulty for the model to make correct predictions, which may ultimately affect performance.

In the experiment, we compared the performance before and after using external knowledge expansion when the tag word length was 2 and 4. Before the extension, the class name was directly used as the label words. In Table 6, we found that when using class names as label words directly, i.e., the MPT column, the performance with length-4 label words is better. This is because more tokens can express more complete semantics,

**Table 5.** Examples of label words using different numbers of masked tokens.

| Class Name | Masked token | Label words |
|---|---|---|
| Therapy or Surgery | 2 | 治疗, 手术, 化疗, 放疗, 透析, 针灸, 穿刺, 移植, 搭桥... |
| | | Therapy, Surgery, Chemotherapy, Radiotherapy, Dialysis, Acupuncture, Puncture, Transplant, Heart bypass, ... |
| | 4 | 治疗方法,手术治疗,化学治疗,放射治疗,血液透析,针灸治疗,穿刺治疗, 器官移植,心脏搭桥, ... |
| | | Therapy, Surgery, Chemotherapy, Radiotherapy, Dialysis, Acupuncture, Puncture, Transplant, Heart bypass, ... |
| Symptom | 2 | 症状, 感受, 炎症, 头晕, 疼痛, 发热, 胸闷, 腹泻, 外伤,... |
| | | Symptom, Patient feeling, inflammation, Dizziness, Ache, Fever, Chest tightness, Diarrhea, trauma, ... |
| | 4 | 症状描述, 患者感受, 局部炎症, 头晕症状, 身体疼痛, 发热症状, 胸闷气短, ... |
| | | Symptom, Patient feeling, Inflammation, Dizziness, Ache, Fever, Chest tightness, ... |

**Table 6.** F1 scores (%) of different numbers of masked tokens.

| Masked token | MPT | KMPT |
|---|---|---|
| 2 | 78.92 | 81.07 |
| 4 | 80.36 | 80.62 |

and words with more tokens can describe categories better when the set of candidate label words is small. Comparing the MPT's performance with the fourth row in Table 2, we found that even when using the two-character class name as a label word, we can achieve performance close to fine tuning (78.92 for MPT and 79.64 for Fine Tuning). Comparing the performance before and after expanding the label words using external knowledge, we found that after expanding the set of label words, the performance with length-2 label words surpassed that of 4-character words. The performance improvement for length-4 label words before and after expansion is not significant, as the knowledge enhancement compensates for the semantic deficiencies brought by shorter label words. However, using longer or multiple-token label words increases the difficulty for the model to output the correct label words, so the improvement is not obvious.

Considering these factors, the KMPT performance we mentioned earlier is based on label words with two masked tokens. If it's challenging to use external knowledge for expansion and only class names can be used as label words, we recommend choosing multi-character Chinese expressions with more complete semantic representation.

## 5   Conclusion

In this paper, we propose KMPT to address Chinese medical text classification problems. KMPT uses multiple tokens as label words, enabling it to express precise Chinese semantics, thus solving difficulties in applying prompt-tuning to Chinese tasks. Additionally, we expand the set of label words using an external KBs, increasing the coverage of label words and reducing subjective bias to enhance the model's generalization performance and accuracy. Experiments on the CHIP-CTC dataset show that KMPT's performance is significantly superior to existing baseline models, both when using the full dataset and in few-shot scenarios. Moreover, KMPT's faster convergence speed compared to traditional fine tuning better adapts to low-resource training scenarios, providing a reference for applying prompt-tuning to Chinese medical text classification tasks in the future. Going forward, we will further explore methods for Chinese medical domain problems by combining existing prompt-based methods, aiming to reduce the difficulty of applying general-domain methods to specific-domain downstream tasks.

**Acknowledgement.** This work is supported by the Natural Science Foundation of China (No. 61962039).

## References

1. Radford, A., Narasimhan, et al.: Improving language understanding by generative pre-training. OpenAI (2018)
2. Devlin, J., Chang, M-W., et al.: Bert: Pre-training of deep bidirectional transformers for language understanding. In: Proceedings of NAACL-HLT, pp. 4171–4186 (2019)
3. Alsentzer, E., Murphy, J., et al.: Publicly available clinical BERT embeddings. In: Proceedings of the 2nd Clinical Natural Language Processing Workshop, pp. 72–78. Association for Computational Linguistics (2019)

4. Yang, F.: Research on BERT Model for Chinese Clinical Natural Language Processing. Peking Union Medical College (2021)
5. Brown, T.B., et al: Language models are few-shot learners. In: Proceedings of the Annual Conference on Neural Information Processing Systems, pp. 18741–18755 (2020)
6. Schick, T., Schütze, H.: Exploiting cloze-questions for few-shot text classification and natural language inference. In: Proceedings of the 16th Conference of the European Chapter of the Association for Computational Linguistics: Main Volume, pp. 255–269. Association for Computational Linguistics (2021)
7. Wallach, H.M.: Topic modeling: beyond bag-of-words. In: Proceedings of the 23rd International Conference on Machine Learning, pp. 977–984 (2006)
8. Mikolov, T., Chen, K., Corrado, G., Dean, J.: Efficient estimation of word representationsin vector space. arXiv preprint arXiv:1301.3781 (2013)
9. Joulin, A., Grave, E., Bojanowski, P., et al.: Bag of tricks for efficient text classification. arXiv preprint arXiv:1607.01759 (2016)
10. Chen, Y.: Convolutional neural network for sentence classification. UWSpace. http://hdl.handle.net/10012/9592 (2015)
11. Liu, P., Qiu, X., Huang, X.: Recurrent neural network for text classification with multi-task learning. arXiv preprint arXiv:1605.05101 (2016)
12. Lai, S., Xu, L., Liu, K., et al.: Recurrent convolutional neural networks for text classification. In: Proceedings of the AAAI Conference on Artificial Intelligence, vol. 29, no. 1 (2015)
13. Yang, Z., Yang, D., et al.: Hierarchical attention networks for document classification. In: Proceedings of NAACL-HLT, pp. 1480–1489 (2016)
14. Zhou, X., Wan, X., Xiao, J.: Attention-based LSTM network for cross-lingual sentiment classification. In: Proceedings of EMNLP, pp. 247–256 (2016)
15. Liu, P., Yuan, W., et al.: Pre-train, prompt, and predict: a systematic survey of prompting methods in natural language processing. ACM Comput. Surv. 55(9), 1–35 (2023)
16. Zhang, Q., Yuan, Q., et al.: Research on medical text classification based on improved capsule network. Electronics 11(14), 2229 (2022)
17. Yao, L., Zhang, Y., et al.: Traditional Chinese medicine clinical records classification using knowledge-powered document embedding. In: Proceedings of the 2016 IEEE International Conference on Bioinformatics and Biomedicine, pp. 1926–1928 (2016)
18. Hughes, M., Li, I., Kotoulas, S., Suzumura, T.: Medical text classification using convolutional neural networks. Stud. Health Technol. 235, 246–250 (2017)
19. Qing, L., Linhong, W., Xuehai, D.: A novel neural network-based method for medical text classification. Future Internet 11(12), 255 (2019)
20. Zhang, N., Jia, Q., Yin, K., Dong, L., Gao, F., Hua, N.: Conceptualized Representation Learning for Chinese Biomedical Text Mining. arXiv preprint arXiv:2008.10813 (2020)
21. Schick, T., Schmid, H., Schütze, H.: Automatically identifying words that can serve as labels for few-shot text classification. In: Proceedings of COLING, pp. 5569–5578 (2020)
22. Gao, T., Fisch, A., Chen, D.: Making pre-trained language models better few-shot learners. arXiv preprint arXiv:2012.15723 (2020)
23. Shin, T., Razeghi, Y., et al.: Eliciting knowledge from language models using automatically generated prompts. In: Proceedings of EMNLP, pp. 4222–4235. Association for Computational Linguistics (2020)
24. Han, X., Zhao, W., Ding, N., et al.: Ptr: prompt tuning with rules for text classification. AI Open 3, 182–192 (2022)
25. Li, X.L., Liang, P.: Prefix-tuning: Optimizing continuous prompts for generation. arXiv preprint arXiv:2101.00190 (2021)
26. Lester, B., Al-Rfou, R., Constant, N.: The power of scale for parameter-efficient prompt tuning. arXiv preprint arXiv:2104.08691 (2021)

27. Ding, N., Hu, S., Zhao, W., et al.: Openprompt: An open-source framework for prompt-learning. arXiv preprint arXiv:2111.01998 (2021)
28. Hambardzumyan, K., Khachatrian, H., May, J.: WARP: word-level adversarial reprogramming. In: Proceedings of the 59th Annual Meeting of the Association for Computational Linguistics and the 11th International Joint Conference on Natural Language Processing, pp. 4921–4933. Association for Computational Linguistics (2021)

# TabMentor: Detect Errors on Tabular Data with Noisy Labels

Yaru Zhang[1], Jianbin Qin[1(✉)], Yaoshu Wang[1], Muhammad Asif Ali[2], Yan Ji[1], and Rui Mao[1]

[1] Shenzhen Institute of Computing Sciences, Shenzhen University, Shenzhen, China
`zhangyaru2020@email.szu.edu.cn`, {`qinjianbin,jy197541,mao`}`@szu.edu.cn`,
`yaoshuw@sics.ac.cn`
[2] King Abdullah University of Science and Technology, Thuwal, Saudi Arabia
`muhammadasif.ali@kaust.edu.sa`

**Abstract.** Existing supervised methods for error detection require access to clean labels in order to train the classification models. This is difficult to achieve in practical scenarios. While the majority of the error detection algorithms ignore the effect of noisy labels, in this paper, we design effective techniques for error detection when both data and labels contain noise. Nevertheless, we present TabMentor, a novel deep-learning model for error detection on tabular data with noisy training labels. TabMentor introduces a deep model for the prediction, i.e., Tab-classifier that suggests the most salient features for the decision step, enabling efficient learning. For feature extraction, it uses existing error detection algorithms, along with some raw features from the datasets. To reduce the negative effect of noisy training labels on the model, TabMentor uses another deep model, i.e., Teachernet, to supervise the training of Tabclassifier. During the training process, both Teachernet and Tab-classifier dynamically learn curriculum from data, allowing Tabclassifier to focus more on clean labeled samples. Performance evaluation using five different data sets shows that the TabMentor excels over the best baseline error detection system by 0.05 to 0.11 in terms of F1 scores.

**Keywords:** Error Detection · Data Quality · Noisy Labels

## 1 Introduction

In data-driven applications, the process of ensuring that tabular data meets the required quality and integrity constraints, known as *data cleaning*, is a time-consuming processing task [30]. Traditionally, data curators manually perform error detection tasks based on prior knowledge and experience. Given the importance of high-quality data, the need to identify and fix incorrect, missing, and duplicate values has become prominent.

Much existing research for error detection was proposed. A few traditional methods rely on violations of integrity constrains [9,12,41] or patterns [14,31] and outliers [28]. These approaches usually require some prior knowledge of data and use predefined detection schemes. For non-expert users, it is difficult to configure error detection algorithms properly. A different approach is learning-based error detection. Such methods leverage machine learning techniques to represent labeled errors and apply the learned models to detect erroneous values [12,19,23,26]. They focus on improving the efficiency of data curators by automating or semi-automating error detection tasks. However, learning-based error detection systems rely on the correctness of the training data. Heterogeneous tabular data are collected from multiple sources and often suffer from erroneous attributes and labels. Noisy features and labels negatively impact the classification model's performance, while noise in labels is more harmful than features [44]. The processes used to construct data often involve some degree of automatic labeling or crowdsourcing, they are inherently error-prone. Large datasets with noisy labels have become increasingly common [36].

The simple approach to erroneous training labels is to correct or delete the corrupted labels. Relabeling may inadvertently introduce more noise, while filtering may remove actual correct data [35]. Therefore, learning with noisy data and effectively improving the model's accuracy is a non-trivial task. Noisy labels will be a burden for the user to train the model. On the other hand, DNN(deep neural networks) have achieved great success in various domains, including text, audio, and image datasets. While in practical applications in many fields, the most common data type is tabular data, including medicine, finance, climate science, etc. For homogeneous image data, text, audio, and video data, DNN for corrupted labels perform well. But on heterogeneous tabular data is still underexplored, especially when there are both erroneous values and labels in the dataset.

To cope with label corruption in training data and support DNN models on tabular data error detection, we propose a new architecture, called TabMentor. Considering the performance shortcomings of deep learning on tabular data, and the previous learning-based error detection model neglecting training label noise, we integrate deep learning on tabular data and the method of overcoming noisy training labels to reduce the negative performance impact of corrupted labels for error detection. We will train two deep models collaboratively. One is for learning and predicting whether the unseen values are erroneous or not. The other is trained to compute the weights of samples to identify whether the labels are noisy or not.

TabMentor introduces a new structure for error detection on tabular data. Inspired by curriculum learning [3], this paper trains a deep classification model Tabclassifier, which can use the sequential attention mechanism to select which key features are used in each decision step to reason and predict whether the value is erroneous or not. We create a more informative representation for each cell. Not only do we extract features from raw datasets but the four existing error detection algorithms to construct feature engineering. In addition, we also train another deep model, named Teachernet, to supervise the training of Tabclassifier. Teachernet learns the weight of each sample which symbolizes the probability that the label is noise. To reduce the influence of noisy labels, only samples with clean labels are used to update Tabclassifier.

**Contribution.** The main contribution of this paper is the new DNN architecture for error detection tasks, TabMentor, which can effectively reduce the negative impact of noisy labels in training data and detect erroneous values accurately. Our solution proposes two key technologies: (i). An informative representation for values that do not only use four existing error detection algorithms but extract features on raw datasets to identify multiple types of actual errors. (ii). We train two collaborative networks Tabclassifier and Teachernet. Tabclassifier uses sequential attention and learnable mask to select the salient features, deciding which features to use for inference at each decision step. Teachernet supervises Tabclassifier, and calculates a weight for each sample, indicating whether the label is correct or not. These two networks interact with each other, updating the model iteratively to reduce the negative effect of corrupted labels.

## 2    Related Works

**No Learning for Error Detection.** Different error detection methods are often tailored for different types of data errors. [6] divides data errors into two main types: *i* quantitative: errors are numeric data that exceed frequency thresholds or distance thresholds [13]. DBoost [28] is a relatively complete framework that applies with outlier detection algorithms: histogram and Gaussian models [5]. *ii* qualitative: which treats erroneous values as violations of predefined rules or patterns, such as integrity constraints [7]. The rule-based system, such as NADEEF [9], GDR [40], Holoclean [33], can run integrity constraint rules to identify errors. There are also systems that detect values that do not conform to data patterns, such as Wrangler [17], commercial systems with Trifacta, and OpenRefine, which help users identify potential quality issues with predefined patterns. These approaches require users to provide some configuration parameters for error detection, and only the correct configuration can perform well.

**Learning-Based Error Detection.** Naturally, learning-based methods rely on pairs of dirty and clean records as training examples, utilizing semi-supervised strategies to avoid the need to manually provide examples [25]. ED2 [26], Raha [23], and Metadata-Driven Error Detection [39] use the ensembling approach to identify the exact location of erroneous values in the tuple. Furthermore, Holodetect [12] uses expression models to learn rich representations and run data augmentation from noisy datasets in a weakly supervised manner. In order to make downstream tasks work properly, there are already some data-cleaning systems that serve ML. Picket [22], DataWig [4], Activeclean [19] do not rely on any external labels but use self-supervised deep learning to learn error detection models that can be applied during training or testing. These methods do not take into account the real existence of noisy labels. TabMentor aims to solve the issues of supervised approaches. It does not require complex feature engineering, which reduces the cost of domain knowledge. TabMentor overcomes the disadvantage of deep models on tabular data and selects salient features for prediction in the decision-making step. In addition, TabMentor solves the adverse effects of noisy labels in the training data.

**Noisy Label Processing.** For noisy labels, an obvious solution is to identify suspicious labels and correct them to the corresponding ground truth. In the presence of a clean subset of data, it is possible to train a network to relabel noisy labels with the prediction of a network trained on the fully clean data [42]. An alternative method to correcting noisy labels is to remove them. There are two methods of data pruning. The first option is to completely remove noisy samples and train a classifier [37]. The second option is to remove the labels and transform the data into two subsets of labeled and unlabeled data. A semi-supervised learning algorithm is then used on the resulting data [21]. Correcting labels or removing data may introduce more noise, resulting in the missing information. Another widely used method to overcome noisy labels is to operate on the input of the classifier. Guiding the network by selecting the correct instances can help the classifier achieve better performance in noisy labels. One is curriculum learning, which starts from clean examples and passes through noisier samples as the classifier improves. [16] implements a Teacher-student approach, where the teacher selects clean samples for the students, and the classifier is continuously updated on the student's performance. In addition, multiple classifiers can be used to help each other choose the next batch of data to train on, which is different from the teacher-student approach. Instead of one network supervising another, they help each other. networks can correct each other's mistakes due to differences in learned representations [11,24]. To overcome the negative effect of noisy labels in the training data, we present TabMentor which learns the order of training samples from data. TabMentor supervises the training of the basic classifier Tabclassifier and adjusts the learning scheme according to data feedback.

## 3    Error Detection with Noisy Labels

In this section, we define the problem of detecting errors on tabular data with noisy training labels, and discuss our approach to solving the problem effectively.

### 3.1    Problem Definition

In this paper, we focus on detecting errors on tabular data in an automatic way using deep learning. Manual labeling is externally difficult, and the cost of domain knowledge is expensive. The labels that we used usually contain a certain proportion of noise. So how to figure out the problem of error detection with noisy training labels? We can define the problem as:

**Definition 1.** (Error Detection with Noisy Labels) Given a dirty training dataset $d$ with $m$ tuples as $T = \{t_1, t_2, ..., t_m\}$ and $n$ attributes as $A = \{a_1, a_2, ..., a_n\}$, it's corresponding labels of each cell as $y = \{l_{d[1,1]}, l_{d[1,2]}, ..., l_{d[i,j]} ..., l_{d[m,n]} | l \in \{0 \text{ or } 1\}\}$, what we need to do is to train a classification model $M$ to predict the correctness of a data cell $d[i,j]$ effectively. So that $l_{d[i,j]} = 0$ if $d[i,j]$ is an erroneous value and $l_{d[i,j]} = 1$ is a correct value. The model takes the dirty training dataset $(X_{train}, Y_{train})$ where both the feature $X_{train}$ and labels $Y_{train}$ may be erroneous. The test dataset $(X_{test}, Y_{test})$ where the features $X_{test}$ might be incorrect however the labels $Y_{test}$ are correct. Although the training labels contain noise, the test labels must be clean to ensure that the metric of the classification model is not affected by corrupted labels.

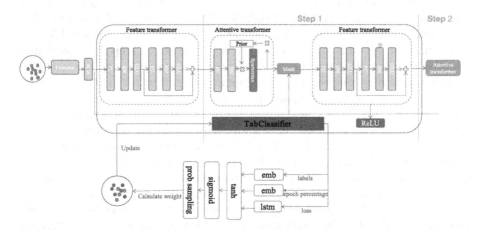

**Fig. 1.** The TabMentor architecture used in our experiments. *Feature transformer* and *Attentive transformer* is the main components of the Tabclassifier. As the role of Studentnet, Tabclassifier is responsible for predicting whether the value is erroneous or not. It outputs losses, epoch percentage, etc. to Teachernet. The Teachernet receives the features of Tabclassifier and calculates the weight of each data.

Using the above definition, we can see that error detection can be regarded as a binary classification problem, what we need to do is training a model to identify erroneous values in the presence of noisy training labels. Our goal is to train an error detection model with good prediction accuracy, such as precision, recall, F1-score, making it close to the model trained on clean data.

## 3.2   Overall Approach

The main purpose of TabMentor is to detect erroneous values effectively when the training dataset contains noisy labels. In TabMentor, it can be achieved by using TabNet [1] and another DNN network as Tabclassifier, Teachernet. It can be treated as a student network and a teacher network. Tabclassifier is responsible for making predictions and computing the cross-entropy loss for each example. The goal of Teachernet is to calculate the weight of each sample. Samples containing noisy labels are given low weights, and correct labels are given high weights. Clean labels are used to retrain the student network Tabclassifier. These two networks restrict each other to reduce the negative impact of noisy labels on deep models.

As shown in Fig. 1, the error detection problem can be treated as a binary classification problem. In the figure, color-coded circles represent normal (blue) and erroneous (orange) cells. While it is not clear whether the label of the training data is noisy or not. For each mini-batch, the system inputs training data with erroneous values and labels. The system uses four existing error detection algorithms and raw datasets to generate feature vectors for each cell, which is effective for handling multiple types of errors. Then together with noisy training labels as the input of TabMentor, which is trained with two deep networks as Tabclassifier and Teachernet. For Tabclassifier, the feature is subjected to a feature transformer by BN (Batch Normalization) layer to obtain an initial feature representation. At each step, the feature of the previous step is masked after attentive transformer. After the input features are passed through the mask, the feature selection is completed and the unimportant features are filtered. Then the features are split into two parts through a feature transformer. One part is calculated after ReLU, and the other part is input to the attentive transformer of the next step to learn the mask. In this way, n-steps decision-making is repeated, and the i-th round is affected by the i-1 round. The trained Tabclassifier is used to calculate the loss and makes predictions on the unseen dataset. Labels and training epoch percentages are encoded through two separate embedding layers. For a sample, its loss can be encoded by a bidirectional LSTM network. The output from LSTM and embedding layers are fed to two connection layers, *tanh* and *sigmoid*. The *sigmoid* activation layer is used to ensure that the weights of the output are between 0 and 1. The probabilistic sampling layer is used to realize the dropout process. The green circles in Fig. 1 are samples with higher weight, representing the labels are correct. Finally, the model is updated only using examples with clean labels.

### 3.3  Tabular Learning for Detection

Most deep neural network approaches on tabular data are hybrid models. They transform data and fuse successful machine learning methods (usually DTs) with neural networks, such as NODE [29], Net-DNF [18]. Due to the great success of attention-based methods, such as text transformer and visual data, researchers proposed to use a deep attention mechanism to deal with heterogeneous tabular data. TabNet [1] is a transformer-based tabular data model. Similar to decision trees, TabNet's architecture consists of multiple sub-networks that are processed in a method of sequential attention. Unlike TabNet, we consider that the labels used for model training contain noise. What we need to do is reduce the impact of erroneous labels on the classification model, and improve the accuracy of detecting errors.

**Feature Extraction.** In order to improve the representation ability of feature vectors for each cell, we not only aggregate the four existing error detection algorithms to form the feature vectors but use the information of raw data as part of vectors. We use rule-based violation, pattern-based violation, outlier, and knowledge-based violation to form multiple error detection strategies and generate feature vectors for each cell. In addition, we also extract features with the raw values using a simple method such as one-hot. We combine the two features to create a more informative representation of the dataset.

For rule-based violation detection, we only detect conflicting data values between columns and mainly focus on rules of the FD form as $\{X \rightarrow Y$, both $X$ and $Y$ is a single attribute value$\}$. In this paper, we generate all attribute pairs as possible functional dependencies for detection strategies. Regarding each attribute pair $\forall_{a \neq a'} \in A$, the detection strategy $s_{a \rightarrow a'}$ marks the value of $d[i, j]$ that violates the functional dependency as erroneous values. Pattern-based violation detection evaluates data values based on compatibility with predefined data schemas and treats values that violate a specific schema as errors. We generate a set of character-checking strategies $s_{ch}$ to check the validity of each cell. In this paper, the generation of outlier policies will be implemented using statistical and distance-based methods [34]. For statistical method with *histogram model*, the resulting strategy $s_o$ treats data cells with frequency less than the certain threshold as errors. For a particular column, the distance-based approach, named *Gaussian model*, builds a Gaussian distribution based on the magnitude of data values. Data in the knowledge base is usually stored in the form of entity relationship, such as Name *isDirectorOf* Movie, where Name and Movie are entity types, and *isDirectorOf* is the relationship between entities. We use entity-relationships in [2] as error detection strategies $s_r$ similarly and identify values that conflict with each entity-relationship $r$.

Run each detection strategy $s \in S_{col}$ on the dataset of a single column, which either marks a data cell $d[i, j]$ as *error* or *correct*. Specifically expressed as:

$$s\,(d[i,j]) = \begin{cases} False, & \text{s marks } d[i,j] \text{ as error} \\ True, & \text{otherwise} \end{cases} \quad (1)$$

For a specific column, we run each error detection strategy $s \in S_{col}$ on data cell $d[i, j]$ respectively where $S_{col} = \{s_1, s_2, ..., s_{|S|}\}$. The output of detection is converted to a logical value into a binary integer. Finally, the total output of all detection strategies constitutes the feature vector of one cell.

$$V_{cell}(d[i,j]) = \{int(s(d[i,j])) \mid \forall_s \in S\} \quad (2)$$

We generate strategies column-based. Since the feature dimensions of each column are not exactly the same, we need to generate features for each column separately. Each column in the data has a different number of strategies. In order to generate an efficient representation of feature vectors in the data, corresponding feature vectors $V_{col}$ are generated for a particular data column $j$.

$$V_{col}(j) = \{v_{cell}(d[i,j]) \mid 1 \leq i \leq m\} \quad (3)$$

For each column, we combine vectors $V_{col}$ with features generated using raw data values directly as input to Tabclassifier. The features we extracted not only from the dataset but from single-column and multiple-column information. It can capture more information to detect erroneous values effectively.

**Classifier Training.** The overall architecture of our classification model is shown in Fig. 1. In this paper, we overcome the negative influence of noisy labels in the training data and train the basic classification model Tabclassifier based on curriculum learning [3]. We use TabNet [1] as our base classification model and train another deep model Teachernet to supervise its training process. Tabclassifier uses the features generated in the previous section and noisy labels as input. After features enter the BN layer, they participate in the calculation of each subsequent step and serve as the input for each step. In each step, there are two important components, feature transformer, and attentive transformer, both of which are composed of BN, fully-connected layer FC, etc. After the feature transformer and attentive transformer, the classification model can select the most silent features to effectively identify correct or erroneous values on unseen data.

For the feature transformer, it includes $FC \rightarrow BN \rightarrow GLU$ which is used to characterize the input. GLU represents the gate linear unit. The output of a feature transformer is divided into $n_d$ ($n$ features for decision) and $n_a$ ($n$ features for attention). $n_a$ is regarded as part of the shared parameters, which will be used to pass the next step as the input of the attentive transformer, and will have an impact on each step of the entire network. As an independent part of each step, $n_d$ will only affect the output of this step. The parameters are processed independently. The output of this step is generated through the ReLu layer to achieve the effect of feature selection. The data output after ReLu will still be summed once when it is transmitted downwards. The output of $n_a$ enters the attentive transformer. The expanded features enter the BN layer and then multiply it with priors to send it into sparsemax. Sparsemax is used for feature selection. The priors here are the previous scale items, indicating how much each feature has been used before. After the attentive transformer, the mask operation is performed to select silent features. Tabclassifier inputs the filtered features into the feature transformer and executes the split as the input of the next step.

## 3.4   Learning Curriculum

Most deep CNNs models perform well on clean training data. Once labels are corrupted, the performance of the model will be affected severely [43]. Inspired by the cognitive process of humans and animals, CL(Curriculum Learning) [3] pointed out that the training gradually includes simple to complex examples in the process of learning. The curriculum specifies a schedule for studying, and samples are learned in a pre-determined order. CL can provide meaningful supervision on the learning of samples and help the model overcome the corrupted labels in the training data.

Inspired by MentorNet [16], our work learns another network Teachernet on a training dataset containing noisy labels to supervise the basic deep model Tabclassifier. Considering the feedback from Tabclassifier, the curriculum can be learned and adjusted dynamically from the data during training. This is a collaborative learning model between students and teachers. Similar to [15], let $f_s(x_i, w)$ denotes Tabclassifier, $L(y_i, f_s(x_i, w))$ represents the cross-entropy loss between the ground truth label $y$ and the Tabclassifier estimated label $y^{'}$ as $L = -[y \log y^{'} + (1-y) \log(1-y^{'})]$, $w$ represents the parameters of $f_s$. Our goal is jointly learn model parameters $w$ and weight variables $v$, and minimize the objective function:

$$\mathbb{F} = \arg \min_{w,v} \frac{1}{n} \sum_{i=1}^{n} v_i^T L(y_i, f_s(x_i, w)) + G(v; \lambda) \tag{4}$$

$v$ represents the weight variable of samples, it can be learned from a network $f_m(z_i; \theta)$ parameterized by $\theta$, where $z_i$ is the input feature to the network. $G(v, \lambda)$ represents the learning scheme, $\lambda$ controls the learning pace. A curriculum can be described as:

**Definition 2.** (Mathematical order curriculum) For the training sets $X = \{x_1, x_2, x_3, \ldots, x_n\}$, an order curriculum can be expressed as:

$$S : X \rightarrow 1, 2, 3, \ldots, n \tag{5}$$

It can be simply regarded as a mathematical sorting function, where $S(x_i) < S(x_j)$ indicates $x_i$ should be learned earlier than $x_j$ in training.

**Example 1.** A predefined curriculum determines the learning scheme for the model to learn new samples. We use a binary schema of predefined curriculum $G(v; \lambda) = -\lambda ||v||_1 = -\lambda \sum_{i=1}^{n} v_i$ as example, which is also used in [20]. Given a fixed $w$, we can define the objective function as $\mathbb{F}_w(v) = \sum_{i=1}^{n} f_{v_i}$, where $f_{v_i} = v_i l_i - \lambda v_i$. The minimum value is obtained by deriving $\nabla_v \mathbb{F}_w(v) = 0$, that is to calculate $\frac{\partial f}{\partial v}$. We have:

$$v_i^* = \begin{cases} 1 & L(y_i, f_s(x_i, \theta)) < \lambda \\ 0 & otherwise \end{cases} \tag{6}$$

Samples with a loss less than a certain threshold $\lambda$ are considered as "easy" $v^* = 1$ and will be selected during training, otherwise they will not be selected. Then the base model Tabclassifier is only trained on the selected "easy" samples.

In this paper, we learn the curriculum from the data dynamically. It is jointly determined by Tabclassifier and Teachernet. We can decide whether to update the learning scheme. If it needs to be updated, $\theta$ is learned from another dataset $D^{'} = \{(x_i, y_i, w), v_i^*\}$, where $D^{'}$ is sampled from $D$. We use binary labels to $v_i^*$, where $v_i^* = 1$ indicates that $y_i$ is a correct label. $\theta$ is learned by minimizing the cross-entropy of $v_i^*$ and $f_m(z_i; \theta)$. We update $w$ and $v$ using an

**Table 1.** A summary of datasets. The error types are missing value(M), typo(T), data format issue(F), and violated functional dependency(VFD)

| Name | Size | Error Rate | Error Types |
|---|---|---|---|
| Flights | 2376 * 7 | 0.30 | M,F,VFD |
| Rayyan | 1000 * 11 | 0.09 | M,T,F,VFD |
| Movies | 7390 * 17 | 0.06 | M,F |
| Hospital | 1000 * 20 | 0.03 | T,VFD |
| Soccer | 20000 * 10 | 0.11 | T,F |

alternating minimization algorithm [8]. After initializing $w$ and $v$, we calculate $z_i = \varphi(x_i, y_i, w^t)$ for each sample of a mini-batch. For step $t$, $v$ is updated by $v^t = f_m(\varphi(Z); \theta)$, where $Z = \{(x_i, y_i, w^t)|1 \leq i \leq n\}$. The weights are computed by the learned model Teachernet. Then we update $w$ using gradient descent as $w^t = w^{t-1} - \alpha \nabla_w \mathbb{F}(w^{t-1}, v^t)$, where $\alpha$ is the learning rate. The difference from existing curriculum learning methods [3,20] is that for each update, the curriculum is learned rather than specified by human experts. It can be based on the basic classification model's feedback to change the curriculum.

## 4  Evaluation

### 4.1  Experimental Setup

We evaluate the effectiveness of our approach on 5 datasets that are described in Table 1. (1) Flights is a real-world dataset that we can obtain from previous research [33]. Data errors occur due to inconsistencies in the time of the same flight in the data source. The cells about 30% in the flight data are erroneous values. (2) Rayyan is a dataset that describes literature information. Rayyan contains about 9% incorrect data cells. Almost all error types are covered in Rayyan [27]. (3)Movies is a dataset taken from the Magellan repository [10]. 6% of erroneous cells exist in this dataset. (4) Hospital is a real-world dataset that we can obtain from Raha [23]. Errors are artificially introduced by injecting typos. This is a benchmark dataset that contains about 3% errors. (5) Soccer is a synthetic dataset that describes the information of soccer players and their teams. It can be obtained from [32]. There exist about 11% erroneous values.

We evaluate TabMentor against ActiveClean [19], Naive-Mix(NM), Naive-sampling(NS), and Raha [23]. ActiveClean is a data cleaning system that updates the model using gradient descent iteratively. NM and NS sampled a batch of candidate data to be cleaned randomly, using the full dataset or examples to be cleaned so far for retraining. We evaluate TabMentor against the following approaches. We also include a configuration-free system that uses multiple error detection algorithms to constitute feature vectors for training [23].

In the evaluation, we report precision, recall, and F1-score to evaluate the performance of error detection. Since TabMentor is trained on every column dependently, we report the performance as the average of full columns. We insert 20% noisy labels on the training dataset randomly. To ensure the accuracy of evaluation, we set the labels of the test dataset to be reliable.

## 4.2   Performance Evaluation

Table 2 shows that TabMentor outperforms the baseline on five datasets in terms of F1-score and recall. Specifically, TabMentor outperforms the best baseline by 0.05 to 0.11 in terms of F1 score and outperforms the best baseline by 0.08 to 0.13 in terms of recall. Since TabMentor considers noisy labels, it can minimize the negative impact of label noise in the training data, thereby improving the performance of the model's error detection. For precision, TabMentor is slightly lower than Raha on the flights dataset, while we have a comparable result with the best excellent baseline on other datasets. The superior performance of Tab-Mentor is strongly supported by our high recall and F1-score. Due to TabMentor not only extracting features from error detection algorithms but from the raw dataset, which can generate a more informative representation for each cell, TabMentor can receive superior recall and is better at detecting errors.

**Table 2.** Comparison with error detection baseline

| Method | Flights | | | Rayyan | | | Movies | | | Hospital | | | Soccer | | |
|---|---|---|---|---|---|---|---|---|---|---|---|---|---|---|---|
| | P | R | $F_1$ | P | R | $F_1$ | P | R | $F_1$ | P | R | $F_1$ | P | R | $F_1$ |
| ActiveClean | 0.61 | 0.69 | 0.65 | 0.76 | 0.81 | 0.78 | 0.84 | 0.84 | 0.84 | 0.85 | 0.86 | 0.85 | 0.84 | 0.84 | 0.84 |
| NS | 0.64 | 0.74 | 0.69 | 0.80 | 0.84 | 0.82 | 0.83 | 0.83 | 0.83 | **0.87** | 0.87 | 0.87 | 0.85 | 0.85 | 0.85 |
| NM | 0.68 | 0.70 | 0.69 | 0.79 | 0.82 | 0.80 | 0.85 | 0.85 | 0.85 | **0.87** | 0.84 | 0.85 | 0.85 | 0.86 | 0.85 |
| Raha | **0.93** | 0.57 | 0.70 | 0.75 | 0.82 | 0.78 | 0.22 | 0.87 | 0.35 | 0.13 | 0.60 | 0.21 | 0.06 | 0.61 | 0.11 |
| TabMentor | 0.77 | **0.86** | **0.81** | **0.84** | **0.92** | **0.89** | **0.86** | **0.98** | **0.92** | 0.86 | **1.0** | **0.92** | **0.86** | **0.99** | **0.92** |

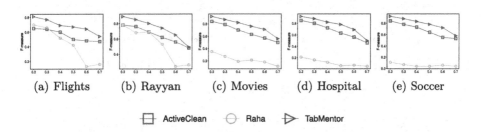

(a) Flights      (b) Rayyan      (c) Movies      (d) Hospital      (e) Soccer

⊞ ActiveClean      ⊖ Raha      ▷ TabMentor

**Fig. 2.** System effectiveness with different proportion of training noisy label

### 4.3 Model Analysis

In this experiment, we evaluate TabMentor's F1-score with three different tree models such as XGBoost, Random Forest, and Decision Tree. Each tree-based classifier is implemented using the scikit-learn library. We use cross-validation to choose the optimal depth. Tree ensemble models such as XGBoost are often recommended for classification and regression problems on tabular data [38]. As shown in Table 3, TabMentor outperforms other tree classifiers on flights and movies. On the other experimental datasets, TabMentor can achieve comparable performance with the tree classifier in Table 3. Although the deep model is not suitable for tabular data, TabMentor can also achieve relatively good performance to the tree model. When comparing the three tree models, Random Forest outperforms XGBoost and Decision Tree slightly.

**Table 3.** Performance comparison of different classifier

| Model | Flights | | | Rayyan | | | Movies | | | Hospital | | | Soccer | | |
|---|---|---|---|---|---|---|---|---|---|---|---|---|---|---|---|
| | P | R | $F_1$ | P | R | $F_1$ | P | R | $F_1$ | P | R | $F_1$ | P | R | $F_1$ |
| XGBoost | 0.82 | 0.74 | 0.78 | 0.84 | **0.94** | **0.89** | 0.85 | 0.95 | 0.90 | 0.85 | 1.0 | 0.92 | **0.86** | 0.99 | 0.92 |
| Random Forest | **0.83** | 0.80 | **0.81** | 0.84 | **0.94** | **0.89** | 0.80 | 0.94 | 0.86 | 0.85 | 1.0 | 0.92 | 0.85 | **1.0** | 0.92 |
| Decision Tree | 0.82 | 0.74 | 0.78 | 0.84 | **0.94** | **0.89** | **0.86** | 0.95 | 0.90 | 0.85 | 1.0 | 0.92 | **0.86** | 0.99 | 0.92 |
| TabMentor | 0.77 | **0.86** | **0.81** | **0.85** | 0.92 | 0.88 | **0.86** | **0.98** | **0.92** | **0.86** | 1.0 | 0.92 | **0.86** | 0.99 | 0.92 |

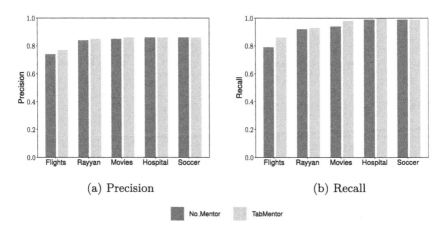

(a) Precision　　　　　　　　　　　(b) Recall

No_Mentor　　TabMentor

**Fig. 3.** Performance comparison with TabMentor and No-mentor

### 4.4 System Effectiveness Analysis

In this section, the effectiveness of TabMentor is evaluated. We analyze the influence of noisy training labels, the impact of Tabclassifier, and analyze the influence of the number of features used for the decision step.

**Influence of Noisy Labels.** In this experiment, we study the impact of different proportions of training noise labels on TabMentor and compare it with Raha, ActiveClean. As shown in Fig. 2, TabMentor outperforms Raha and ActiveClean on all five datasets, cause it takes into account the influence of noisy labels and only uses clean samples to update the model during training. The F1-score of Raha and ActiveClean on the three datasets of Movies, Hospitals, and Soccer has always been lower than TabMentor. Raha and TabMentor's performance is superior to other methods on datasets with multi-column errors such as Flights and Rayyan. Since they both consider errors in more than just a single column. But when the ratio of noisy labels in the training data increases to 0.5, the performance of Raha decreases significantly. While TabMentor outperforms Raha when noisy labels take up a larger proportion.

**Analysis of Mentor.** In this experiment, we analyze the implementation of Teachernet in TabMentor. We use precision and recall as our metric to evaluate the efficiency of the trained model. We run Tabclassifier with no supervise, named No-Mentor which trains a model with noisy training labels. As shown in Fig. 3, TabMentor outperforms No-Mentor in terms of recall especially on Flights and Movies. The precision of TabMentor on the two datasets of Flights and Rayyan is superior to No-Mentor. Since we use another network Teachernet to learn a curriculum from data, it controls Tabclassifier to train with clean samples which can prevent the model from learning the corrupted information. If we directly use the noisy training labels as No-Mentor, the model Tabclassifier will be more prone to overfitting to noisy training labels which induced a negative influence on the model. Therefore, TabMentor gives superior performance compared to using noisy training labels without processing.

**Table 4.** Performance comparison of different $\sigma$ values

| $\sigma$ | Flights | | | Rayyan | | | Movies | | | Hospital | | | Soccer | | |
|---|---|---|---|---|---|---|---|---|---|---|---|---|---|---|---|
| | P | R | $F_1$ | P | R | $F_1$ | P | R | $F_1$ | P | R | $F_1$ | P | R | $F_1$ |
| $\sigma = 8$ | **0.83** | 0.80 | **0.81** | 0.83 | **0.93** | 0.81 | 0.86 | 0.98 | 0.92 | **0.86** | 1.0 | **0.92** | 0.86 | 0.99 | 0.92 |
| $\sigma = 16$ | 0.77 | **0.86** | **0.81** | 0.84 | 0.92 | **0.89** | 0.86 | 0.98 | 0.92 | **0.86** | 1.0 | **0.92** | 0.86 | 0.99 | 0.92 |
| $\sigma = 32$ | 0.74 | 0.81 | 0.77 | **0.85** | 0.92 | 0.88 | 0.86 | 0.98 | 0.92 | 0.85 | 0.99 | 0.91 | 0.86 | **1.0** | 0.92 |
| $\sigma = 64$ | 0.77 | 0.83 | 0.80 | 0.81 | **0.93** | 0.86 | 0.86 | 0.98 | 0.92 | **0.86** | 1.0 | **0.92** | 0.86 | 0.99 | 0.92 |

**Propagation Experiment.** In this experiment, we investigate the effect of different values of $\sigma$ in TabMentor, where $\sigma$ is the number of features used for decision steps. It is decided by the values of $n_d$ and $n_a$ as $\sigma = n_a - n_d$. A very high value of $n_d$ and $n_a$ may suffer from overfitting and yield poor generation. Table 4 shows that changing values of $\sigma$ has a minimal effect on the performance of TabMentor, especially in Movies and Soccer. With the value of $\sigma$ increasing, it is also worth noting that the performance of TabMentor reported in Table 4 becomes worse. The values of $\sigma$ indicate the magnitude of the difference between $n_d$ and $n_a$. In experiments, larger values of $\sigma$ will lead to lower performance.

## 5  Conclusion

We presented TabMentor, a learning-based approach with deep model for error detection on tabular data that can effectively maintain excellent performance in the case of noise in both attributes and training labels. TabMentor contains a model Tabclassifier that allows input raw datasets without any preprocessing. When making decisions, a learnable mask is used to select salient features so that the learning ability of decision step is not wasted on irrelevant features. In addition, it cooperates with another network teachernet to learn curriculum from data dynamically and reduces the negative impact of noisy labels on classification models. TabMentor achieves better performance on five experimental datasets when compared to previous state-of-art systems.

**Acknowledgements.** National Key R&D program of China 2021YFB3301500, Guangdong Provincial Natural Science Foundation 2019A1515111047, Shenzhen Colleges and Universities Continuous Support Grant 20200811104054002, Guangdong "Pearl River Talent Recruitment Program" under Grant 2019ZT08X603, the 14th "115" Industrial Innovation Group (Project 4) of Anhui Province, NSFC 62072311, U2001212, Guangdong Project 2020B1515120028, and Shenzhen Project JCYJ20210324094402008.

## References

1. Arik, S.Ö., Pfister, T.: Tabnet: attentive interpretable tabular learning. In: AAAI (2021)
2. Auer, S., Bizer, C., Kobilarov, G., Lehmann, J., Cyganiak, R., Ives, Z.: DBpedia: a nucleus for a web of open data. In: Aberer, K., et al. (eds.) ASWC/ISWC -2007. LNCS, vol. 4825, pp. 722–735. Springer, Heidelberg (2007). https://doi.org/10.1007/978-3-540-76298-0_52
3. Bengio, Y., Louradour, J., Collobert, R., Weston, J.: Curriculum learning. In: Proceedings of the 26th Annual International Conference on Machine Learning (2009)
4. Biessmann, F., et al.: Datawig: missing value imputation for tables. J. Mach. Learn. Res. **20**, 1–6 (2019)
5. Chandola, V., Banerjee, A., Kumar, V.: Anomaly detection: a survey. ACM Comput. Surv. (CSUR) **41**, 1–58 (2009)
6. Chu, X., Ilyas, I.F., Krishnan, S., Wang, J.: Data cleaning: overview and emerging challenges. In: Proceedings of the 2016 International Conference on Management of Data (2016)
7. Chu, X., Ilyas, I.F., Papotti, P.: Holistic data cleaning: putting violations into context. In: ICDE. IEEE (2013)
8. Csiszár, I.: Information geometry and alternating minimization procedures. Stat. Decis. **1**, 205–237 (1984)
9. Dallachiesa, M., et al.: Nadeef: a commodity data cleaning system. In: 2013 ACM SIGMOD (2013)

10. Das, S., Doan, A., Psgc, C.G., Konda, P., Govind, Y., Paulsen, D.: The magellan data repository (2015)
11. Han, B., et al.: Co-teaching: robust training of deep neural networks with extremely noisy labels. Adv. Neural Inf. Process. Syst. **31** (2018)
12. Heidari, A., McGrath, J., Ilyas, I.F., Rekatsinas, T.: Holodetect: few-shot learning for error detection. In: Proceedings of the 2019 International Conference on Management of Data (2019)
13. Hellerstein, J.M.: Quantitative data cleaning for large databases. UNECE (2008)
14. Huang, Z., He, Y.: Auto-detect: data-driven error detection in tables. In: Proceedings of the 2018 International Conference on Management of Data (2018)
15. Jiang, L., Meng, D., Zhao, Q., Shan, S., Hauptmann, A.G.: Self-paced curriculum learning. In: Twenty-Ninth AAAI Conference on Artificial Intelligence (2015)
16. Jiang, L., Zhou, Z., Leung, T., Li, L.J., Fei-Fei, L.: Mentornet: learning data-driven curriculum for very deep neural networks on corrupted labels. In: International Conference on Machine Learning. PMLR (2018)
17. Kandel, S., Paepcke, A., Hellerstein, J., Heer, J.: Wrangler: interactive visual specification of data transformation scripts. In: Proceedings of the Sigchi Conference on Human Factors in Computing Systems (2011)
18. Katzir, L., Elidan, G., El-Yaniv, R.: Net-dnf: effective deep modeling of tabular data. In: International Conference on Learning Representations (2020)
19. Krishnan, S., Wang, J., Wu, E., Franklin, M.J., Goldberg, K.: Activeclean: interactive data cleaning for statistical modeling. In: PVLDB (2016)
20. Kumar, M., Packer, B., Koller, D.: Self-paced learning for latent variable models. Adv. Neural Inf. Process. Syst. (2010)
21. Li, J., Socher, R., Hoi, S.C.: Dividemix: learning with noisy labels as semi-supervised learning. arXiv preprint arXiv:2002.07394 (2020)
22. Liu, Z., Zhou, Z., Rekatsinas, T.: Picket: self-supervised data diagnostics for ml pipelines. arXiv (2020)
23. Mahdavi, M., et al.: Raha: a configuration-free error detection system. In: SIGMOD (2019)
24. Malach, E., Shalev-Shwartz, S.: Decoupling "when to update" from "how to update". Adv. Neural Inf. Process. Syst. (2017)
25. Neutatz, F., Chen, B., Abedjan, Z., Wu, E.: From cleaning before ml to cleaning for ml. IEEE (2021)
26. Neutatz, F., Mahdavi, M., Abedjan, Z.: Ed2: two-stage active learning for error detection-technical report. arXiv (2019)
27. Ouzzani, M., Hammady, H., Fedorowicz, Z., Elmagarmid, A.: Rayyan-a web and mobile app for systematic reviews. Syst. Rev. **5**, 1–10 (2016)
28. Pit-Claudel, C., Mariet, Z., Harding, R., Madden, S.: Outlier detection in heterogeneous datasets using automatic tuple expansion (2016)
29. Popov, S., Morozov, S., Babenko, A.: Neural oblivious decision ensembles for deep learning on tabular data. arXiv preprint arXiv:1909.06312 (2019)
30. Rahm, E., Do, H.H.: Data cleaning: problems and current approaches. IEEE (2000)
31. Raman, V., Hellerstein, J.M.: Potter's wheel: an interactive data cleaning system. In: VLDB (2001)
32. Rammelaere, J., Geerts, F.: Explaining repaired data with cfds. In: VLDB (2018)
33. Rekatsinas, T., Chu, X., Ilyas, I.F., Ré, C.: Holoclean: holistic data repairs with probabilistic inference. arXiv (2017)
34. Ridzuan, F., Zainon, W.M.N.W.: Diagnostic analysis for outlier detection in big data analytics. Procedia Comput. Sci. **197**, 685–692 (2022)

35. Rosales, R., Fung, G., Tong, W.: Automatic discrimination of mislabeled training points for large margin classifiers. In: Proceedings of Snowbird Machine Learning Workshop. Citeseer (2009)
36. Sambasivan, N., Kapania, S., Highfill, H., Akrong, D., Paritosh, P., Aroyo, L.M.: "everyone wants to do the model work, not the data work": data cascades in high-stakes AI. In: proceedings of the 2021 CHI Conference on Human Factors in Computing Systems (2021)
37. Sharma, K., Donmez, P., Luo, E., Liu, Y., Yalniz, I.Z.: NoiseRank: unsupervised label noise reduction with dependence models. In: Vedaldi, A., Bischof, H., Brox, T., Frahm, J.-M. (eds.) ECCV 2020. LNCS, vol. 12372, pp. 737–753. Springer, Cham (2020). https://doi.org/10.1007/978-3-030-58583-9_44
38. Shwartz-Ziv, R., Armon, A.: Tabular data: deep learning is not all you need. Inf. Fusion **81**, 84–90 (2022)
39. Visengeriyeva, L., Abedjan, Z.: Metadata-driven error detection. In: SSDBM (2018)
40. Yakout, M., Elmagarmid, A.K., Neville, J., Ouzzani, M., Ilyas, I.F.: Guided data repair. arXiv (2011)
41. Yan, J.N., Schulte, O., Zhang, M., Wang, J., Cheng, R.: Scoded: statistical constraint oriented data error detection. In: 2020 ACM SIGMOD (2020)
42. Yuan, B., Chen, J., Zhang, W., Tai, H.S., McMains, S.: Iterative cross learning on noisy labels. In: 2018 IEEE Winter Conference on Applications of Computer Vision (WACV). IEEE (2018)
43. Zhang, C., Bengio, S., Hardt, M., Recht, B., Vinyals, O.: Understanding deep learning (still) requires rethinking generalization. Commun. ACM **64**, 107–115 (2021)
44. Zhu, X., Wu, X.: Class noise vs. attribute noise: a quantitative study. Artif. Intell. Rev. **22**, 177–210 (2004)

# Label-Aware Hierarchical Contrastive Domain Adaptation for Cross-Network Node Classification

Peng Xue[1], Mengqiu Shao[1], Xi Zhou[2], and Xiao Shen[1(✉)]

[1] School of Computer Science and Technology, Hainan University, Haikou, China
{pxue,mengqiushao,xshen}@hainanu.edu.cn
[2] College of Tropical Crops, Hainan University, Haikou, China
xzhou@hainanu.edu.cn

**Abstract.** Cross-network node classification (CNNC) has gained a great deal of attention recently, which aims to transfer the knowledge from a label-rich source network to accurately classify nodes for a different but related unlabeled target network. To tackle the problem of network shift, the existing CNNC algorithms combine graph neural networks (GNNs) and domain adaptation (DA) to solve the problem. However, GNNs are vulnerable to network structure noises, and the traditional DA methods mainly focus on matching the marginal distributions and cannot guarantee the alignment of the class-conditional distributions of different networks. To remedy these deficiencies, we propose a novel label-aware hierarchical contrastive domain adaptation (LHCDA) model to address CNNC. On one hand, we use multi-head graph attention network (GAT) to learn noise-resistant node embeddings. On the other hand, a label-aware hierarchical contrastive domain adaptation module is designed to align the class-conditional distributions across networks at both node-node level and node-class level. Since target labels are unavailable, we use K-means clustering to generate pseudo-labels and employ the prediction confidence to reduce the noises. Extensive experimental results on six CNNC tasks demonstrate that the proposed LHCDA model is superior than previous state-of-the-art CNNC methods.

**Keywords:** Cross-network Node Classification · Contrastive Domain Adaptation · Graph Neural Networks

## 1 Introduction

Node classification has been a popular research topic in the field of graph machine learning. Traditional node classification methods are designed for a single network, which assume that both training nodes and test nodes are sampled from the same network [1]. However, in practical applications, node classification tasks can be conducted across different networks, and the shortage of label information in the target network will reduce the effectiveness of classification [2]. In this context, cross-network node classification (CNNC) [3] is proposed to classify the nodes in the unlabeled target network by leveraging the rich label information of the source network.

X. Yang et al. (Eds.): ADMA 2023, LNAI 14178, pp. 183–198, 2023.
https://doi.org/10.1007/978-3-031-46671-7_13

Recently, CNNC has received increasing attention in the field of graph machine learning. The existing CNNC [3–11] methods typically combine graph neural networks (GNNs) and domain adaptation (DA), however, such combination also suffer from the following limitations: firstly, the existing CNNC methods usually adopt graph convolutional network (GCN) [1] to learn node embeddings, which can easily cause over-smoothing issue, i.e., node embeddings are indistinguishable [12, 13]. Secondly, the existing CNNC methods generally adopt the traditional DA methods [14–17] to align the marginal distributions between the source and the target networks, while neglecting the class information. As a result, the alignment of the class-conditional distributions across networks cannot be guaranteed.

To address these limitations, we propose a label-aware hierarchical contrastive domain adaptation (LHCDA) model for CNNC. On one hand, we employ the multi-head graph attention network (GAT) [18] as the GNN encoder to learn node embeddings, since GAT with adaptive edge weights is more resistant to network structure noise. On the other hand, we propose a label-aware hierarchical contrastive domain adaptation module to achieve the class-conditional alignment at both node-node and node-class levels. For node-node contrastive domain adaptation, given an anchor node from one network, we randomly select nodes from the other network that belong to the same class as the anchor to form positive pairs. And for node-class contrastive domain adaptation, given an anchor node from one network, the positive pairs are formed as the class prototype embeddings with the same class-label from the other network. We minimize the distance of all positive pairs (i.e. same class across networks) while maximize the distance of all negative pairs (i.e. different classes across networks), as a result, the intra-class domain discrepancy is minimized and the inter-class domain discrepancy is maximized. Since target labels are unavailable, we use K-means clustering to produce pseudo-labels and employ the prediction confidence [19] to reduce the noises. Thus, the proposed label-aware hierarchical contrastive domain adaptation module can overcome the limitation of the traditional DA methods that cannot guarantee the alignment of class-conditional distributions between different networks. Extensive experiments on six CNNC tasks demonstrate that the proposed LHCDA method achieves state-of-the-art performance. The contributions of this work are summarized as follows:

1. Unlike previous CNNC methods that generally employ GCN as the GNN encoder to learn node embeddings, we choose multi-head GAT as the GNN encoder, which can ease the network structure noise and over-smoothing issue.
2. We propose a label-aware hierarchical contrastive domain adaptation module to align the class-conditional distributions across networks at both node-node and node-class levels, which overcomes the shortcoming of most existing CNNC methods that ignores the alignment of the class-conditional distributions.
3. Extensive experiments on real-world datasets verify the proposed LHCDA achieves the state-of-the-art CNNC performance.

# 2  Related Work

## 2.1  Contrastive Domain Adaptation

In recent years, contrastive learning has been regarded as a very important part of self-supervised learning and has been widely used in computer vision (CV) [20] and natural language processing (NLP) [21]. Instead of relying on manual labels, self-supervised learning directly uses the data itself as the supervised signals, and aims to learn representations that maximize the mutual information between different contrastive views. Inspired by this, researchers have applied contrastive learning techniques to graph-structured data and proposed a line of graph contrastive learning methods [22], aiming to learn graph representations from unlabeled graph data to alleviate the problem of lack of labels.

Inspired by the remarkable performance of contrastive learning on unsupervised representation learning tasks, researchers have recently explored the application of contrastive learning techniques in the context of domain adaptation. CAN [23] proposed a new metric based on contrastive domain discrepancy, aiming to minimize intra-class domain discrepancy and maximize inter-class domain discrepancy. CDCL [24] employed contrastive learning to achieve domain alignment, and for a given anchor image, it treats images from different domains with the same category of the anchor as positive samples, and images from different domains with different categories of the anchor as negative samples. CLDA [25] aligns features in the source and target domains at both the category level and the sample level. CDA [26] explores a novel self-supervised domain adaptation problem where labels are not available for both source and target domains, and uses an unsupervised approach to define positive and negative pairs so that the learnable parameters obtained from pre-training are robust to domain discrepancy. It should be noted that compared to conventional domain adaptation methods based on statistical matching or adversarial learning, the research on contrastive domain adaptation is still at an early stage and there is still much room for exploration. In addition, only a few contrastive domain adaptation methods are designed for CV, while developing the contrastive domain adaptation methods for graph-structured data remain unexplored.

## 2.2  Cross-Network Node Classification

CNNC receives increasing attention in graph machine learning that aims to accurately predict the node labels of the target network using the labeled data from the source network and unlabeled data from the target network. To generate pseudo-labels for target network, CDNE [3] trains a logistic regression classifier based on the node attributes and observed node labels of the source network, and then uses this classifier to infer fuzzy labels for unlabeled nodes of the target network. In ACDNE [5], the deep network embedding module employs two feature extractors to jointly maintain node attribute affinity and topological proximities, a node classifier is also added to distinguish different node labels, and an adversarial domain adaptative technique based on gradient reversal layer (GRL) [14] is used to make the node representations network invariant. DM-GNN [4] improves on ACDNE by incorporating a label-propagation node classifier and the conditional adversarial domain adaptation technique. AdaGCN [6] uses GCN to learn node embeddings and Wasserstein distance-based guided adversarial domain adaptation to reduce the distribution differences between different networks. ASN [9] employs two domain-private encoders to extract domain-specific features in each network and employs a domain-shared encoder to learn domain-invariant features across networks. UDAGCN [7] uses GRL-based adversarial domain adaption and constructs dual-GCN capturing local consistency and global consistency during feature aggegation. Based on UDAGCN, GCLN [8] introduces a node-graph based graph contrastive learning method to maximize the mutual information between node embedding and graph embedding.

The proposed LHCDA model innovatively incorporates the contrastive domain adaptation technique to address CNNC, which is distinct from other CNNC methods in three aspects. Firstly, the proposed LHCDA uses multi-head GAT with adaptive edge weights for neighbor aggregation, which are more resistant to network structure noise. Meanwhile, compared to most CNNC methods, which only consider to match marginal distribution across networks, the proposed LHCDA sets up a hierarchical label-aware graph contrastive domain adaptation at both node-node and node-class levels, where the class information is taken into account to better match the class-conditional distributions across networks. Finally, the proposed LHCDA uses both the traditional domain adaptation method and the hierarchical graph contrastive domain adaptation, which can not only match the marginal distribution, but also match the class-conditional distribution across networks.

## 3   Problem Definition

Let $\mathcal{G} = (\mathcal{V}, \mathcal{E}, A, X, Y)$ denote a network, where $\mathcal{V} = (v_1, v_2, \ldots, v_n)$ represents a set of nodes and $\mathcal{E} \subseteq \mathcal{V} \times \mathcal{V}$ represents a set of edges in $\mathcal{G}$. $A \in \{0, 1\}^{n \times n}$ is the adjacency matrix of $\mathcal{G}$, where $n$ is the number of nodes in $\mathcal{G}$, and $A_{ij} = 1$ if $(v_i, v_j) \in \mathcal{E}$, otherwise $A_{ij} = 0$. $X \in \mathbb{R}^{n \times \omega}$ is the node attribute matrix, where $\omega$ is the number of node attributes, and $x_i \in \mathbb{R}^{\omega}$ is the node attribute vector of $v_i$. $Y \in \{0, 1\}^{n \times c}$ is the node label matrix, where $c$ is the number of label categories, and $Y_{ip} = 1$ if $v_i$ is labeled with category $p$, otherwise, $Y_{ip} = 0$. Suppose that we have a source network $\mathcal{G}^s = (\mathcal{V}^s, \mathcal{E}^s, A^s, X^s, Y^s)$ with label information and a target network $\mathcal{G}^t = (\mathcal{V}^t, \mathcal{E}^t, A^t, X^t)$ without label information, the goal of CNNC is to leverage the label information of the source network to assist node classification in a completely unlabeled target network.

## 4   Proposed Model

In this section, we present our model *Label-aware Hierarchical Contrastive Domain Adaptation* (LHCDA) to address CNNC.

### 4.1   Overview of Model Framework

As shown in Fig. 1, the proposed LHCDA model consists of five main components, namely a GNN encoder $f_h$, a node classifier $f_y$, a domain discriminator $f_d$, a negative learning module, and a label-aware hierarchical contrastive domain adaptation module.

Specifically, the GNN encoder $f_h$ is applied to learn each network's node embeddings, the node classifier $f_y$ is trained to make node representations label-discriminative, and the domain discriminator $f_d$ is employed in an adversarial manner to align marginal distributions across networks. Besides, the negative learning module is employed to guide the selection process for pseudo-labeling, and the label-aware hierarchical contrastive domain adaptation module aims to align the class-conditional distributions across networks.

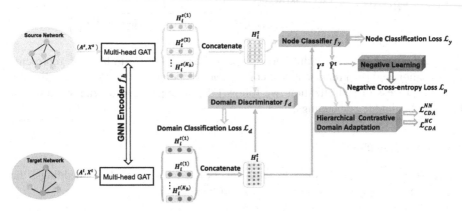

**Fig. 1.** The overview framework of the proposed LHCDA model. It consists of a GNN encoder, a node classifier, a domain discriminator, a negative learning module, and a hierarchical contrastive domain adaptation module.

## 4.2 Graph Neural Network Encoder

GNNs have demonstrated excellent performance in graph representation learning. GNNs learn node embeddings by fusing the graph structure and node features, the conventional GNNs include GCN [1], GAT [18], GraphSAGE [27] and so on [28, 29]. In the CNNC problem, a series of methods [6, 8] choose to use GCN as the GNN encoder to learn node embeddings. However, GCN uses fixed edge weights during neighborhood aggregation and causes the over-smoothing issue easily. Thus, instead of GCN, we adopt multi-head GAT [18] as the GNN encoder, which can learn more robust node embeddings by taking advantage of adaptive edge weights.

We adopt a mini-batch training strategy. Specifically, we randomly sample a batch of nodes $\mathbb{B} = \{v_i\}_{i=1}^{b}$ from a given network $\mathcal{G}$. The embedding of each node can be updated at each GNN layer using the embeddings of its neighboring nodes. After $l$ layer's aggregation, the embedding of each node would capture the structural information of the $l$-hop neighborhood. The process of this update can be described as:

$$a_i^{(l-1)} = \text{AGGREGATE}^{l-1}\left(\left\{h_u^{(l-1)}, \hat{A}_{iu}^{(l-1)} : u \in \mathcal{N}_i^{\mathbb{B}}\right\}\right)$$
$$h_i^l = \text{COMBINE}^{l-1}\left(\left\{h_i^{(l-1)}, a_i^{(l-1)}\right\}\right) \tag{1}$$

where $\mathcal{N}_i^\mathbb{B}$ is a set of first-order neighbors of $v_i$ in the batch $\mathbb{B}$, $a_i^{(l-1)}$ denotes the aggregated neighborhood embeddings of $v_i$ among $\mathcal{N}_i^\mathbb{B}$ at $(l-1)$-th layer, and $h_i^l$ denotes the embedding of $v_i$ at $l$-th layer.

During neighborhood aggregation of $v_i$ at the $(l-1)$-th layer, the adaptive edge weight of $(v_i, v_j)$, which reflects the importance degree of $v_j$ towards $v_i$ is learned by GAT [18], as:

$$\hat{A}_{iu}^{(l-1)} = \frac{\exp\left(\text{LeakyReLU}\left(t^{(l-1)^T}\left[W^{(l-1)}h_i^{(l-1)} \| W^{(l-1)}h_u^{(l-1)}\right]\right)\right)}{\Sigma_{j\in\mathcal{N}_i^\mathbb{B}\cup\{i\}}\exp\left(\text{LeakyReLU}\left(t^{(l-1)^T}\left[W^{(l-1)}h_i^{(l-1)} \| W^{(l-1)}h_j^{(l-1)}\right]\right)\right)} \tag{2}$$

where $t^{(l-1)}$ and $W^{(l-1)}$ are the learnable parameters at $(l-1)$-th layer, $\cdot^T$ represents transposition and $[\cdot\|\cdot]$ represents the concatenation operation. In order to stabilize the learning process of self-attention, we follow [18] to employ multi-head attentions to learn the final node embeddings.

### 4.3  Node Classifier

Here, we use a node classifier to learn label-discriminative node embeddings, the classifier is constructed by a single-layer multi-head GAT,

$$\hat{y}_i = f_y(h_i; \theta_h) \tag{3}$$

where $h_i$ represents the final embedding of $v_i$, $\theta_h$ is the learnable parameters in the node classifier. Based on the observed node labels of the source network and the node label probabilities predicted by the node classifier in Eq. (3), the node classification loss is defined as:

$$\mathcal{L}_y = -\frac{1}{|\mathbb{B}^s|}\sum_{i=1}^{|\mathbb{B}^s|}\sum_{p=1}^{c} Y_{ip}^s \log \hat{Y}_{ip}^s + \left(1 - Y_{ip}^s\right)\log\left(1 - \hat{Y}_{ip}^s\right) \tag{4}$$

where $\widehat{Y}_{ip}^s$ denotes the predicted probability of $v_i^s$ belonging to category $p$, and $|\mathbb{B}^s|$ denotes the size of nodes in the source mini-batch.

### 4:4  Domain Discriminator

To reduce the distribution discrepancy between the source network and the target network, we use an adversarial domain adaptation method based on GRL [14] to train GNN encoder and domain discriminator. Here, we employ a multi-layer perceptron (MLP) to construct the domain discriminator $f_d$, which can be defined as:

$$\hat{d}_i = f_d(h_i; \theta_d) \tag{5}$$

where $h_i$ denotes the final embedding of $v_i$ and $\theta_d$ is the learnable parameters in the domain discriminator, $\hat{d}_i$ is the predicted probability of $v_i$ from the target network. Based on the ground-truth domain labels of the nodes in the source network and the target network, and the predicted domain probabilities of the domain discriminator in Eq. (5), the domain classification loss is defined as:

$$\mathcal{L}_d = -\frac{1}{|\mathbb{B}^s| + |\mathbb{B}^t|} \sum_{i=1}^{|\mathbb{B}^s|+|\mathbb{B}^t|} d_i \log \hat{d}_i + (1 - d_i) \log \left(1 - \hat{d}_i\right) \tag{6}$$

where $d_i$ is the ground-truth domain label of $v_i$, and $d_i = 1$ if $v_i \in \mathbb{B}^t$, and $d_i = 0$ if $v_i \in \mathbb{B}^s$. To achieve simultaneously adversarial training of the GNN encoder and the domain discriminator, we insert a GRL between the two components instead of training them alternately.

### 4.5 Pseudo-labeling for Target Network

In the CNNC problem, ground-truth node labels from the source network are available, while from the target network are unavailable. However, the proposed label-aware hierarchical contrastive domain adaptation scheme requires to define positive and negative pairs based on the label information from different networks. To address this, we follow [23] to leverage K-means clustering to generate pseudo-labels. Specifically, the initial cluster centroid of each class is set as the class prototype (i.e., $c_k = \frac{\sum_{i=1}^{n^s} Y_{ik}^s h_i^s}{\sum_{i=1}^{n^s} Y_{ik}^s}$) in the source network, and the number of clusters are set as the number of label categories (i.e. $c$). Then, by assigning each target node to the closest centroid, we can obtain the pseudo-labels of target network, according to the clustering results. These clusters can represent the distribution of different classes in the target network. To make the cluster-labels more robust, we remove the ambiguous nodes far from its assigned cluster centroids.

In addition, we follow [19] to select a subset of nodes with extremely low prediction confidence for negative learning, as:

$$g_q^i = 1 \left[ \hat{Y}_{iq}^t \le \mathcal{T}_\gamma \right] \tag{7}$$

where $\hat{Y}_{iq}^t$ denotes the prediction confidence of a target node $v_i^t$ belonging to class $q$, and $\mathcal{T}_\gamma$ is the threshold of prediction confidence. Note that if $v_i^t$ is selected for negative learning w.r.t. category $q$, then it reflects that $v_i^t$ is very unlikely to belong to class $q$. Then, taking the selected nodes for negative learning to iteratively re-train the model, the negative cross-entropy loss is defined as:

$$\mathcal{L}_P = -\frac{1}{S^i} \sum_{q=1}^{c} g_q^i \log \left(1 - \hat{Y}_{iq}^t\right) \tag{8}$$

where $S^i = \sum_q g_q^i$ denotes the number of pseudo-labels assigned to $v_i^t$. Minimizing Eq. (8) makes $\hat{Y}_{iq}^t$ tend to be 0 if $\hat{Y}_{iq}^t \le \mathcal{T}_\gamma$. That is, if $v_i^t$ is unlikely to belong to class $q$, then the model would progressively decrease $\hat{Y}_{iq}^t$ to be 0, resulting in an increase in the probability values of other classes.

### 4.6  Label-Aware Hierarchical Contrastive Domain Adaptation

Next, we propose a label-aware hierarchical contrastive domain adaptation approach to minimize the intra-class domain discrepancy and maximize the inter-class domain discrepancy, thus matching class-conditional distributions of the source and target networks.

- Definition of positive and negative pairs at node-node level: Considering $\mathcal{G}^s$ and $\mathcal{G}^t$ as two contrastive views, sample a source batch $\mathbb{B}^s = \{v_i^s\}_{i=1}^b$ from $\mathcal{G}^s$, and a target batch $\mathbb{B}^t = \{v_i^t\}_{j=1}^b$ from $\mathcal{G}^t$. Given the node embedding $h_i^s$ in $\mathbb{B}^s$ as the anchor, the positives are defined as the node embeddings from different networks belonging to the same class. And the negatives of the anchor $h_i^s$ are defined as the node embeddings from different networks belonging to different classes. Similar definition can be given if a node embedding in the target network is selected as the anchor. Figure 2 illustrates the idea.

Based on the above definition of positive and negative pairs, the node-node contrastive domain adaptation (CDA) loss associated with the anchor $h_i^s$ is defined as:

$$\ell_{\text{CDA}}^{\text{NN}}\left(h_i^s\right) = -\log \frac{\frac{1}{|\mathcal{P}_i^s|}\sum_{v_j^t \in \mathcal{P}_i^s}\exp\left(h_i^s \cdot h_j^t / \tau\right)}{\underbrace{\sum_{v_j^t \in \mathcal{P}_i^s}\exp\left(h_i^s \cdot h_j^t / \tau\right)}_{\text{positive pairs}} + \underbrace{\sum_{v_i^t \notin \mathcal{P}_i^s}\exp\left(h_i^s \cdot h_l^t / \tau\right)}_{\text{negative pairs}}} \tag{9}$$

where $\mathcal{P}_i^s = \left\{v_j^t | v_j^t \in \mathcal{V}^t, \bar{\bar{Y}}_{jp}^t = Y_{ip}^s = 1, \exists p \in \{1, \cdots, c\}\right\}$ represents a set of positives associated with $h_i^s$, in which $Y_{ip}^s = 1$ indicates that the source network node $v_i^s$ has an observed label of category $p$ and $\bar{\bar{Y}}_{jp}^t = 1$ indicates that the target network node $v_j^t$ has a pseudo-label of category $p$, $\tau$ is a temperature parameter, and $\langle \cdot \rangle$ is the inner product operation. Minimizing Eq. (9) will pull positive pairs (i.e. cross-network nodes with the same class-label) close, and push negative pairs (i.e. cross-network nodes with different class-labels) away.

Similarly, choosing the node embedding $h_j^t$ in $\mathbb{B}^t$ as an anchor, its associated node-node CDA loss can be defined similar to Eq. (9). Choosing the embedding of each node in $\mathbb{B}^s$ and $\mathbb{B}^t$ as anchors, the total node-node CDA loss is defined as:

$$\mathcal{L}_{\text{CDA}}^{\text{NN}} = \frac{1}{|\mathbb{B}^s|}\sum_{i=1}^{|\mathbb{B}^s|}\ell_{\text{CDA}}^{\text{NN}}\left(h_i^s\right) + \frac{1}{|\mathbb{B}^t|}\sum_{j=1}^{|\mathbb{B}^t|}\ell_{\text{CDA}}^{\text{NN}}\left(h_j^t\right) \tag{10}$$

- Definition of positive and negative pairs at node-class level: Considering $\mathcal{G}^s$ and $\mathcal{G}^t$ as two contrastive views, sample a source batch $\mathbb{B}^s = \{v_i^s\}_{i=1}^b$ from $\mathcal{G}^s$, and a target batch $\mathbb{B}^t = \{v_i^t\}_{j=1}^b$ from $\mathcal{G}^t$. Given the node embedding $h_i^s$ in $\mathbb{B}^s$ as the anchor, the positives are defined as the class prototype embeddings with the same class-label from different networks. The negatives of anchor $h_i^s$ are defined as class prototype embeddings from different networks with different class. Similar definition can be given if a node embedding in target network is selected to be the anchor. Figure 3 illustrates the idea.

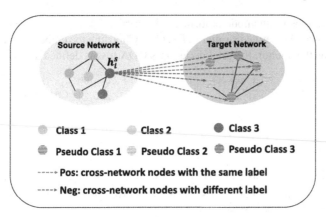

**Fig. 2.** Illustration of positives and negatives in node-node contrastive domain adaptation.

Based on the above definition of positive and negative pairs, given $h_i^s$ as an anchor, its associated node-class CDA loss is defined as:

$$\ell_{\text{CDA}}^{\text{NC}}\left(h_i^s\right) = -\log \frac{\sum_{k=1}^{c} \mathbb{I}\left(Y_{ik}^s = 1\right)\exp\left(h_i^s \cdot c_k^t / \tau\right)}{\underbrace{\sum_{k=1}^{c} \mathbb{I}\left(Y_{ik}^s = 1\right)\exp\left(h_i^s \cdot c_k^t / \tau\right)}_{\text{positive pairs}} + \underbrace{\sum_{q=1}^{c} \mathbb{I}\left(Y_{iq}^s = 0\right)\exp\left(h_i^s \cdot c_q^t / \tau\right)}_{\text{negative pairs}}} \tag{11}$$

where $c_k^t = \frac{\sum_{i=1}^{n^t} \bar{\bar{Y}}_{ik}^t h_i^t}{\sum_{i=1}^{n^t} \bar{\bar{Y}}_{ik}^t}$ represents the class prototype embedding of the target network corresponding to category $k$, and $\mathbb{I}(\cdot)$ is an indicator. Minimization Eq. (11) makes each anchor to be close to the corresponding class prototype of different networks while far away from the different class prototypes of different networks.

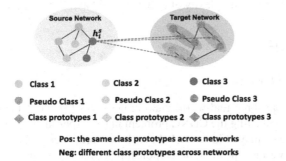

**Fig. 3.** Illustration of positives and negatives in node-class contrastive domain adaptation.

Similarly, choosing the node embedding $h_j^t$ in $\mathbb{B}^t$ as an anchor, its associated node-class CDA loss can be defined similar to Eq. (11). Choosing the embedding of each node in $\mathbb{B}^s$ and $\mathbb{B}^t$ as the anchor, the total node-class CDA loss is defined as:

$$\mathcal{L}_{CDA}^{NC} = \frac{1}{|\mathbb{B}^s|} \sum_{i=1}^{|\mathbb{B}^s|} \ell_{CDA}^{NC}(h_i^s) + \frac{1}{|\mathbb{B}^t|} \sum_{j=1}^{|\mathbb{B}^t|} \ell_{CDA}^{NC}(h_j^t) \tag{12}$$

### 4.7 Overall Objectives

The overall objective consists of five parts: the node classification loss $\mathcal{L}_y$, the domain classification loss $\mathcal{L}_d$, the node-node CDA loss $\mathcal{L}_{CDA}^{NN}$, the node-class CDA loss $\mathcal{L}_{CDA}^{NC}$ and the negative learning loss $\mathcal{L}_P$, as:

$$\min_{\theta_h \theta_y} \left\{ \mathcal{L}_y + \lambda \max_{\theta_d}\{-\mathcal{L}_d\} + \mu \mathcal{L}_{CDA}^{NN} + \varphi \mathcal{L}_{CDA}^{NC} + \eta \mathcal{L}_P \right\} \tag{13}$$

where $\theta_h$, $\theta_y$ and $\theta_d$ represent the learnable parameters of the GNN encoder $f_h$, the node classifier $f_y$ and the domain discriminator $f_d$, respectively. In addition, $\lambda$, $\mu$, $\varphi$ and $\eta$ denote the weight of $\mathcal{L}_d$, $\mathcal{L}_{CDA}^{NN}$, $\mathcal{L}_{CDA}^{NC}$ and $\mathcal{L}_P$ respectively to balance the importance of different loss terms.

### 4.8 Time Complexity

The time complexity of GAT encoder on $\mathbb{B}^s$ and $\mathbb{B}^t$ is $\mathcal{O}(bwf + (|\mathcal{E}^{\mathbb{B}^s}| + |\mathcal{E}^{\mathbb{B}^t}|)f)$, where $b$ is the batch size, $w$ is the number of node attributes, $f$ is the number of embedding dimensions, $|\mathcal{E}^{\mathbb{B}^s}|$ and $|\mathcal{E}^{\mathbb{B}^t}|$ are the number of edges in $\mathbb{B}^s$ and $\mathbb{B}^t$. The time complexity of node-node CDA on $\mathbb{B}^s$ and $\mathbb{B}^t$ is $\mathcal{O}(b^2 f)$, and the time complexity of node-class CDA on $\mathbb{B}^s$ and $\mathbb{B}^t$ is $\mathcal{O}(bcf)$, where $c$ denotes the number of label categories. Because $|\mathcal{E}^{\mathbb{B}^s}| \ll b^2$ and $|\mathcal{E}^{\mathbb{B}^t}| \ll b^2$, the time complexity of LHCDA on $\mathbb{B}^s$ and $\mathbb{B}^t$ is $\mathcal{O}(bwf + b^2 f + bcf)$.

## 5 Experiments

### 5.1 Experimental Setup

**Datasets.** We adopt three real-world citation networks [3] (i.e. Citationv1, DBLPv7 and ACMv9) for evaluating the CNNC problem in our experimental studies, the details of the datasets are displayed in Table 1. In each network, each node represents a paper and each edge represents the citation relationship between two papers. Each node has a vector of attributes, which are the keywords extracted from the paper titles. Each node has a category label, which represents the research area to which the paper belongs. We alternate two networks as the source or the target network for CNNC tasks. For example, we use A→C to denote that ACMv9 is the source network while Citationv1 is the target network. Six CNNC tasks are conducted, namely A→C, A→D, C→A, C→D, D→A and D→C.

**Table 1.** Statistics of the Datasets.

| Dataset | # Nodes | # Edges | # Attributes | # Labels |
|---------|---------|---------|--------------|----------|
| Citationv1 | 8935 | 15113 | 6775 | 5 |
| DBLPv7 | 5484 | 8130 | | |
| ACMv9 | 9360 | 15602 | | |

**Baselines.** We compared the proposed model with the following state-of-the-art baselines:

- Domain Adaptation: MMD [30] is a statistical matching DA method, and GRL [14] is an adversarial learning DA method.
- Graph Neural Networks: GCN [1] and GAT [18] are the most well-known GNNs for single-network node classification. GCN conducts graph convolutions on the given network structure and node attributes for node classification. GAT incorporates an attention mechanism to determine the importance of each node to the central node.
- Cross-network Node Classification: CDNE [3], ACDNE [5], ASN [9], AdaGCN [6] and UDAGCN [7] are the state-of-the-art CNNC algorithms.

**Implementation Details.** We adopted PyTorch [31] to implement our proposed method, and for other baselines, we adopted their open-source implementations. To train our method, we adopted the Adam optimizer with 100 epochs over shuffled mini batches with the batch size $b$ of 4000. A $\ell_2$-norm regularization was applied on the trainable parameters to prevent overfitting with the weight decay of 0.01. The initial learning rate $lr_0$ was set as 0.01 and the learning rate was decayed as $\frac{lr_0}{(1+10p)^{0.75}}$, where $p$ linearly changes from 0 to 1 during training. The domain discriminator was constructed by an MLP with two hidden layers, with the dimensionality of each layer as 32. We set the weight of $\mathcal{L}_{CDA}^{NC}$ (i.e., $\varphi$) and the weight of $\mathcal{L}_{CDA}^{NN}$ (i.e., $\mu$) as 1, and the weight of $\mathcal{L}_P$ (i.e., $\eta$) as 1. We set the temperature parameter $\tau$ as 1. In addition, we set the number of attention heads in the GNN encoder as 4, and the number of attention heads in the node classifier as 2. Besides, we set the number of hidden layers of the GNN encoder $\mathbb{L}$ as 2,3 or 4 for different tasks, for A→C, $\mathbb{L}$ is 2, for C→A and D→A, $\mathbb{L}$ is 3, others are set as 4. Besides, the number of embedding dimensions $f$ of each attention head was set as 16. Finally, we set the threshold $\delta$ to filter out noisy cluster-labels as 0.1, and the threshold $\mathcal{T}_y$ of negative learning as 0.2.

## 5.2 Performance Comparison

To evaluate the CNNC performance, we follow [3–6] to adopt Micro-F1 and Macro-F1 as two evaluation metrics. Table 2 shows the performance comparisons of different methods for CNNC. Firstly, it can be found that our method is significantly better than the DA methods (i.e. MMD and GRL) and the GNNs (i.e. GCN and GAT). This is because the DA methods do not consider network topology, while the GNNs do not consider the discrepancy between different networks.

**Table 2.** Micro-F1 and Macro-F1 scores of CNNC. The best results among all comparing algorithms are shown in bold.

| Tasks | F1(%) | MMD | GRL | GCN | GAT | CDNE | ACDNE | ASN | AdaGCN | UDAGCN | **LHCDA** |
|---|---|---|---|---|---|---|---|---|---|---|---|
| A→C | Micro | 54.48 | 56.73 | 73.56 | 77.44 | 78.91 | 83.27 | 80.15 | 80.16 | 81.44 | **83.89** |
|  | Macro | 52.01 | 54.92 | 70.03 | 75.03 | 77.00 | 81.66 | 76.89 | 78.05 | 79.79 | **82.15** |
| A→D | Micro | 54.48 | 55.35 | 68.22 | 72.04 | 72.03 | 76.57 | 76.88 | 75.42 | 76.40 | **78.84** |
|  | Macro | 51.16 | 52.49 | 64.13 | 69.25 | 69.78 | 74.31 | 74.37 | 72.25 | 74.77 | **76.88** |
| C→A | Micro | 54.16 | 55.53 | 71.32 | 71.72 | 77.52 | 79.56 | 74.76 | 76.01 | 76.75 | **80.04** |
|  | Macro | 51.15 | 53.45 | 69.19 | 70.41 | 76.79 | 78.88 | 74.53 | 75.64 | 75.88 | **79.74** |
| C→D | Micro | 57.01 | 57.85 | 71.24 | 73.18 | 74.15 | 77.35 | 76.84 | 76.60 | 78.25 | **80.11** |
|  | Macro | 53.58 | 55.15 | 68.12 | 72.61 | 71.71 | 76.09 | 73.77 | 74.83 | 77.00 | **78.94** |
| D→A | Micro | 51.43 | 53.11 | 66.83 | 67.77 | 76.59 | 76.34 | 69.16 | 73.81 | 73.10 | **78.60** |
|  | Macro | 46.51 | 50.07 | 62.91 | 65.21 | 75.91 | 76.09 | 70.38 | 71.80 | 72.17 | **77.24** |
| D→C | Micro | 53.40 | 56.27 | 71.63 | 74.34 | 76.91 | 82.09 | 78.58 | 80.54 | 77.47 | **82.52** |
|  | Macro | 49.62 | 54.13 | 67.19 | 73.10 | 78.05 | 80.25 | 74.82 | 78.76 | 74.88 | **80.59** |
| Avg | Micro | 54.16 | 55.81 | 70.47 | 72.75 | 76.02 | 79.20 | 76.06 | 77.09 | 77.24 | **80.67** |
|  | Macro | 50.67 | 53.06 | 66.93 | 70.94 | 74.87 | 77.88 | 74.13 | 75.22 | 75.75 | **79.26** |

Secondly, LHCDA exhibits better performance than other CNNC baselines, for example, LHCDA outperforms the second-best method ACDNE by an average of 1.47% and 1.38% in terms of F1-micro and F1-macro. This is mainly because other CNNC methods usually use the traditional DA methods, which only consider the cross-network marginal distribution alignment while ignoring the cross-network class-conditional distribution alignment. However, in LHCDA, we propose a label-aware hierarchical contrastive domain adaptation, which can align the class-conditional distributions across networks effectively. On the other hand, previous CNNC baselines generally choose GCN as the GNN encoder, in contrast, we choose GAT as the GNN encoder. The outperformance of GAT over GCN can also verify that the adaptive edge weights are beneficial for CNNC than the fixed edge weights.

### 5.3 Ablation Study

As shown in Table 3, we conducted ablation studies to demonstrate the effectiveness of the different model designs in LHCDA. We can find that in the absence of node classifier or domain discriminator, the performance drops sharply. This is because in DA, the target risk is upper bounded by the source risk and domain discrepancy [32]. In addition, the performance becomes significantly worse if without both $\mathcal{L}_{CDA}^{NN}$ and $\mathcal{L}_{CDA}^{NC}$. This reflects that the proposed hierarchical contrastive domain adaptation module is indeed effective for cross-network class-conditional alignment. Moreover, without either $\mathcal{L}_{CDA}^{NN}$ or $\mathcal{L}_{CDA}^{NC}$, the F1 scores are lower than LHCDA, which means that contrastive domain adaptation at both node-node and node-class levels are helpful for cross-network class-conditional alignment. Finally, we also find that if without negative learning loss $\mathcal{L}_P$, the result would

**Table 3.** Model variants of LHCDA on six tasks.

| Model variants | F1(%) | A→C | A→D | C→A | C→D | D→A | D→C |
|---|---|---|---|---|---|---|---|
| w/o node classifier | Micro | 64.47 | 58.66 | 55.78 | 54.95 | 56.18 | 59.94 |
| | Macro | 56.29 | 49.17 | 50.52 | 45.54 | 52.99 | 52.51 |
| w/o domain discriminator | Micro | 59.38 | 65.58 | 67.25 | 73.81 | 65.19 | 71.82 |
| | Macro | 51.14 | 52.94 | 55.83 | 67.85 | 54.84 | 60.59 |
| w/o $\mathcal{L}_{CDA}^{NN}$ | Micro | 83.01 | 78.54 | 78.47 | 79.47 | 76.58 | 81.58 |
| | Macro | 81.22 | 76.25 | 76.83 | 77.47 | 74.66 | 78.53 |
| w/o $\mathcal{L}_{CDA}^{NC}$ | Micro | 83.16 | 77.96 | 78.14 | 79.19 | 76.12 | 82.21 |
| | Macro | 80.44 | 75.51 | 73.76 | 76.60 | 71.52 | 78.77 |
| w/o $\mathcal{L}_{CDA}^{NN}$ and $\mathcal{L}_{CDA}^{NC}$ | Micro | 81.32 | 77.93 | 75.16 | 78.47 | 72.51 | 80.97 |
| | Macro | 75.79 | 75.36 | 68.66 | 75.44 | 64.16 | 76.05 |
| w/o $\mathcal{L}_P$ | Micro | 82.61 | 78.06 | 78.55 | 77.84 | 77.82 | 78.99 |
| | Macro | 77.26 | 76.45 | 78.14 | 73.17 | 76.51 | 76.36 |
| LHCDA (Ours) | Micro | 83.89 | 78.84 | 80.04 | 80.11 | 78.60 | 82.52 |
| | Macro | 82.15 | 76.88 | 79.74 | 78.94 | 77.24 | 80.59 |

reduce significantly, which means that negative learning is beneficial for pseudo-labeling in our model.

## 5.4 Parameter Sensitivity

Here, we study the impacts of the hyper-parameters $b$, $K_h$, $f$ and $\mathbb{L}$ on the task A→C, since during the experiments we found that the performance is insensitive to parameters $\mu$, $\varphi$ and $\eta$. From Fig. 4(a), we found that as the batch size $b$ increases, the results are better, probably because in label-aware contrastive domain adaptation, no positive pairs can be formed if $b$ is too small. The parameter $K_h$ represents the number of attention heads in the GNN encoder, as shown in Fig. 4(b), when $K_h$ equals 1, the result is the worst, and when it equals 4, the best result is achieved. For the number of embedding dimensions, we can find that $f$=8 achieves the worst result, while the performance is insensitive to the values of 16, 32 and 64 in Fig. 4. (c). Finally in Fig. 4. (d), the GNN encoder with the number of hidden layers $\mathbb{L} \in \{2, 3\}$ can obviously improve the performance over a single-layer GNN, this is because aggregating the information from high-order neighbors is beneficial for node classification [33]. While $\mathbb{L}$ equals 4 or 5, there is a significant performance drop, because too deep GNNs can cause over-smoothing and over-fitting easily.

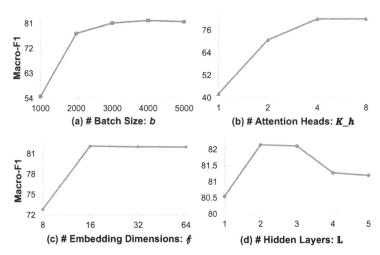

**Fig. 4.** Sensitivities of the hyper-parameters $b$, $K_h$, $f$, and $\mathbb{L}$ on the task A → C.

## 6   Conclusion

In this paper, we propose a novel LHCDA model for CNNC, by exploiting a label-aware hierarchical contrastive domain adaptation module to align the class-conditional distributions across networks at both node-node and node-class levels. Specifically, the node-node contrastive domain adaptation can pull together nodes of the same class across networks while pushing apart nodes from different classes across networks. Besides, the node-class contrastive domain adaptation makes nodes close to the corresponding class prototype in different networks while far away from the different class prototypes in different networks. The experiments on the real-word benchmark datasets demonstrate that the proposed method surpasses the strong CNNC baselines.

**Acknowledgments.** This research was supported in part by Hainan Provincial Natural Science Foundation of China (No. 322RC570), National Natural Science Foundation of China (No. 62102124), and the Research Start-up Fund of Hainan University (No. KYQD(ZR)-22016).

## References

1. Kipf, T.N., Welling, M.: Semi-supervised classification with graph convolutional networks. In: International Conference on Learning Representations (2017)
2. Shen, X., Mao, S., Chung, F.-L.: Cross-network learning with fuzzy labels for seed selection and graph sparsification in influence maximization. IEEE Trans. Fuzzy Syst. **28**(9), 2195–2208 (2020)
3. Shen, X., Dai, Q., Mao, S., Chung, F.-L., Choi, K.-S.: Network together: Node classification via cross-network deep network embedding. IEEE Trans. Neural Netw. Learn. Syst. **32**(5), 1935–1948 (2021)
4. Shen, X., Pan, S., Choi, K.-S., Zhou, X.: Domain-adaptive message passing graph neural network. Neural Netw. **164**, 439–454 (2023)

5. Shen, X., Dai, Q., Chung, F.I., Lu, W., Choi, K.S.: Adversarial deep network embedding for cross-network node classification. In Proceedings of the AAAI Conference on Artificial Intelligence, vol. 34, no. 03, pp. 2991-2999 (2020)
6. Dai, Q., Wu, X.-M., Xiao, J., Shen, X., Wang, D.: Graph transfer learning via adversarial domain adaptation with graph convolution. IEEE Trans. Knowl. Data Eng. **35**(5), 4908–4922 (2023)
7. Wu, M., Pan, S., Zhou, C., Chang, X., Zhu, X.: Unsupervised domain adaptive graph convolutional networks. In: Proceedings of The Web Conference 2020, pp. 1457-1467 (2020)
8. Wu, M., Pan, S., Zhu, X.: Attraction and repulsion: unsupervised domain adaptive graph contrastive learning network. IEEE Trans. Emerg. Top. Comput. Intell. **6**(5), 1079–1091 (2022)
9. Zhang, X., Du, Y., Xie, R., Wang, C.: Adversarial separation network for cross-network node classification. In Proceedings of the 30th ACM International Conference on Information & Knowledge Management, pp. 2618–2626 (2021)
10. Xiao, J., Dai, Q., Xie, X., Dou, Q., Kwok, K.W., Lam, J.: Domain adaptive graph infomax via conditional adversarial networks. IEEE Trans. Netw. Sci. Eng. **10**(1), 35–52 (2022)
11. Zhang, Y., Song, G., Du, L., Yang, S., Jin, Y.: DANE: domain adaptive network embedding. In: International Joint Conference on Artificial Intelligence, pp. 4362–4368 (2019)
12. Dai, E., Jin, W., Liu, H., Wang, S.: Towards robust graph neural networks for noisy graphs with sparse labels. In Proceedings of the Fifteenth ACM International Conference on Web Search and Data Mining, pp. 181–191 (2022)
13. Li, Q., Han, Z., Wu, X.M.: Deeper insights into graph convolutional networks for semi-supervised learning. In: Proceedings of the AAAI Conference on Artificial Intelligence, vol. 32, no. 1 (2018)
14. Ganin, Y., et al.: Domain-adversarial training of neural networks. J. Mach. Learn. Res. **17**(1), 2096–2030 (2016)
15. Long, M., Cao, Y., Wang, J., Jordan, M.: Learning transferable features with deep adaptation networks. In: International Conference on Machine Learning, PMLR pp. 97–105 (2015)
16. Ganin, Y., Lempitsky, V.: Unsupervised domain adaptation by backpropagation. In: International Conference on Machine Learning, pp. 1180–1189. PMLR (2015)
17. Saenko, K., Kulis, B., Fritz, M., Darrell, T.: Adapting visual category models to new domains. In: Daniilidis, K., Maragos, P., Paragios, N. (eds.) ECCV 2010. LNCS, vol. 6314, pp. 213–226. Springer, Heidelberg (2010). https://doi.org/10.1007/978-3-642-15561-1_16
18. Veličković, P., Cucurull, G., Casanova, A., Romero, A., Lio, P., Bengio, Y.: Graph attention networks. In: International Conference on Learning Representations (2018)
19. Rizve, M.N., Duarte, K., Rawat, Y.S., Shah, M.: In defense of pseudo-labeling: an uncertainty-aware pseudo-label selection framework for semi-supervised learning. In: International Conference on Learning Representations (2021)
20. Chen, T., Kornblith, S., Norouzi, M., Hinton, G.: A simple framework for contrastive learning of visual representations. In: International Conference on Machine Learning, pp. 1597–1607. PMLR (2020)
21. Aberdam, A., et al.: Sequence-to-sequence contrastive learning for text recognition. In: Proceedings of the IEEE/CVF Conference on Computer Vision and Pattern Recognition, pp. 15302–15312 (2021)
22. Xie, Y., Xu, Z., Zhang, J., Wang, Z., Ji, S.: Self-supervised learning of graph neural networks: a unified review. IEEE Trans. Pattern Analy. Mach. Intell. 2412–2429 (2022)
23. Kang, G., Jiang, L., Yang, Y., Hauptmann, A.G.: Contrastive adaptation network for unsupervised domain adaptation. In: Proceedings of the IEEE/CVF Conference on Computer Vision and Pattern Recognition, pp. 4893–4902 (2019)
24. Wang, R., Wu, Z., Weng, Z., Chen, J., Qi, G.J., Jiang, J.G.: Cross-domain contrastive learning for unsupervised domain adaptation. IEEE Trans. Multimedia, 1665–1673 (2022)

25. Singh, A.: CLDA: contrastive learning for semi-supervised domain adaptation. Adv. Neural. Inf. Process. Syst. **34**, 5089–5101 (2021)
26. Thota, M., Leontidis, G.: Contrastive domain adaptation. In: Proceedings of the IEEE/CVF Conference on Computer Vision and Pattern Recognition, pp. 2209–2218 (2021)
27. Hamilton, W., Ying, Z., Leskovec, J.: Inductive representation learning on large graphs. Adv. Neural. Inf. Process. Syst. **30**, 1025–1035 (2017)
28. Wu, F., Souza, A., Zhang, T., Fifty, C., Yu, T., Weinberger, K.: Simplifying graph convolutional networks. In: International Conference on Machine Learning, pp. 6861–6871: PMLR (2019)
29. Gasteiger, J., Bojchevski, A., Günnemann, S.: Predict then propagate: Graph neural networks meet personalized pagerank. In: International Conference on Learning Representations (2019)
30. Gretton, A., Borgwardt, K., Rasch, M., Schölkopf, B., Smola, A.: A kernel method for the two-sample-problem. Adv. Neural Inf. Process. Syst. **19** (2006)
31. Paszke, A., et al.: Pytorch: an imperative style, high-performance deep learning library. Adv. Neural Inf. Process. Syst. **32** (2019)
32. Ben-David, S., Blitzer, J., Crammer, K., Kulesza, A., Pereira, F., Vaughan, J.W.: A theory of learning from different domains. Mach. Learn. **79**, 151–175 (2010)
33. Shen, X., Chung, F.L.: Deep network embedding with aggregated proximity preserving. In: Proceedings of the 2017 IEEE/ACM International Conference on Advances in Social Networks Analysis and Mining 2017, pp. 40–43 (2017)

# Semi-supervised Classification Based on Graph Convolution Encoder Representations from BERT

Jinli Zhang[1(✉)], Zongli Jiang[1], Chen Li[2], and Zhenbo Wang[1]

[1] Beijing University of Technology, Pingleyuan Street, Beijing, China
lz73798@gmail.com, jiangzl@bjut.edu.cn, wangzhenbo@emails.bjut.edu.cn
[2] Graduate School of Informatics, Nagoya University, Chikusa, Nagoya 464-8602,
Japan
li.chen.z2@a.mail.nagoya-u.ac.jp

**Abstract.** Attention-based models have attracted crazy enthusiasm both in natural language processing and graph processing. We propose a novel model called Graph Encoder Representations from Transformers (GERT). Inspired by the similar distribution between vertices in graphs and words in natural language, GERT utilizes the equivalent of sentences-vertices obtained from truncated random walks to learn the local information of vertices. Then, GERT combines the strengths of local information learned from random walks and long-distance dependence obtained from transformer encoder models to represent latent features. Compared to other transformer models, the advantages of GERT include extracting local and global information, being suitable for homogeneous and heterogeneous networks, and possessing stronger strengths in extracting latent features. On top of GERT, we integrate convolution to extract information from the local neighbors and obtain another novel model Graph Convolution Encoder Representations from Transformers (GCERT). We demonstrate the effectiveness of proposed models on six networks DBLP, BlogCatalog, CiteSeerX, CoRE, Flickr, and PubMed. Evaluation results show that our models improve $F_1$ scores of current state-of-the-art methods up to 10%.

**Keywords:** Bert · graph neural networks · network embedding

## 1 Introduction

There are many types of graphs in the real world, such as social graphs, protein-protein graphs, citation graphs, knowledge graphs, etc. It is well recognized that graph data are sophisticated and challenging, especially due to the high computational complexity, low parallelizability and inapplicability of current machine learning methods. To tackle these challenges, recent studies advanced graph embedding techniques that provide effective and efficient solutions. Graph embeddings vectorize a graph by preserving the structural information and representing a graph with low dimensional vectors. The embeddings capture the graph topology, the relationships of local neighborhoods and other relevant information corresponding to the graph. When the embeddings pack node properties into a vector with a small dimension, it is easier to employ machine learning methods. Besides, the vector operations are simpler and faster than comparable operations on graphs. Recently, universal representations of graphs learned by graph embedding techniques have been successfully applied in various graph analysis tasks, such as link prediction, node classification [25], node clustering, recommendation [5], network alignment and community detection, etc.

The approaches to embedding graphs in low-dimensional space have been improved greatly over recent decades. Learning from language models for a graph has been proven to be an effective approach to capturing structural features by generating the context of nodes. Deepwalk [17] first uses random walks to generate contextual sequences of nodes from graphs, then the skip-gram algorithm is used to update the objective function. Node2vec [7] extends the idea by modifying Deepwalk with breadth-first and depth-first sampling in a random walk. Recently, a notable natural language model BERT is proposed [3], which has been the state-of-the-art method. The essence cell of BERT is the transformer block [20] that is the transduction model relying entirely on self-attention to compute representations of input and output. The outstanding performance of BERT inspires us to utilize the essence of BERT to analyze graph data.

In this paper, we address the semi-supervised classification by proposing a novel framework GERT: Graph Encoder Representations from Transformers. To our best knowledge, GERT is the first model to apply BERT in graphs. Using random walk on graphs to obtain the sequence of randomly generated nodes. GERT takes the sequences of nodes in a graph as input and generates a latent representation as output. In this way, GERT can be used in both homogeneous networks and heterogeneous networks.

Recently, applications of deep neural networks in graph analysis are showing strong performances, especially convolutional operations in graphs. Many studies on graph convolutional networks have emerged [23,24]. The advantages of convolutional operations on graphs are summarized as follows: First, graph convolutions are more efficient and convenient to composite with other traditional methods in graphs; Second, graph convolutions aggregate information of local neighbors. To integrate information from local neighbors, we make a modification to GERT by adding convolutional operations. Then, we obtain another novel model GCERT. To demonstrate our models, we evaluate their performances on semi-surprised classification tasks in six real-world networks. Experimental results show that our models outperform the state-of-the-art baselines. The embeddings from our models are general and can be applied to other tasks. For the scope of our study, we only report the performances of classification tasks. Our major contributions are summarized as follows:

- Our models GERT and GCERT fuse BERT to graph data. Through random walks on a graph, our proposed models take the sequences of nodes as input, then utilize self-attention mechanisms to draw global dependencies between adjacent nodes in the structure of the graph. Furthermore, our models can preserve the topological structure of graphs.

- Our models can handle various tasks in both homogeneous networks and heterogeneous networks.
- Our models are extensively evaluated on semi-surprised classification tasks on five homogeneous networks and one heterogeneous network. Results show that our proposed models significantly outperform state-of-the-art baselines. Strong performances show that our methods improve the ability to extract information from graph data.

## 2    Related Work

### 2.1   Network Embedding

Network embedding, also called network representation, is a way to incorporate information about the structure of graphs into machine learning models. Network embedding learns to preserve network topology structure, vertex content, and other side information to map nodes into a low-dimensional vector space. The feature representations of nodes can be easily applied to downstream tasks using machine learning algorithms.

From the algorithmic perspective, commonly used models are usually summarized into three categories:

- **Matrix factorization based methods.** An adjacency matrix represents the topological structure of graphs, where the values in the matrix represent the relationships between relevant nodes. We can simply conduct matrix factorization on the adjacency matrix to obtain the embeddings. Matrix factorization is very dependent on the matrix properties. One of the earliest methods is Laplacian eigenmaps, such as Laplacian Eigenmaps (LE) technique. Another type of method is the inner-product one, which is based on a pairwise, inner-product decoder. For example, High-Order Proximity preserved Embedding (HOPE for short) [15] utilizes an inner-product decoder to obtain efficient node embeddings.
- **Random walk approaches.** Random walk based methods are among the most successful models, which capture structural relationships between vertices. They preserve the structure of graphs by mapping the co-occur on truncated random walks into similar low-dimensional space. These methods have superior performances on some downstream tasks by using node embeddings. Deepwalk is the pioneer job in using the random walk to generate node embeddings in graphs. DDRW [13], GENE [1], and Tri-DNR [16] are the extensions of preserving structures by using random walk.
- **Deep neural networks based methods.** Deep neural network based methods can extract complex structural features and learn deep, highly nonlinear node embeddings, which has shown huge successes in other fields. SiNE [21] and SDAE are deep neural network methods in homogeneous network embedding. HAN [22] is the deep learning method applied in heterogeneous network embedding.

## 2.2 Attention Models on Graph

Incorporating an attention mechanism can address problems such as the large and complex patterns and noise data in graphs. In particular, the three main advantages of attention on graphs can be listed as follows [12]: First, attention improves the ability to handle noisy parts of graphs, further enhancing the signal-to-noise ratio. Second, attention assigns the highlight scores to the most relevant neighbors in graphs. Third, attention can make the results more interpretable. According to the types of attention, there are two major categories for attention on graphs: learn attention weights and similarity-based attention. Among the methods, GAM [11] utilizes random walks to sample nodes of a graph. The difference between our models and GAM is that GAM encodes information from visited nodes using an RNN to construct a subgraph embedding. Our models fuse BERT to generate node embeddings. The second difference is that GAM is based on a subgraph, while our models are based on nodes.

## 2.3 Graph Convolution

Inspired by the success of CNN in the computer vision field, many studies have been proposed to apply the re-definition of convolution into network data. These methods can be divided into two main streams [24]: spectral based methods and spatial based methods. Graph Convolutional Networks (GCNs) [2,10] and their variants have obtained significant attention and become popular methods for learning network embedding.

# 3 Problem Definition and Framework

A graph is represented as $\mathcal{G} = (\mathcal{V}, \mathcal{E}, \mathcal{X})$, where $\mathcal{V}$ represents node set in graph, $\mathcal{E} \subseteq \mathcal{V} \times \mathcal{V}$ are the edge set between the nodes and $\mathcal{X}$ consist content features associated with each node. The graph contains various types of nodes and edges, denoted as two mapping functions $\phi : \mathcal{V} \to \mathcal{T}$ and $\psi : \mathcal{E} \to \mathcal{R}$, where $\mathcal{T}$ and $\mathcal{R}$ are node type and edge type, respectively. If $|\mathcal{T}| + |\mathcal{R}| > 2$, the graph can be regarded as heterogeneous graph. When $|\mathcal{T}| + |\mathcal{R}| = 2$, the graph is homogeneous graph. The topological structure of graphs is usually expressed by the adjacency matrix $\mathbf{A}$, in which 0 denotes that no edge between the two nodes, and 1 represents two nodes connected in a graph.

Given a graph, our objective is to map each node into a low-dimensional space $Z_{\mathcal{E}} \in \mathbb{R}^{|V| \times d}$, $d$ is the latent dimension of the node. The $Z$ preserves the global topological structure of a graph, and also represents information of local neighbors with target nodes. Then, we utilize the embedding matrix to help the classification decision.

## 3.1 Overall Framework

Our goal is to learn robust feature representations to perform the classification and build an effective model. To obtain strong features, we leverage the

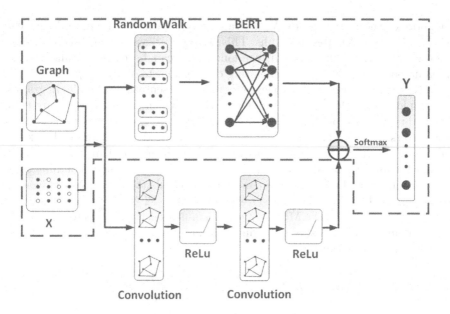

**Fig. 1.** GCERT model and GERT is labeled in the red box. (Color figure online)

transformer encoder model based on random walks to map vertices into low-dimensional space. The overall framework of our model is shown in Fig. 1, which is consisted of two modules: random Walk BERT and graph convolutional layer.

## 4    The Proposed Model

In our model, we aim to extract more effective information on target vertices from both long-distance and local neighbors. Thus, we need to address the following two questions: 1) how to achieve the input of BERT from graphs? 2) how to integrate the information of local neighbors to obtain the latent embeddings?

### 4.1    Random Walks on BERT

Inspired by Deepwalk [17], we utilize random walks to integrate language models and graphs. In a graph $G$ given a node $v_i$, a random walk can be denoted as $\mathcal{S}_{v_i}$. The random walks of nodes $v_i$ are random variables $\mathcal{S}_{v_i}^{1}, \mathcal{S}_{v_i}^{2}, \cdots, \mathcal{S}_{v_i}^{k}$, where $\mathcal{S}_{v_i}^{j}$ is randomly chosen from the neighbors of node $v_{j-1}$. We can set the length of random walks to control the connecting neighbors with nodes $v_i$. The sequences by random walks correspond to sentences in natural language processing.

Using random walks as the fundamental method to extract local features of target nodes has two main advantages. First, it is to preserve the local structure of graphs to the maximum extent, as the nodes in the random walks are the neighbors of the target nodes. When randomly selecting walking nodes, it can process all types of nodes in graphs. This strength allows our model to consider heterogeneous networks. Second, random walks can be easily parallelized, as different random walks can be exploring different parts of the graphs.

In this work, we use BERT to learn feature representations of nodes in graphs. The random variables are the input of BERT, which is designed to pre-train deep bidirectional representations by jointly learning the left and right context in all layers. The fundamental structure of BERT is the transformer block. To elaborate, we draw the implementation of Transformer the principle is as in standard Transformer. As shown in Fig. 1, the Transformer block contains three parts: Multi-headed self-Attention mechanism (MulAtt), Layer Normalization (LNorm), and Position-wise Feed-forward Network (PFN). Given a sequence of input vector $\mathbf{S} \in \mathbb{R}^{n \times d}$, the MulAtt mechanism utilizes a query vector $\mathbf{Q} \in \mathbb{R}^{n \times d}$ ($n$ is the number of nodes, $d$ is the dimension of queries) as input to soft choose the relevant information.

$$MulAtt(\mathbf{Q}, \mathbf{K}, \mathbf{V}) = softmax(\frac{\mathbf{Q}\mathbf{K^T}}{\sqrt{d}})\mathbf{V} \tag{1}$$

where $\mathbf{K} = \mathbf{S}W_i^K, \mathbf{V} = \mathbf{S}, W_i^V \ W_i^Q, W_i^K$, and $W_i^V$, are head-specific projections (parameters matrices) for $Q$, $K$, and $V$, respectively.

Similar to Transformer, we also use multi-head attention with $k$ heads to get more useful information.

$$H_i = MulAtt(\mathbf{Q}W_i^Q, \mathbf{S}W_i^K, \mathbf{S}W_i^V), i \in 1, 2, \ldots, k \tag{2}$$

Suppose it has $s$ parallel attention layers, we concatenate $k$ attention heads to achieve $MulAtt$.

$$MulAtt(Q, K, V) = (H_1 \oplus H_2 \oplus \ldots \oplus H_s)W^M \tag{3}$$

where $W^M$ is the learnable parameter, and $\oplus$ is concatenation operation.

A layer normalization $LN$ is connected with the multi-head attention layer in each transformer layer $l$.

$$\tilde{m}^l = MulAtt(h^{l-1}) \tag{4}$$

$$m^l = LNorm(\tilde{m}^l + h^{l-1}) \tag{5}$$

After a layer normalization $LNorm$, there is a fully connected feed-forward network, which ensures each position is separately and identically. What is more, the normalization network contains two linear transformation layers connected by the ReLU activation function.

$$\tilde{h}^l = PFN(m^l) = ReLU(m^l) \tag{6}$$

By alternatively updating the random walks nodes, the BERT based on random walks can finally study the non-local information.

## 4.2    Graph Convolutional Layer

Random walks on BERT can represent long-distance features of neighborhoods. To obtain local information of target nodes, we introduce the graph convolutional layer to convolute local information. Graph convolutional operation is flexible and effective for efficient information propagation on graphs. In this work, we use one convolutional layer to extract features of local neighbors.

Let $\mathbf{D}$ is the diagonal matrix of the adjacency matrix $\mathbf{A}$, where $\mathbf{D}(i, i) = \sum_{j=1}^{N} a_{ij}$, and $a_{ij}$ is the element of $ith$ row and $jth$ column in $\mathbf{A}$. $\mathbf{X}$ is a node feature matrix where each row is a feature vector, then in the convolutional layer, the graph convolution operator $g$ can be computed as

$$g \bullet \mathbf{X} := \mathbf{D}^{-\frac{1}{2}} \mathbf{A} \mathbf{D}^{-\frac{1}{2}} \mathbf{X} \tag{7}$$

where $\bullet$ denotes as a convolution operator. To manage the convolution operator, it can compute convoluted features of each node as the weighted average of features of neighbors. For each layer, the output $\mathbf{H}^{(l)}, i = 1, 2, \ldots$ is defined as

$$\mathbf{H}^{(l)} = activation(g \star \mathbf{H}^{(l-1)} \mathbf{W}^{(l)}) \tag{8}$$

$$\mathbf{H}^{(0)} = \mathbf{X} \tag{9}$$

where $\mathbf{W}$ is the weight matrix, $activation$ is conventionally chosen $Relu$ function.

As illustrated in Fig. 1, our first model GERT takes the sequence of nodes obtained from the random walk as input, then learns the representations based on BERT. Finally, GERT utilizes softmax to obtain the classification results. GCERT concatenates the embeddings between the embedding of GERT and the embedding of the graph convolutional layer to address the semi-supervised classification task.

## 5    Experiments Setup

In this section, we evaluate our model on six real-world datasets. We thoroughly evaluate GCERT on semi-surprised classification tasks and analyze the sensibility influenced by several parameters.

**Table 1.** Statistics of datasets

| Datasets | ♯ Nodes | ♯ Edges | ♯ Node Types | ♯ Labels |
|---|---|---|---|---|
| CoRA | 2,708 | 5,429 | 1 | 7 |
| Blogcatalog | 5,196 | 171,743 | 1 | 6 |
| CiteSeerXX | 3,312 | 4,660 | 1 | 6 |
| PubMed | 19,717 | 44,338 | 1 | 3 |
| Flickr | 7,575 | 239,738 | 1 | 9 |
| DBLP | 41,936 | 534,009 | 3 | 20 |

## 5.1 Datasets

We provide an overview of datasets in this section. Our model can deal with homogeneous networks and heterogeneous networks on classification tasks. We choose five real-world homogeneous networks. As most heterogeneous networks lack prior knowledge about classification tasks, we use the benchmark dataset, DBLP, to evaluate our model. The statistics of graphs in our experiments are provided in Table 1.

- **CoRA, CiteSeerX, PubMed** [18] The CoRA, CiteSeerX, and PubMed graphs are homogeneous networks, which are all citation networks. In citation networks, the nodes denote papers, and the relationships between nodes are edges that express the citation links.
- **DBLP** [9] is a heterogeneous network. It contains three types of nodes: authors, papers, and conferences. There are "writing" relationships between authors and papers, and "publishing" relationships between papers and conferences. In this experiment, we perform node classification *w.r.t.* on the given labels.
- **BlogCatalog** [19] is a network of social relationships provided by bloggers from the BlogCatalog website, where the labels represent the topic categories provided by the authors.
- **Flickr** [14] is a social network in which users share their photos with friends. The nodes denote users and the relationships between users are friendships. The labels are the interest groups of users.

## 5.2 Baselines

We compare GCERT with two categories of methods: homogeneous network methods and heterogeneous network methods. As described in related works, the Deepwalk method has utilized random walks mechanism; GCNs convolute the information of local neighbors.

- **GraphSAGE** [8] is a general inductive framework that leverages node feature to represent node embeddings. GraphSAGE can efficiently sample and aggregate information of the local neighborhoods.
- **ANRL** [26] is a network representation learning method. ANRL utilizes the information of neighborhoods by a neighbor enhancement autoencoder and skip-gram models.

For the heterogeneous network, we select three popular heterogeneous network baselines: HIN2Vec, metapath2vec, and HeGAN.

- **HIN2Vec** [6] learns heterogeneous networks node embeddings. HIN2Vec is a neural network model, designed to capture the rich semantics embedded in heterogeneous networks by exploiting different types of relationships among nodes.

- **metapath2vec** [4] develops scalable representation models to learn embeddings of heterogeneous networks. Metapath2ve model formalizes meta-path based random walks to construct the heterogeneous neighborhood of a node and then leverages a heterogeneous skip-gram model to perform node embeddings.
- **HeGAN** [9] is inspired by generative adversarial networks that train both a discriminator and a generator in a minimax game. The discriminator and generator of HeGAN are designed to be relation-aware to capture the rich semantics in heterogeneous networks.

### 5.3 Experiment Settings

To facilitate the comparisons between our models and relevant baselines, we set up the same criterion to ensure fairness. Specifically, we randomly sample 10% data as a testing dataset and the rest of the data is as training data. We repeat this process 10 times and report the average performance. The parameters are listed below. In addition, we also modify each of the parameters and control the rest to examine the parameter sensitivity of the proposed methods.

- The number of walks per node $w$: 1
- The walk length $l$: 8
- The vector dimension $d$: 32
- The size of negative samples: 5
- Learning rate: le-5
- Epoch: 20
- Batch-size: 32

## 6  Results and Analysis

In this section, we report our experiment results. First, we thoroughly evaluate our models on a number of classification tasks in homogeneous networks. Then, we present an experimental analysis of our models on heterogeneous networks. Finally, we analyze the parameter sensitivity of our framework.

### 6.1 Node Classification in Homogeneous Network

To evaluate the effectiveness of our models, we perform node classification on the homogeneous network. Similar to previous studies, we select $Macro - F1$ Score and $Micro - F1$ Score to measure the performance of GERT and GCERT. The results are shown in Table 2. Overall, our proposed models GERT and GCERT consistently and significantly outperform all baselines in terms of both metrics in all homogeneous networks.

**Table 2.** Node classification performance on homogeneous networks. The best performance runs per metric per dataset are marked in boldface.

| Method | Core(%) | | BlogCatalog(%) | | CiteSeerX(%) | | Flickr(%) | | PubMed(%) | |
|---|---|---|---|---|---|---|---|---|---|---|
| | Macro-F1 | Micro-F1 | Macro-F1 | Micro-F1 | Macro-F1 | Micro-F1 | Macro-F1 | Micro-F1 | Macro-F1 | Micro-F1 |
| GCN | 70.6 | 72.4 | 62.7 | 63.3 | 66.7 | 66.9 | 53.6 | 55.4 | 76.5 | 78.3 |
| deepwalk | 62.7 | 63.9 | 60.8 | 62.9 | 58.6 | 66.8 | 35.7 | 43 | 75.1 | 78.6 |
| GraphSAGE | 74.5 | 76.3 | 62.7 | 64.2 | 59.3 | 62.8 | 64.7 | 65.3 | 81.1 | 81.8 |
| ANRL | 77.4 | 63.5 | 65.7 | 67.9 | 67.1 | 71.5 | 67.3 | 69.7 | 84.6 | 84.7 |
| GERT | **95.2** | 93 | 87.2 | **89.1** | 70.1 | 76.5 | 73.2 | 75.8 | 86.2 | 88.4 |
| GCERT | 94.3 | **94.5** | **88.3** | 88.7 | **72.4** | **77.3** | **74.9** | **76.7** | **87.6** | **89.2** |

**CoRA, CiteSeerX, BlogCatalog.** From Table 2, both GERT and GCERT perform consistently better than other baselines when predicting for paper categories on CoRA, BlogCatalog and CiteSeerX network. Specifically, GERT outperforms the baselines from 17% to 30% on the CoRA network. The methods which integrate the local neighbors, such as GCN, GraphSAGE and ANRL perform better than the random walk methods by at least 8% with respect to both $F1$ score. After we integrate both methods together and add an attention mechanism to get our models, the performance improves a lot.

The $Micro-F1$ performance of GERT and GCERT improve 10% more than baselines on CiteSeerXX network. As convolution the local neighbors' information, GCERT performs better than GERT by 2% $Macro-F1$ score.

BlogCatalog network is considerably larger than both previous citation networks we have experimented on. However, our models still perform better than other baselines from 20% to 30%. Additionally, GERT performs 20% more than the best baseline on both $F1$ scores.

**Flickr.** When predicting the interest of groups of users, GERT and GCERT achieve the best performance among all the baselines on $Macro-F1$ Score and $Micro-F1$ Score. ANRL performs 67.3% on $Micro-F1$ score, which is the best $Micro-F1$ score among the baselines. Additionally, GCERT outperforms ANRL 7% on $Micro-F1$ score. In other words, GERT and GCERT consistently perform quite well in $Macro-F1$ and $Micro-F1$.

**PubMed.** PubMed is also a larger network that is much closer to a real-world graph than networks we have previously considered. From Table 2, our models perform better than other baselines from 2% to 10%. What is more, GCERT improves 1.4% than GERT concerning the $Macro-F1$ score.

**Table 3.** Author Node classification performance on the heterogeneous network. The best performance runs per metric per dataset are marked in boldface.

|  | Macro-F1(%) | Micro-F1(%) |
|---|---|---|
| metapath2vec | 93 | 92.9 |
| HeGAN | 93.8 | 93.8 |
| HIN2Vec | 91.1 | 91.4 |
| GERT | 95.1 | 95.5 |
| GCERT | **96** | **96.2** |

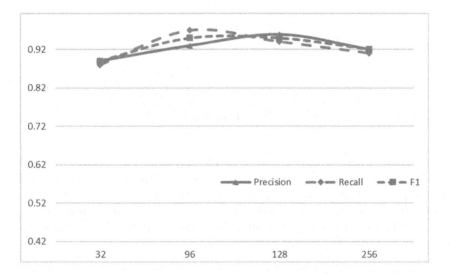

**Fig. 2.** Dimension

## 6.2    Node Classification in Heterogeneous Network

We conduct experiments on the DBLP network which is a heterogeneous network. From Table 3, we can see that our proposed model GCERT outperforms the baseline from 0.9% to 4.9% on $Macro - F1$. Additionally, GERT performs 1.7% than the best baselines HeGAN on $Macro - F1$. As to integrating the convolutional information, GCERT has a stronger performance than GERT.

## 6.3    Parameter Sensitivity

To investigate the impact of parameters on our proposed models, we conduct experiments on the classification task, using the CoRA network as a reference. In the interest of brevity, we keep other parameters the same when we vary the number of embedding dimensions (d), the number of epochs, s and the number of batch sizes.

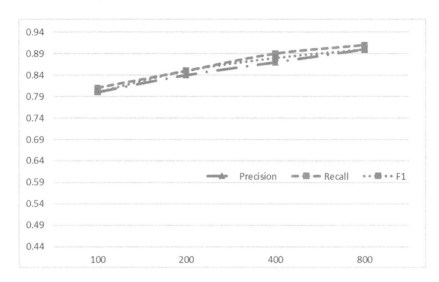

**Fig. 3.** Epoch

**Dimension.** Figure 2 shows the effect of increasing the number of embedding dimensions available to the proposed methods. The performance is slightly increased as the number of dimensions $d$ is increasing until $d = 128$. Overall the results show that our model's performance is relatively stable. In other words, our model is not sensitive to embedding dimensions. Nevertheless, when $d = 128$, our model obtains the optimal performance.

**Epoch.** We vary the number of epochs in 100, 200, 400, and 800 to evaluate our models' performance. From Fig. 3, when $epoch = 100$, our model can obtain 80% with respect to precision. As adding the number of epochs, the performance is moderately increasing. The best performance is obtained when epochs are 800. In conclusion, our model is generally stable around the number of epochs.

**Batch Size.** We observe the performance of our model when the number of batch sizes varies on 12, 24, 32, 48, and 64. From Fig. 4, Our model achieves optimal precision when $batchsize = 64$. When $batchsize = 48$, the recall of our model is the best around all the values. In general, the performance is stable when varying the number of batch sizes.

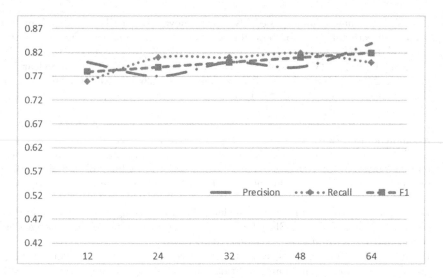

**Fig. 4.** Batch Size

## 7    Conclusions

We propose a novel framework, which can effectively learn the latent representations of nodes in graphs. We are among the first to analyze graph data using
BERT. Our model GERT to learn node embeddings by using the information
from truncated random walks as the input of BERT. Furthermore, to better
leverage, the information of local neighbors, another model GCERT we propose
integrates the convolution operation into GERT. In addition, our model can also
learn the latent representations of nodes in heterogeneous networks. Moreover,
our experiments are conducted on a variety of different networks, which not
only contain homogeneous networks but also include heterogeneous networks.
The semi-supervised classification results demonstrate the effectiveness of our
models.

In future work, we will further explore the attention mechanism in graphs.
Besides, we will strengthen the theoretical justifications of the algorithms.

**Acknowledgements.** This work was supported by the National Natural Science
Foundation of China, 92267107.

## References

1. Chen, J., Zhang, Q., Huang, X.: Incorporate group information to enhance net-
   work embedding. In: Proceedings of the 25th ACM International on Conference on
   Information and Knowledge Management, pp. 1901–1904. ACM (2016)
2. Chen, Yu., Wu, L.: Graph neural networks: graph structure learning. In: Graph
   Neural Networks: Foundations, Frontiers, and Applications, pp. 297–321. Springer,
   Singapore (2022). https://doi.org/10.1007/978-981-16-6054-2_14

3. Devlin, J., Chang, M.W., Lee, K., Toutanova, K.: BERT: pre-training of deep bidirectional transformers for language understanding. arXiv preprint arXiv:1810.04805 (2018)
4. Dong, Y., Chawla, N.V., Swami, A.: metapath2vec: scalable representation learning for heterogeneous networks. In: Proceedings of the 23rd ACM SIGKDD International Conference on Knowledge Discovery and Data Mining, pp. 135–144. ACM (2017)
5. Fan, S., et al.: Metapath-guided heterogeneous graph neural network for intent recommendation (2019)
6. Fu, T.Y., Lee, W.C., Lei, Z.: HIN2Vec: explore meta-paths in heterogeneous information networks for representation learning. In: Proceedings of the 2017 ACM on Conference on Information and Knowledge Management, pp. 1797–1806. ACM (2017)
7. Grover, A., Leskovec, J.: node2vec: scalable feature learning for networks. In: Proceedings of the 22nd ACM SIGKDD International Conference on Knowledge Discovery and Data Mining, pp. 855–864. ACM (2016)
8. Hamilton, W., Ying, Z., Leskovec, J.: Inductive representation learning on large graphs. In: Advances in Neural Information Processing Systems, pp. 1024–1034 (2017)
9. Hu, B., Fang, Y., Shi, C.: Adversarial learning on heterogeneous information networks (2019)
10. Kipf, T.N., Welling, M.: Semi-supervised classification with graph convolutional networks. arXiv preprint arXiv:1609.02907 (2016)
11. Lee, J.B., Rossi, R., Kong, X.: Graph classification using structural attention. In: Proceedings of the 24th ACM SIGKDD International Conference on Knowledge Discovery & Data Mining, pp. 1666–1674. ACM (2018)
12. Lee, J.B., Rossi, R.A., Kim, S., Ahmed, N.K., Koh, E.: Attention models in graphs: A survey. arXiv preprint arXiv:1807.07984 (2018)
13. Li, J., Zhu, J., Zhang, B.: Discriminative deep random walk for network classification. In: Proceedings of the 54th Annual Meeting of the Association for Computational Linguistics (Volume 1: Long Papers), pp. 1004–1013 (2016)
14. Meng, Z., Liang, S., Bao, H., Zhang, X.: Co-embedding attributed networks. In: Proceedings of the Twelfth ACM International Conference on Web Search and Data Mining, pp. 393–401. ACM (2019)
15. Ou, M., Cui, P., Pei, J., Zhang, Z., Zhu, W.: Asymmetric transitivity preserving graph embedding. In: Proceedings of the 22nd ACM SIGKDD International Conference on Knowledge Discovery and Data Mining, pp. 1105–1114. ACM (2016)
16. Pan, S., Wu, J., Zhu, X., Zhang, C., Wang, Y.: Tri-party deep network representation. Network **11**(9), 12 (2016)
17. Perozzi, B., Al-Rfou, R., Skiena, S.: DeepWalk: online learning of social representations. In: Proceedings of the 20th ACM SIGKDD International Conference on Knowledge Discovery and Data Mining, pp. 701–710. ACM (2014)
18. Sohn, K., Lee, H., Yan, X.: Learning structured output representation using deep conditional generative models. In: Advances in Neural Information Processing Systems, pp. 3483–3491 (2015)
19. Tang, L., Liu, H.: Relational learning via latent social dimensions. In: Proceedings of the 15th ACM SIGKDD International Conference on Knowledge Discovery and Data Mining, pp. 817–826. ACM (2009)
20. Vaswani, A., et al.: Attention is all you need. In: Advances in Neural Information Processing Systems, pp. 5998–6008 (2017)

21. Wang, S., Tang, J., Aggarwal, C., Chang, Y., Liu, H.: Signed network embedding in social media. In: Proceedings of the 2017 SIAM International Conference on Data Mining, pp. 327–335. SIAM (2017)

22. Wang, X., et al.: Heterogeneous graph attention network. In: The World Wide Web Conference, pp. 2022–2032. ACM (2019)

23. Wu, L., et al.: Graph neural networks for natural language processing: A survey. arXiv preprint arXiv:2106.06090 (2021)

24. Wu, Z., Pan, S., Chen, F., Long, G., Zhang, C., Yu, P.S.: A comprehensive survey on graph neural networks. arXiv preprint arXiv:1901.00596 (2019)

25. Zhang, J., Jiang, Z., Li, T.: CHIN: classification with META-PATH in heterogeneous information networks. In: Florez, H., Diaz, C., Chavarriaga, J. (eds.) ICAI 2018. CCIS, vol. 942, pp. 63–74. Springer, Cham (2018). https://doi.org/10.1007/978-3-030-01535-0_5

26. Zhang, Z., et al.: ANRL: attributed network representation learning via deep neural networks. In: Proceedings of the Twenty-Seventh International Joint Conference on Artificial Intelligence (IJCAI). vol. 18, pp. 3155–3161 (2018)

# Global Balanced Text Classification for Stable Disease Diagnosis

Zhuoyang Xu[✉], Xuehan Jiang, Siyue Chen, Yue Zhao, Gang Hu,
Xingzhi Sun, and Guotong Xie

Ping An Healthcare Technology, Beijing, China
xuzhuoyang534@pingan.com.cn

**Abstract.** Disease diagnosis plays an important role in the application of clinical decision system. When applying artificial intelligence models in disease diagnosis, one should be aware of the spurious correlations learned from the data. Such correlations could be introduced by the biased distribution of the data and would be reinforced during the optimization process. The spurious correlations will cause the model ultra-sensitive to some trivial variations in the input, and even introduce some significant errors when applied in non-independent and identically distributed (non-i.i.d) data with the training data. In this paper, we addressed this problem by applying the global balancing method to adjust the distribution balance in the learned representation space of the text via the assignment of global sample weights. By learning the global sample weights that minimized the balance loss, the text representation space had been iteratively optimized by attenuating the strength of the spurious correlations introduced by the training data. Our algorithm, TCGBR (Text Classification with the Global Balance Regularizer), showed the reduced sensitivity to the trivial variations of the input text and increased diagnosis accuracy when tested on the non-i.i.d data. Furthermore, we built the causal graph to investigate the causal relationships in established diagnostic models and found the diagnostic logic of TCGBR was more in line with the clinical common sense. The presented architecture can be used by practitioners to improve the stability of diagnostic models and the technique of global balancing can be naturally generalized to other representation-based deep learning models.

**Keywords:** Disease diagnosis · Stable learning · Global balance · Causal discovery · Linguistic causal effect

## 1 Introduction

Machine learning and deep learning are promising in the healthcare domain, e.g., the application of the clinical decision system [1,23,28]. Artificial intelligence greatly benefits the efficiency and the accessibility of high-quality healthcare services [8]. AI model is widely used to reinforce the correlations between the inputs and the outputs. Nevertheless, it should be aware of the existence of the

© The Author(s), under exclusive license to Springer Nature Switzerland AG 2023
X. Yang et al. (Eds.): ADMA 2023, LNAI 14178, pp. 214–228, 2023.
https://doi.org/10.1007/978-3-031-46671-7_15

spurious correlations, which could be introduced by the selection bias of the training data. This problem is especially critical in the disease diagnosis task where severe consequences may occur once the spurious correlations mislead the model in non-independent and identically distributed (non-i.i.d) data. As the reduction of diagnostic error has been regarded as an urgent task by health care organizations (HCOs) [6], there is a need to reduce spurious correlations when developing diagnostic models.

To remove the spurious correlations, one possible way is to learn the causal relationship between the input and the output. The causal relationship is stable to transfer to the non-i.i.d data [25]. There are typically two paradigms to learn the causal relationship from data. The first one is the structural causal model (SCM) [5], which carries out the causal inference based on the causal graph. The second one is the potential outcome framework [17] that evaluates the causal coefficient between $X$ and $Y$ by adjusting the distribution of the confounders.

In SCMs, the causal graph is constructed by the domain experts which is infeasible to consider all implicit relationships. Although there are algorithms for causal discovery from the data, such as the PC algorithm [21], the hypothesis that the data should be infinity is hard to be satisfied especially in electronic health records (EHRs) where there are limited records for the infrequent diseases. Besides, the nodes in the causal graph usually represent some entities, limiting the application of the SCMs to the non-structure data.

The potential outcome framework is more feasible in the real application. By counterbalancing the distributions of the confounders in the treatment group and the control group divided by $X$, the coefficient between $X$ and $Y$ represents the causal contribution of $X$ to $Y$ according to the causal inference theory [5]. Based on this theory, many methods have been proposed to counterbalance the effect of the confounders, such as the matching-based methods [2] and the weighted-based methods [4]. Shen et al. proposed to generalize this weighted-based method to high dimensions and brought up the global balance regularizer in [18]. When the global balance regularizer is implemented, the image classification model learns more goal-related features than environmental features. For example, the image is classified as a dog due to the features of the dog instead of the grass which is frequently co-occurred with dogs.

In the implementation of the diagnostic model using TextCNN [9], the result is ultrasensitive to trivial variations of the inputs. For example, adding a period at the end of the input does not change semantics, but the model may give a different diagnosis, which reflects some spurious correlations was learned during the training process of TextCNN. Therefore, it is necessary to introduce causal inference techniques in the modeling of the disease diagnosis models. Jonathan G. Richens et al. are the first to propose the use of causal machine learning to improve diagnostic accuracy. They came up with a twin network containing the factual graph and the counterfactual graph [15]. They used a three-layer noisy-OR Bayesian network to represent the pairwise causal relationship between the risk factors with the diseases and the diseases with the symptoms. Their results show that this counterfactual algorithm achieves expert clinical accuracy, suggesting that causal inference is critical to the application of machine learning in healthcare.

Nevertheless, EHRs are text-riches data. Constructing the causal graph from the text data is expert resource-dependent. Using the potential outcome framework, Reid Pryzant et al. proposed adding reversal layers in the language representation network to remove the effect of confounders in lexicon interpretation [14]. Yao et al. brought up a framework to estimate the treatment effect with text covariates [24]. The authors combined the generative adversarial network with the potential outcome framework to remove instrumental information that only related to the treatment variable in the original input. However, all the published works are based on the definition of the con-founder variables in the text, and are only designed for some specific scenarios. This limits the generalization of their models in the wide application of causal inference with textual data.

Inspired by the works from Cui's group and his collaborators [10,11,18], we designed and implemented the global balance regularizer in the TextCNN model, which was a widely used NLP algorithm to learn the representation of text data. The global balance regularizer measured the distribution bias of other dimensions when evaluating the coefficient between the anchor dimension and the output label. By minimizing the global balance regularizer, the bias introduced by the distribution of other dimensions was minimized, therefore the spurious correlation had been attenuated. To evaluate what had been learned in the NLP model with and without a global balancing regularizer, we also interpreted the results with a model-agnostic interpretation approach proposed in [7]. We further built the causal graph for the model with and without a global balancing regularizer in i.i.d and non-i.i.d data. By comparing the causal graphs, we found the balanced NLP model revealed more causal relationships between the disease and symptoms than the NLP model without the balancing regularizer. Besides, such causal relationships were more stable when transferred from i.i.d data to non-i.i.d data for the balanced model than the model without the balancing regularizer.

Our main contribution includes the following three aspects:

1. We proposed an approach to integrate the global balance with the representation learning-based NLP model for the diagnosis classification task and conduct extensive experiments concerning different real-world datasets to show the better diagnosis performance of our TCGBR model in the non-i.i.d dataset.
2. The effect of the balanced TCGBR model was evaluated extensively. The spurious correlations, measured in the fraction of punctuation as important lexicons, decreased for the balanced TCGBR model.
3. We proposed a framework to discover the causal relationships for text input and constructed the causal graphs for TextCNN and TCGBR models in different datasets. The causal graph illustrated the TCGBR found causal relationships in agreement with the clinical perception. Besides, the causal graph constructed from the TCGBR was more stable in the non-i.i.d dataset.

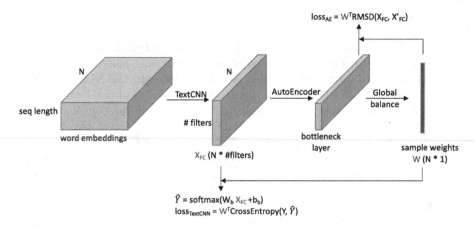

**Fig. 1.** The TCGBR model we proposed in the current study.

## 1.1 Problem Definition

We focused on the spurious correlations learned in the NLP models in disease classification. Those spurious correlations were introduced by the selection bias of training data and led to the ultra-sensitivity of the model to trivial variations of the input. To attenuate the spurious correlations and evaluate the result, we defined our problem as to how to develop and evaluate unbiased disease classification models from text data.

## 2 Methodology

The methodology part included the algorithm of the global balancing for text classification proposed in the current study and the framework of causal discovery for text data. The algorithm of the global balancing for text classification consisted of three modules as shown in Fig. 1. The first module was an NLP model that vectorized the input tokens into a condensed representation network. The second module was used to reduce the dimensions of the text representation. The last module was the global balance part, which was used to update the global sample weights. These three modules shared the global sample weights.

The framework for causal discovery also contained three steps as shown in Fig. 2. First, the importance of each word in a text sample was obtained using the method proposed in [7]. By tokenizing the sentence, the importance of each token was calculated. Second, the whole label space was clustered based on the confusion matrix of the classification. Diseases in the same cluster were those that were difficult to differentiate according to the classification model. Therefore, inter-class diseases were independent, and the causal graph was built separately for each disease cluster. The last step was to construct the causal graph for each disease cluster using the FCI algorithm [22].

We will introduce the TCGBR algorithm and the causal discovery framework in detail in the following subsections after the Preliminaries.

## 2.1  Preliminaries

TextCNN (Text Convolutional Neural Network) is an efficient and widely used network architecture for text representation learning and classification [9]. It was a revolution to introduce the idea of convolution into the NLP area. The input of TextCNN was a tensor with the size of (dimensions of word embedding, sequence length of the input text, batch size). By applying the convolution and pooling operation, the tensor was converted to a vector with batch size. Multiple convolution kernels could be used with different sizes, thus various features could be extracted from the original text embedding. And after the convolution and pooling operation, the text was represented with a vector of size $n$, where $n$ was the number of filters. Finally, the vector with size $n$ was used for classification or prediction.

To interpret what the model had learned, we utilized a model-agnostic approach proposed in [7]. The author proposed a way to quantify the information of a specific word learned by the neural network (NN) concerning the task of the NN. It was realized via the perturbation approximation. The original initial embeddings of the input text were represented as $X$. With a random perturbation $\Delta$ being added to $X$, the permutated representation was also fed into the NN with the same parameters $\varnothing$. By minimizing the difference between $\varnothing(X + \Delta)$ and $\varnothing(X)$ while maximizing $\Delta$ simultaneously, the effect of important words to the model output becomes stable.

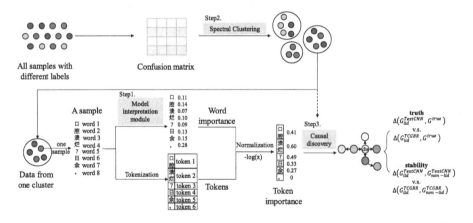

**Fig. 2.** The framework for causal discovery for the NLP model.

## 2.2  The Global Balance Regularizer

The potential outcome framework is a popular paradigm to estimate the causal relationship between $X$ and $Y$ in the existence of the confounder $Z$. As shown in Fig. 3, the arrows indicated the causal relationships between variables. Confounders are defined as a variable that is related to both the input variable and the output variable. For the sake of simplicity, let us assume all the variables are binary. Instances with $X = 0$ are grouped as the control group, while instances with $X = 1$ belong to the treatment group. The potential outcome framework suggested balancing the distribution of $Z$ between the control group and the treatment group would estimate the direct causal contribution of $X$ to $Y$ under the assumptions of causal inference.

**Fig. 3.** The potential outcome framework.

Based on this theory, researchers proposed the inverse of the propensity weighting (IPW) strategy [16]. The propensity referred to the probability of the instance being allocated in the treatment group for $Z$. Therefore, according to the IPW, samples weighted by the inverse of the propensity would ensure balancing distribution of $Z$ in the treatment and control group. Thus, samples weighted by the propensity score would estimate the direct causal contribution of $X$ to $Y$ according to the potential outcome framework. Shen et al. proposed the global balancing method, which was a generalization of confounder balance in high dimensions [18]. The global balance regularizer was formulated as the following:

$$L_{global\_balance} =$$
$$\lambda_0 \sum_{j=1}^{p} \| \frac{X_{-j}^T (W \odot I_j)}{W^T I_j} - \frac{X_{-j}^T (W \odot (1 - I_j))}{W^T (1 - I_j)} \|_2^2 + \lambda_1 \|W\|_2^2 \qquad (1)$$

where the first term quantifies the overall difference of the first momentum of the distribution between the treatment group and the control group, the second term is the constrain to punish the unreasonable large sample weights. In the equation, $X$ represented the input matrix with the size $(N, p)$ where $N$ was the number of samples, $p$ was the dimensions of features. $X_{-j}^T$ was the transpose of $X$ with the $j - th$ row equaled to zero. $W$ represented the sample weight vector which had positive values. $I$ was the indicator matrix and $I_j$ indicated whether the sample belonged to the treatment group in the $j - th$ column. The first term within $\| \cdot \|_2^2$ gave the distribution difference between the treatment group and the control group for the $j - th$ column.

Optimized with the global balance regularizer would update the global sample weights which minimize the bias introduced by the confounders when learning the causal relationship between the input and the output.

**Table 1.** Module summary in TCGBR.

| Module | Input | Output | Loss |
|---|---|---|---|
| TextCNN | original text, embeddings, $W$ | $X_{FC}$ | weighted CE[a] |
| AutoEncoder | $X_{FC}$ , $W$ | bottleneck layer | weighted RMSE[b] |
| Global balance | bottleneck layer | $W$ | balance loss |

[a]Cross-entropy.
[b]Root mean square error.

### 2.3   The Network Architecture and Optimization of TCGBR

In the current study, we proposed to implement the global balance regularizer in the text classifier based on representation learning.

As shown in Fig. 1, our network contained three modules. The input and output of each module were listed in Table 1. First, input with the original text and the word embeddings, TextCNN converted the text in the representation space by minimizing the cross-entropy of classification. $X_{FC}$ was used to represent the fully connected layer just before the classification layer in TextCNN. Then, $X_{FC}$ was input into AE and the representation got dimension reduced. The last module was the global balance which updated the sample weights by minimizing the balance loss.

The total loss of the network TCGBR is defined as the following:

$$J(W, \Phi, \Psi) = L(\Phi, W)_{TextCNN} + \lambda_{AE} L(\Psi, W)_{AE}$$
$$+ L(W)_{global\_balance} \tag{2}$$

where $\Phi$ and $\Psi$ are the parameters for TextCNN and AE respectively. $W$ is the global sample weights shared by all three modules.

The total loss was composed of three parts and was optimized separately. For each batch, we first initialized the global sample weights as vectors of 1. Then through forwarding calculation, we got the $X_{FC}$ and the bottleneck layer. And the balance loss was calculated based on the bottleneck layer. Then the global sample weights were updated until converged or the maximum steps were reached. Then $\Phi$ was updated with the new global sample weights. Finally, $\Psi$ was updated with the new global sample weights.

### 2.4   The Evaluation Methodology

In addition to the diagnostic accuracy on i.i.d data and non-i.i.d data, to verify whether TCGBR could make more reasonable inferences, we also performed the following two evaluations. First, we calculated the fraction of punctuation ranked as the top 3 important tokens in the data. As the punctuation brought little information in the diagnosis classification, the fraction of punctuation ranked as the top 3 important tokens measured the spurious relationship leaned by the model. We expected that TCGBR had a lower fraction of punctuation in the top

3 important tokens. Second, we evaluated the causal relationship learned by different NLP models. As calibrated the spurious relationship, the TCGBR should reveal more reasonable causal relationships than TextCNN. And the difference of the causal relationship revealed by the TCGBR i.i.d data and the TCGBR non-i.i.d data should be smaller than the difference of that in the TextCNN model, as the TCGBR was more stable.

We calculated the word-level importance using the perturbation-based approximation method proposed in [7]. The main idea for this method could be found in the preliminaries. The smaller the value was, the higher importance of a word. Since the causal relationship should be on the token level, we further tokenized the text and calculated the token-level importance with the minimum word-level importance of the token. It should be noted the word-level importance was calculated separately for each instance, therefore we normalized the importance value within each instance by dividing the maximum importance value of all the tokens in an instance. Furthermore, the negative logarithm of the normalized importance value was calculated, therefore the importance value was converted to a positive value, and the larger the value, the more important the token.

The efficiency of constructing a causal graph depended on the node considered in the graph. Therefore, we sub-grouped all the diagnoses such that diseases from two different subgroups would be independent. We applied community detection and used the confusion matrix to represent the affinity between each node. The number of the cluster was determined by the largest gap of eigenvalues of the affinity matrix according to [26].

To obtain the causal relationship, we applied the algorithms of causal discovery implemented in the causal-learn toolkits[1]. We used the algorithm called fast causal inference (FCI) [22], which revealed causal relationships based on the Markov equivalence class. The FCI algorithm identified the causal relationship based on the independent test between the different combinations of variables. Besides, the FCI algorithm relaxed the constraint of unobserved confounders which suited the electronic health records.

# 3 Experiments and Results

## 3.1 Datasets

We used visiting data both from the primary hospitals of several cities and the top three hospitals in one city in China. The data contained the chief complaint (cc) with the true diagnosis label, and some of the data also contained the history of present illness (hpi). The statistic of the datasets was listed in Table 2. The dataset name contained the hospital levels, the city, and the information for diagnosis. We had constructed the TextCNN model and TCGBR using all the datasets listed in Table 2. All the datasets were randomly split into training, validation, and i.i.d-test set in a ratio of 8:1:1.

---

[1] https://github.com/cmu-phil/causal-learn.

**Table 2.** Overview of the datasets.

| Dataset | Sample size | # of class | Avg. length | Non-i.i.d test-set1, testset2 |
|---|---|---|---|---|
| Primary-CityA-cc | 298839 | 2000 | 9.6 | Primary-CityA-cc_hpi, Primary-CityB-cc |
| Primary-CityA-cc_hpi | 182971 | 2000 | 44.2 | Primary-CityA-cc, TopHsp3-cc_hpi |
| TopHsp3-cc | 213117 | 365 | 6.8 | TopHsp3-cc_hpi, TopHsp1-cc |
| TopHsp3-cc_hpi | 150827 | 365 | 59.0 | TopHsp3-cc, TopHsp1-cc_hpi |

## 3.2 Experiments Setups

**Implementation Details of TextCNN.** We implemented TextCNN in Python 3.6.9 and TensorFlow 1.15.2. We used kernel size of (2,3,4) in the convolution and each kernel size with 256 filters. Therefore, the dimension for $X_{FC}$ was 768. The words embeddings had the dimension of 1024 and were initialized randomly. We trained for 100 epochs with the learning rate of 0.0003 and the batch size of 2560 until the ACC@1 in the validation dataset had not been improved for several epochs.

**Implementation Details of AutoEncoder.** We implemented AE using a three layers NN to reduce the dimension. From the input layer to the bottleneck layer, there were 256, 128, and 16 neurons in each layer. The reconstruction NN was just symmetric with the dimension reduction NN.

**Implementation Details of Global Balance.** For each batch optimization, the optimization was terminated after (4*epoch*epoch+100) steps if not converged. This served as a self-adapted mechanism that enabled fast representation learning at the initial stage. Then the representation was fine-tuned with the well-balanced sample weights.

**Implementation Details of FCI.** Fisher test was used in the FCI algorithm, and the alpha was set to 0.05 as the default. The algorithm was implemented in the causal-learn toolkit.

## 3.3 The Evaluation Process

We compared TCGBR with TextCNN in the following five aspects:

1). The ACC@1, ACC@3, and ACC@5 for text classification in the test datasets. These test datasets were i.i.d with the training data.

2). The generalization of the model to other data resources, here and after noted as "non-i.i.d test data". We used ACC@1, ACC@3, and ACC@5 to evaluate the model performance in different non-i.i.d test data. The higher performance indicated a more generalized model had been learned.

3). The fraction of punctuation being the top 3 important tokens in the test dataset: as punctuation was more likely to be unrelated to the semantics, therefore correlations between the punctuation and the output may be spurious correlations. The lower the fraction of punctuation indicated the more causal relationships had been learned.

4). The truthfulness of the causal graphs built on the i.i.d test dataset on the two models.

5). The stability of the causal graphs built on the two models: the stability was calculated as the difference between the causal graphs built on the i.i.d dataset and the non-i.i.d dataset for the same NLP model.

**Table 3.** Performance of all models in the i.i.d test data.

| Train data | Network | ACC@1 | ACC@3 | ACC@5 |
|---|---|---|---|---|
| Primary-CityA-cc | TextCNN | 0.631 | **0.817** | **0.870** |
|  | TCGBR | **0.632** | 0.815 | 0.869 |
| Primary-CityA-cc_hpi | TextCNN | **0.682** | **0.851** | 0.894 |
|  | TCGBR | 0.681 | 0.850 | 0.894 |
| TopHsp3-cc | TextCNN | 0.635 | 0.873 | **0.937** |
|  | TCGBR | 0.635 | **0.874** | 0.936 |
| TopHsp3-cc_hpi | TextCNN | **0.722** | **0.926** | **0.966** |
|  | TCGBR | 0.721 | 0.924 | 0.965 |

## 3.4 Result Analysis

**Performance of TextCNN and TCGBR Trained in Multiple Datasets.**
The ACC@1, ACC@3, and ACC@5 for all models in the i.i.d and non-i.i.d test sets were listed in Table 3 and Table 4, respectively. The better results for the same training dataset were noted in bold.

From Table 3, we found the diagnosis performance for TextCNN and TCGBR had no significant difference for the same dataset.

From Table 4, we found for most cases, the TCGBR network had an increase of about 2% in the non-i.i.d test datasets compared with TextCNN, suggesting the TCGBR network could learn more causal relationships.

**Table 4.** Performance of all models in the non-i.i.d test data. The testset1 and testset2 are listed in Table 2

| Train data | Network | ACC@1/ACC@3/ACC@5 on testset1 | ACC@1/ACC@3/ACC@5 on testset2 |
|---|---|---|---|
| Primary-CityA-cc | TextCNN | 0.519/0.738/0.811 | 0.317/0.535/0.624 |
| | TCGBR | **0.534/0.758/0.823** | **0.324/0.556/0.645** |
| Primary-CityA-cc_hpi | TextCNN | 0.471/0.706/0.792 | 0.266/0.497/0.595 |
| | TCGBR | **0.487/0.724/0.813** | **0.286/0.528/0.617** |
| TopHsp3-cc | TextCNN | 0.489/0.746/0.849 | 0.507/0.752/0.876 |
| | TCGBR | **0.496/0.757/0.862** | **0.525/0.765/0.890** |
| TopHsp3-cc_hpi | TextCNN | 0.549/0.772/0.851 | 0.517/0.776/0.887 |
| | TCGBR | **0.566/0.787/0.865** | **0.535/0.784/**0.887 |

**Table 5.** The number of samples had punctuation as the top 3 important words in each model.

| | i.i.d test data (Primary-CityA-cc) | non-i.i.d test data (Primary-CityA-cc_hpi) |
|---|---|---|
| Test data size | 29884 | 182971 |
| TextCNN | 6395 (21.4%) | 42449 (23.2%) |
| TCGBR | 5200 (17.4%) | 32020 (17.5%) |

**The Fraction of Punctuation as the Top 3 Important Tokens.** We calculated the number of samples that ranked punctuation as the top 3 important tokens in models trained with the Primary-CityA-cc data. The results were listed in Table 5. The spurious correlation of the punctuation with the disease labels learned by the model was reduced by 4% in the i.i.d test data and 5.7% in the non-i.i.d test data when using the TCGBR network. It was interesting to note that this type of spurious correlation had increased from the i.i.d test data to the non-i.i.d test data for the model trained with TextCNN, but remained almost the same for the model trained with TCGBR.

**The Quantitative Comparison of the Stability.** We calculated the difference of confusion matrix between i.i.d dataset and non-i.i.d dataset. The difference was measured using Manhattan distance. The distance for TCGBR was 4.8% lower than the distance for TextCNN.

**The Causal Graphs for TextCNN and TCGBR on Different Test Datasets.** With spectral clustering, the diseases were grouped into independent communities. To better compare the truthfulness and the stability of the causal graphs, we chose a disease cluster that had the same cluster members in TextCNN and TCGBR. It was found that hyperthyrea and primary hyperthyrea were clustered in the same community both in TextCNN and TCGBR, therefore we constructed the causal graphs for this community.

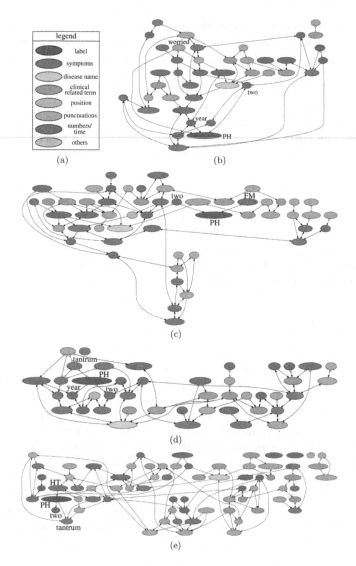

**Fig. 4.** The causal graphs to distinguish hyperthyrea from primary hyperthyrea for (b) the TextCNN model in the i.i.d test data, (c) the TextCNN model in the non-i.i.d data, (d) the TCGBR model in the i.i.d data, and (e) the TCGBR model in the non-i.i.d data. The node was colored by the type of the token as illustrated in the legend (a). The names of the label node and its adjacent nodes were specified. PH: Primary hyperthyroidism; FM: Four months; HT: Hand trembling.

As shown in Fig. 4, four causal graphs were built for two NLP models on two test datasets. The nodes in the causal graphs were colored based on the category of the nodes, such as symptoms that were closely related with the disease diagnosis were colored in red while the punctuation was colored in light blue. The edges between nodes in the causal graphs represent the relationship discovered based on the dataset. The arrow endpoint of the edge represented the causal relationship, while the hollow ones represented the undetermined causal relationship.

For the causal graphs built for TextCNN, the label node (primary hyperthyrea) was mostly connected with nodes of numbers or times (Fig. 4b and Fig. 4c). While the label node in the causal graphs built for TCGBR (Fig. 4d and Fig. 4e) was also connected with some symptoms, and was surrounded by other symptom nodes. Therefore, the causal graph built on TCGBR was more consistent with clinical perception.

By comparing Fig. 4b (TextCNN i.i.d dataset) with Fig. 4c (TextCNN non-i.i.d dataset), the topology of the graph model changed a lot, especially the nodes around the label node (primary hyperthyrea), indicating the instability of the causal relationship found in the TextCNN model. While for TCGBR, the label node was surrounded by symptom nodes in both causal graphs on i.i.d dataset (Fig. 4d) and non-i.i.d dataset(Fig. 4e). Besides, some symptoms seemed stable nearby the label node, such as tantrum.

## 4   Discussion

Diagnosis is one of the most important tasks performed by clinical decision systems. As the 'bottom of the iceberg' of patient-safety, diagnostic errors have been included as a high-priority problem in primary care [13,19,20]. Although a series of works adopted machine learning and deep learning approaches to develop diagnostic models from abundant patient data [3,9,12,27], the spurious correlations will cause the model ultra-sensitive to some trivial variations and introduce diagnostic errors. One possible solution to this kind of error is to the build the model with a 'causal' way, rather than simply rely on associative inference with the biased real-world data. Jonathan G. Richens et al. verified the effectiveness of using causal machine learning to develop diagnostic models with idea of SCM [15]. In this work, we have proposed a extensible algorithm based on deep learning that automatically build the diagnostic models with idea of potential outcome framework, which is more feasible than SCM in the real-world application.

While we have integrated the causal regularizer into a widely used baseline-TextCNN, the regularizer could be generalized to any representation-based deep learning models. Further experiments could be performed to validate the effectiveness of the causal regularizer on other deep learning models in the future. Besides, in addition to our evaluation metrics, further evaluation design could be performed to determine the effectiveness of the algorithm in clinical practice.

## 5   Conclusion

We proposed the TCGBR algorithm to balance the distribution of the confounders in the representation-based NLP model for disease diagnosis. And by attenuating the balance loss, the spurious correlations could be reduced, leading to a stable diagnosis classification to non-i.i.d test data and desensitization of the trivial variation of the input, such as the punctuation. Three tools were applied to validate the effectiveness of the causal learning: the model performance in the non-i.i.d dataset, the fractions of spurious correlations, and the causal graph. All those measurements showed TCGBR captured the causal relationship between the diseases and the chief complaints, making it could be stably transferred to the non-i.i.d dataset. The presented architecture could be used by practitioners to improve the stability of diagnostic models. Besides, our algorithm of integrating the causal regularizer could be generalized to other representation-based deep learning models.

**Acknowledgement.** The work is supported by Shenzhen City's Science and Technology Plan Project (JSGG20210802153806021).

# References

1. Choy, G., et al.: Current applications and future impact of machine learning in radiology. Radiology **288**(2), 318 (2018)
2. Coca-Perraillon, M.: Local and global optimal propensity score matching. In: SAS Global Forum. vol. 185, pp. 1–9 (2007)
3. De Fauw, J., et al.: Clinically applicable deep learning for diagnosis and referral in retinal disease. Nat. Med. **24**(9), 1342–1350 (2018)
4. DuGoff, E.H., Schuler, M., Stuart, E.A.: Generalizing observational study results: applying propensity score methods to complex surveys. Health Serv. Res. **49**(1), 284–303 (2014)
5. Fenton, N.E., Neil, M., Constantinou, A.C.: The book of why: the new science of cause and effect, judea pearl, dana mackenzie. basic books (2018). Artif. Intell. **284**, 103286 (2020)
6. Graber, M.L., Trowbridge, R., Myers, J.S., Umscheid, C.A., Strull, W., Kanter, M.H.: The next organizational challenge: finding and addressing diagnostic error. Joint Comm. J. Qual. Patient Saf. **40**(3), 102–110 (2014)
7. Guan, C., Wang, X., Zhang, Q., Chen, R., He, D., Xie, X.: Towards a deep and unified understanding of deep neural models in NLP. In: International Conference on Machine Learning, pp. 2454–2463. PMLR (2019)
8. Jiang, X., et al.: Characteristics of online health care services from china's largest online medical platform: cross-sectional survey study. J. Med. Internet Res. **23**(4), e25817 (2021)
9. Kim, Y.: Convolutional neural networks for sentence classification. In: Moschitti, A., Pang, B., Daelemans, W. (eds.) Proceedings of the 2014 Conference on Empirical Methods in Natural Language Processing, EMNLP 2014, 25–29 October 2014, Doha, Qatar, A meeting of SIGDAT, a Special Interest Group of the ACL, pp. 1746–1751. ACL (2014)
10. Kuang, K., et al.: Treatment effect estimation via differentiated confounder balancing and regression. ACM Trans. Knowl. Discov. Data (TKDD) **14**(1), 1–25 (2019)
11. Kuang, K., Xiong, R., Cui, P., Athey, S., Li, B.: Stable prediction with model misspecification and agnostic distribution shift. In: Proceedings of the AAAI Conference on Artificial Intelligence. vol. 34, pp. 4485–4492 (2020)

12. Liang, H., et al.: Evaluation and accurate diagnoses of pediatric diseases using artificial intelligence. Nat. Med. **25**(3), 433–438 (2019)

13. Liberman, A.L., Newman-Toker, D.E.: Symptom-disease pair analysis of diagnostic error (spade): a conceptual framework and methodological approach for unearthing misdiagnosis-related harms using big data. BMJ Qual. Saf. **27**(7), 557–566 (2018)

14. Pryzant, R., Shen, K., Jurafsky, D., Wagner, S.: Deconfounded lexicon induction for interpretable social science. In: Proceedings of the 2018 Conference of the North American Chapter of the Association for Computational Linguistics: Human Language Technologies, Volume 1 (Long Papers), pp. 1615–1625 (2018)

15. Richens, J.G., Lee, C.M., Johri, S.: Improving the accuracy of medical diagnosis with causal machine learning. Nat. Commun. **11**(1), 1–9 (2020)

16. Rosenbaum, P.R., Rubin, D.B.: The central role of the propensity score in observational studies for causal effects. Biometrika **70**(1), 41–55 (1983)

17. Rubin, D.B.: Causal inference using potential outcomes: design, modeling, decisions. J. Am. Stat. Assoc. **100**(469), 322–331 (2005)

18. Shen, Z., Cui, P., Kuang, K., Li, B., Chen, P.: Causally regularized learning with agnostic data selection bias. In: Proceedings of the 26th ACM International Conference on Multimedia, pp. 411–419 (2018)

19. Singh, H., Meyer, A.N., Thomas, E.J.: The frequency of diagnostic errors in outpatient care: estimations from three large observational studies involving us adult populations. BMJ Qual. Saf. **23**(9), 727–731 (2014)

20. Singh, H., Schiff, G.D., Graber, M.L., Onakpoya, I., Thompson, M.J.: The global burden of diagnostic errors in primary care. BMJ Qual. Saf. **26**(6), 484–494 (2017)

21. Spirtes, P., Glymour, C., Scheines, R.: Causation, Prediction, and Search, 2nd edn. MIT Press, Adaptive computation and machine learning (2000)

22. Spirtes, P., Meek, C., Richardson, T.S.: Causal inference in the presence of latent variables and selection bias. In: Besnard, P., Hanks, S. (eds.) UAI 1995: Proceedings of the Eleventh Annual Conference on Uncertainty in Artificial Intelligence, Montreal, Quebec, Canada, 18–20 August 1995, pp. 499–506. Morgan Kaufmann (1995)

23. Wong, K.K., Fortino, G., Abbott, D.: Deep learning-based cardiovascular image diagnosis: a promising challenge. Futur. Gener. Comput. Syst. **110**, 802–811 (2020)

24. Yao, L., Li, S., Li, Y., Xue, H., Gao, J., Zhang, A.: On the estimation of treatment effect with text covariates. In: International Joint Conference on Artificial Intelligence (2019)

25. Yu, B.: Three principles of data science: predictability, computability, and stability (PCS). In: Abe, N., et al. (eds.) IEEE International Conference on Big Data (IEEE BigData 2018), Seattle, WA, USA, 10–13 December 2018, p. 4. IEEE (2018)

26. Zelnik-Manor, L., Perona, P.: Self-tuning spectral clustering. In: Advances in Neural Information Processing Systems, vol. 17 (2004)

27. Zhang, Z., et al.: Pathologist-level interpretable whole-slide cancer diagnosis with deep learning. Nat. Mach. Intell. **1**(5), 236–245 (2019)

28. Zhu, R., Tu, X., Huang, J.: Using deep learning based natural language processing techniques for clinical decision-making with EHRs. In: Dash, S., Acharya, B.R., Mittal, M., Abraham, A., Kelemen, A. (eds.) Deep Learning Techniques for Biomedical and Health Informatics. SBD, vol. 68, pp. 257–295. Springer, Cham (2020). https://doi.org/10.1007/978-3-030-33966-1_13

# Graph

# Dominance Maximization in Uncertain Graphs

Atharva Tekawade and Suman Banerjee[(✉)]

Department of Computer Science and Engineering, Indian Institute of Technology
Jammu, Jammu 181221, India
{2018uee0137,suman.banerjee}@iitjammu.ac.in

**Abstract.** An uncertain graph is defined as a collection of nodes with their interconnected edges, where each edge is assigned with its existence probability value. Motivated by many real-world scenarios, in this paper, we define the notion of *dominance* for a given subset of nodes as the sum of its cardinality and the total probability with which the remaining nodes in the graph are dominated by the subset. For a given uncertain graph $\mathcal{G}(\mathcal{V}, \mathcal{E}, \mathcal{P})$, and a positive integer $k$, we introduce the Problem of *Dominance Maximization* where the goal is to choose a $k$-element subset to maximize the dominance in the uncertain graph. First, we show that the dominance function is non-negative, monotone, and submodular. Subsequently, we show that the problem of maximizing dominance in an uncertain graph is NP-hard. Then we propose an incremental greedy strategy based on the marginal gain in dominance computation that leads to $(1 - \frac{1}{e})$-factor approximate solution. With a slight change, we show that this strategy can be made efficient without losing the approximation ratio. Finally, we exploit the submodularity property of the dominance function and construct a combinatorial object called 'Pruned Submodularity Graph' and use this effectively to solve the dominance maximization problem. All the methodologies have been analyzed to understand their time and space requirements. We implement the proposed solution approaches with real-world network datasets and conduct many experiments. We observe that for larger datasets the Pruned Submodularity Graph-based approach takes reasonable computational time and produces better solutions in terms of dominance value compared to the baseline methods.

**Keywords:** Uncertain Graph · Dominance Maximization · Incremental Greedy Strategy · Existence Probability · Submodularity

## 1 Introduction

In many real-world scenarios, obtained data are of the following type: a set of *entities* and *pairwise relationships* among them. Examples include social networks where the users are the entities and 'friendship' is the pairwise relationship, transportation networks where the cities are the entities and *connectivity* is the pairwise relationship, and so on. Such scenarios are essentially captured using

© The Author(s), under exclusive license to Springer Nature Switzerland AG 2023
X. Yang et al. (Eds.): ADMA 2023, LNAI 14178, pp. 231–246, 2023.
https://doi.org/10.1007/978-3-031-46671-7_16

graphs where the entities are represented by the vertices and we put an edge between two vertices if there exists a relationship between the corresponding two entities. In practice, uncertainty occurs due to imprecise information, errors in computation, redundancy in input data, and so on [4]. Hence, in many situations, available data is uncertain in nature. Mining this uncertain data has gained huge attention in the research community since in last one decade [2]. Different kinds of problems have been studied in the arena of uncertain data management and mining such as *pattern mining* [1], *clustering* [12], *outlier detection* [5], *classification* [13] and so on. Many real-world networks are uncertain (i.e., edges of the network are marked with existence probability) in nature such as social networks (where edges are marked with influence probabilities), communication networks (where edges are marked with successful packet delivery probability), protein-protein interaction network (where edge probability signifies the successful docking probability of two proteins) etc. [4].

Due to applications in different domains, analysis and mining of uncertain graphs gained huge attention in the past decade. Several problems have been studied in this domain including structural pattern query problems (clique, reliable subgraph, truss, core, etc.), graph algorithmic problems (e.g., spanning tree, graph clustering, information flow maximization, etc.), different computational problems (reliability computation, conditional reliability computation, simrank computation, etc.), and so on. Look into the survey by Banerjee [4] for more details. The problem of finding a minimum cardinality Dominating Set is a well-studied problem in algorithmic graph theory. Given a simple, undirected graph $G(V, E)$, a subset of its vertex set $\mathcal{S} \subseteq V(G)$ is said to be a dominating set in $G$ if for every vertex $v \in V(G)$ either $v$ is in $\mathcal{S}$ or at least one neighbor of $v$ is in $\mathcal{S}$. Finding dominating set of minimum size in a graph is a well-known NP-hard problem with applications in many areas of computer science and operations research such as *wireless networks* [6], *graph mining* [11], *social network analysis* [11], *facility location* [3], and many more. However, the notion of dominating set has not been generalized in the context of uncertain graphs. In this paper, we bridge this gap by introducing the notion of dominance which correctly generalizes the concept of dominating set.

Many real-world problems can be mapped as the maximization of the dominance function by choosing a limited number of nodes. Consider a social network represented as an uncertain graph where edge probabilities represent the respective influence probability between two users. In this context, the maximization of dominance by selecting $k$ number of users will signify the maximization of influence in one hop. Consider another situation where a sensor network is represented as an uncertain graph where the edge probability signifies the successful data transmission probability. Consider a subset of the sensors as data generator. Then the dominance of these nodes signifies the expected number of sensors in which the sensed data will be reached. These real-life situations motivates us to define the notion of dominance. In particular, our contributions are as follows:

– We introduce the noble Dominance Maximization Problem which correctly generalizes the concept of dominating set problem in the context of uncertain graph.

- We show that the proposed dominance function is non-negative, monotone, submodular, and subadditive.
- We propose three solution approaches to solve this problem. The first one is an $(1 - \frac{1}{e})$-factor approximation algorithm, the second one is an iterative algorithm exploiting the submodularity property of the dominance function, and the third one is the based on Pruned Submodularity Graph.
- All the proposed solution approaches have been analyzed to understand its time and space requirement.
- We conduct a number of experiments with three real-world datasets have been conducted to show the effectiveness of the proposed methodologies

Rest of the paper is organized as follows. Section 2 describes background definitions and defines the problem formally. The proposed solution approaches have been described in Sect. 3. Section 4 describes the experimental evaluation of the proposed solution approaches. Finally, Sect. 5 concludes this study and gives future research directions.

## 2    Preliminaries and Problem Definition

The graphs considered in this paper are finite, simple, and undirected. We denote the graph by $G(V, E)$ where $V(G) = \{v_1, v_2, \ldots, v_n\}$ is the set of $n$ vertices and $E(G) = \{e_1, e_2, \ldots, e_m\}$ is the set of $m$ edges. The number of vertices and edges of graph $G$ is denoted by $n$ and $m$, respectively. For any arbitrary vertex $u \in V(G)$, its neighborhood is defined as the set of vertices that are adjacent to $u$ and denoted as $N(u)$, i.e., $N(u) = \{v : (uv) \in E(G)\}$. The cardinality of the neighborhood of $u$ is defined as its *degree* and denoted by $deg(u)$, i.e., $deg(u) = |N(u)|$. Now, we state the notion of uncertain graphs in Definition 1.

**Definition 1 (Uncertain Graph).**    *We denote an uncertain graph by* $\mathcal{G}(\mathcal{V}, \mathcal{E}, \mathcal{P})$ *where* $\mathcal{V}(\mathcal{G}) = \{v_1, v_2, \ldots, v_n\}$ *are the set of vertices,* $\mathcal{E}(\mathcal{G}) = \{e_1, e_2, \ldots, e_m\}$ *($\mathcal{E}(\mathcal{G}) \subseteq \mathcal{V}(\mathcal{G}) \times \mathcal{V}(\mathcal{G})$) are the set of edges, and* $\mathcal{P}$ *is the edge weight function that assigns each edge to its existence probability, i.e.,* $\mathcal{P} : \mathcal{E}(\mathcal{G}) \longrightarrow (0, 1]$.

So, the deterministic graph (generally, called graphs) is the special case of uncertain graphs, where each edge has the existence probability of 1. For any two vertices $u$ and $v$ if $(uv) \in \mathcal{E}(\mathcal{G})$, we denote its existence probability by $\mathcal{P}(u, v)$. If $(uv) \notin \mathcal{E}(\mathcal{G})$ then $\mathcal{P}(u, v) = 0$. Next, we define the notion of dominating set of a deterministic graph in Definition 2.

**Definition 2 (Dominating Set).**    *Given a graph* $G(V, E)$, *with its vertex set* $V(G)$ *and edge set* $E(G)$, *a subset of its vertex set* $S$ *is said to be the dominating set of* $G$ *if the vertices that are not in* $S$ *has at least one neighbor in* $S$, *i.e.,* $\forall v \in V(G) \setminus S$, $N(v) \cap S \neq \emptyset$.

Given an undirected graph $G(V, E)$, finding its minimum cardinality dominating set is a well-known NP-hard Problem. It can be observed that the problem of dominating set can not be directly applied on uncertain graphs because the edges are marked with the existence probability value. Hence, we generalize the notion of dominating set in the context of uncertain graph. For a given uncertain graph $\mathcal{G}(\mathcal{V}, \mathcal{E}, \mathcal{P})$ and its subset of vertices $\mathcal{S} \subseteq V(\mathcal{G})$, we introduce the notion of 'dominance' in Definition 3.

**Definition 3 (Dominance in Uncertain Graph).** *Given an uncertain graph* $\mathcal{G}(\mathcal{V}, \mathcal{E}, \mathcal{P})$, *and a subset of its vertices* $\mathcal{S}$, *its dominance in* $\mathcal{G}$ *is defined as the total probability with which the nodes in* $\mathcal{V}(\mathcal{G})$ *can be dominated, which can be expressed by Eq. No. 1.*

$$\mathcal{D}(\mathcal{S}) = |\mathcal{S}| + \sum_{u \in V(\mathcal{G}) \setminus \mathcal{S}} 1 - \prod_{v \in N(u) \cap \mathcal{S}} (1 - \mathcal{P}(u, v)) \tag{1}$$

Now, from Eq. No. 1 it can be observed that for any $\mathcal{S} \subseteq \mathcal{V}(\mathcal{G})$, $0 \leq \mathcal{D}(\mathcal{S}) \leq |\mathcal{S}| + |N(\mathcal{S})|$. Also if we consider for all $(uv) \in E(\mathcal{G})$, $\mathcal{P}(u, v) = 1$, from Eq. No. 1 we can verify that the dominance value will be $|\mathcal{S}| + |N(\mathcal{S})|$. In this case, $\mathcal{G}$ will be a deterministic graph and $\mathcal{S}$ will be a dominating set in $\mathcal{G}$. This means that Definition 3 correctly generalizes the concept of dominating set in the context of uncertain graphs. Next, we introduce the Problem of Dominance Maximization.

**Definition 4 ( Dominance Maximization Problem).** *Given an uncertain graph* $\mathcal{G}(\mathcal{V}, \mathcal{E}, \mathcal{P})$ *and a positive integer* $k$, *the Dominance Maximization Problem asks to find out a set* $\mathcal{S} \subseteq V(\mathcal{G})$ *of* $k$ *nodes that maximizes* $\mathcal{D}(\mathcal{S})$. *Mathematically, this problem can be stated using Eq. No. 2.*

$$\mathcal{S}^* \longleftarrow \underset{\mathcal{S} \subseteq V(\mathcal{G}) \ and \ |\mathcal{S}| = k}{argmax} \mathcal{D}(\mathcal{S}) \tag{2}$$

From the computational point of view the Dominance Maximization Problem can be posed as follows:

---

**DOMINANCE MAXIMIZATION PROBLEM**

**Input:** An uncertain graph $\mathcal{G}(\mathcal{V}, \mathcal{E}, \mathcal{P})$, and $k \in \mathbb{Z}^+$.
**Problem:** Find out a set $\mathcal{S} \subseteq V(\mathcal{G})$ with $|\mathcal{S}| = k$ such that $\mathcal{D}(\mathcal{S})$ is maximized.

---

A set function $f$ defined on the ground set $\mathcal{X} = \{x_1, x_2, \ldots, x_n\}$ assigns a positive real number including 0 for each subset of $\mathcal{X}$, i.e., $f : 2^{\mathcal{X}} \longrightarrow \mathbb{R}_0^+$. A set function is said to be normalized if $f(\emptyset) = 0$. A set function can follow several properties such as *non-negativity, monotonicity, submodularity*, etc. These properties are stated in Definitions 5, 6, and 7, respectively. The solution methodologies that we have proposed in this paper exploit some of these properties of the dominance function.

**Definition 5 (Non-negativity of a Set Function).** *A set function $f(.)$ defined on the ground set $\mathcal{X}$ is said to be non-negativity, if for all $\mathcal{Y} \subseteq \mathcal{X}$, $f(\mathcal{Y}) \geq 0$.*

**Definition 6 ( Monotonicity of a Set Function).** *A set function $f(.)$ defined on the ground set $\mathcal{X}$ is said to be monotone, if for all $\mathcal{Y} \subseteq \mathcal{X}$, and $\forall u \in \mathcal{X} \setminus \mathcal{Y}$, $f(\mathcal{Y} \cup \{u\}) \geq f(\mathcal{Y})$.*

**Definition 7 (Sub-modularity of a Set Function).** *A set function $f(.)$ is defined on the ground set $\mathcal{X}$ is said to be submodular, if $\forall \mathcal{Y}_1 \subseteq \mathcal{Y}_2 \subseteq \mathcal{X}$, and $\forall u \in \mathcal{X} \setminus \mathcal{Y}_2$, $f(\mathcal{Y}_1 \cup \{u\}) - f(\mathcal{Y}_1) \geq f(\mathcal{Y}_2 \cup \{u\}) - f(\mathcal{Y}_2)$.*

# 3   Proposed Solution Approach

In this section, we describe the proposed solution approaches. Before that we first establish few important properties of the dominance function. In this section due to space limitation proofs of many lemmas and theorems are omitted and that will appear in a subsequent journal version of this paper.

**Lemma 1.** *The dominance function $\mathcal{D}(.)$ is non-negative, monotone, submodular.*

Now, we introduce the notion of 'Marginal Gain in Dominance' which is sated in Definition 8.

**Definition 8 (Marginal Gain in Dominance).** *Given an uncertain graph $\mathcal{G}(\mathcal{V}, \mathcal{E}, \mathcal{P})$, a subset of the vertices $\mathcal{S} \subseteq V(\mathcal{G})$ and a node $u \in V(\mathcal{G}) \setminus \mathcal{S}$, its marginal gain in dominance is defined as the difference in the dominance value when the node is included in $\mathcal{S}$ and when not. This can be expressed by Eq. No. 3.*

$$\Delta(u|\mathcal{S}) = \mathcal{D}(\mathcal{S} \cup \{u\}) - \mathcal{D}(\mathcal{S}) \tag{3}$$

The proposed approach is based on the marginal dominance gain works as follows. Initially, we start with an empty set, and in each iteration, the node that causes the maximum marginal gain in the dominance value is added to the set. This process is repeated $k$ times. Algorithm 1 describes the procedure. As the dominance function is submodular and we are maximizing this function with respect to the cardinality constraint, according to Nemhauser et al. [10] Algorithm 1 will generate $(1 - \frac{1}{e})$-factor approximate solution. Hence, Theorem 1 holds.

**Theorem 1.** *For a given $k \in \mathbb{Z}^+$, let $\mathcal{S}^A$ be the set obtained in Algorithm 1, and $\mathcal{S}^{opt}$ be the optimal set of size $k$. Then $\mathcal{D}(\mathcal{S}^A) \geq (1 - \frac{1}{e})\mathcal{D}(\mathcal{S}^{opt})$.*

Now, we proceed to analyze this algorithm for time and space requirements. Let, $n$, $m$, and $d^{max}$ denote the number of nodes, edges, and maximum degree of the network. In each iteration of the while loop, computing the marginal gain in dominance for each $u \in V(\mathcal{G}) \setminus \mathcal{S}$ requires $\mathcal{O}(k \cdot d^{max})$ time. As it is looping

---

**Algorithm 1:** Marginal Gain-based Approach for Dominance maximization

---

> **Input** : The Uncertain Graph $\mathcal{G}(\mathcal{V}, \mathcal{E}, \mathcal{P})$, $k \in \mathbb{Z}^+$.
> **Output**: The Set $\mathcal{S}^{\mathcal{A}}$

1  $\mathcal{S}^{\mathcal{A}} \longleftarrow \emptyset$;
2  **while** $|\mathcal{S}^{\mathcal{A}}| < k$ **do**
3  $\quad$ $u \longleftarrow \underset{v \in \mathcal{V}(\mathcal{G}) \setminus \mathcal{S}}{argmax} \; \Delta(v|\mathcal{S}^{\mathcal{A}})$;
4  $\quad$ $\mathcal{S}^{\mathcal{A}} \longleftarrow \mathcal{S}^{\mathcal{A}} \cup \{u\}$
5  **end**
6  *return* $\mathcal{S}^{\mathcal{A}}$

---

for $k$ times, the running time of Algorithm 1 is $\mathcal{O}(k^2 \cdot d^{max})$. $k$ and $d_{max}$ will be of $\mathcal{O}(n)$ in the worst case. Hence the time requirement of Algorithm 1 is of $\mathcal{O}(n^3)$. Also, it is easy to observe that this can be implemented with $\mathcal{O}(1)$ space. Hence Theorem 2 holds.

**Theorem 2.** *Running time and space requirement of Algorithm 1 will be $\mathcal{O}(n^3)$ and $\mathcal{O}(1)$, respectively.*

Now, it is important to observe that, in Algorithm 1 in each iteration of the while loop marginal gain in dominance has been computed for all the nodes that are not in $\mathcal{S}$ and this step penalizes the efficiency in terms of computational time. The proposed methodology can be improved by exploiting the submodularity property of the dominance function.

From the execution time point of view, Algorithm 1 can be improved by exploiting the submodularity property of the dominance function. In the first iteration of the while loop the node with the maximum individual dominance value is included in $\mathcal{S}^{\mathcal{A}}$. Also we sort the nodes based on the obtained marginal gain in the first iteration. It can be observed that for any node $u \in \mathcal{V}(\mathcal{G})$, $\Delta(u|\emptyset) = \mathcal{D}(u)$ as $\Delta(\emptyset) = 0$. By the submodularity property of the dominance function and for any $\mathcal{S} \subseteq \mathcal{V}(\mathcal{G})$ and $u \in \mathcal{V}(\mathcal{G}) \setminus \mathcal{S}$, $\Delta(u|\mathcal{S}) \leq \Delta(u|\emptyset)$. Let, $\mathcal{S}'$ denotes the sorted list of nodes based on the marginal gain in the first iteration and without loss of generality also assume that $i$-th node $u_i$ is included in $\mathcal{S}^{\mathcal{A}}$. Now, in the second iteration we compute the marginal gain of the nodes in the order in which they are placed in $\mathcal{S}'$. Now, assume that in the second iteration for the node $u_j \in V(\mathcal{G}) \setminus \{u_i\}$, the value of $\Delta(u_j|\{u_i\})$ is greater than $\Delta(u_{j+1}|\{u_i\})$, where $u_j$ and $u_{j+1}$ are two consecutive nodes in $\mathcal{S}'$. Now, it is important to observe that even if we compute the marginal gains of the nodes $u_{j+2}, u_{j+3}, \cdots$, with respect to $\{u_i\}$ they can not be more than that of the marginal gain of $u_j$. Hence, $u_j$ is the node with the maximum marginal gain in the second iteration and included in $\mathcal{S}^{\mathcal{A}}$. So, we can the skip the marginal gain computations of the nodes $u_{j+2}, u_{j+3}, \cdots$ are redundant and can be avoided. Initially, this approach was proposed by Leskovec et al. [9]. However, this technique was adopted to solve many real-world problems. Algorithm 2 describes the procedure in pseudo code.

It is important to observe that in the worst case the performance of Algorithm 2 as bad as Algorithm 1. However, in experimentation with real-world datasets, we observe a significant improvement in computational time. Hence, Theorem 3 holds.

---

**Algorithm 2:** Improved Algorithm based o the Marginal Gain for the Dominance Maximization

---

**Input** : The Uncertain Graph $\mathcal{G}(\mathcal{V}, \mathcal{E}, \mathcal{P})$, $k \in \mathbb{Z}^+$.
**Output**: The Set $\mathcal{S}^{\mathcal{A}}$

1   $\mathcal{S}^{\mathcal{A}} \longleftarrow \emptyset$;
2   **while** $\exists u \in \mathcal{V}(\mathcal{G}) \setminus \mathcal{S}^{\mathcal{A}}$ *and* $|\mathcal{S}^{\mathcal{A}}| < k$ **do**
3     **for** *All* $u \in \mathcal{V}(\mathcal{G}) \setminus \mathcal{S}$ **do**
4       $Curr[u] = \text{FALSE}$;
5     **end**
6     **while** *TRUE* **do**
7       $w \longleftarrow \underset{v \in \mathcal{V}(\mathcal{G}) \setminus \mathcal{S}^{\mathcal{A}}}{argmax} \Delta(v | \mathcal{S}^{\mathcal{A}})$;
8       **if** $Cur[w] = TRUE$ **then**
9         $\mathcal{S}^{\mathcal{A}} \longleftarrow \mathcal{S}^{\mathcal{A}} \cup \{w\}$;
10        **break**;
11      **else**
12        Compute $\Delta(w | \mathcal{S}^{\mathcal{A}})$;
13        $Cur[w] = \text{TRUE}$;
14      **end**
15     **end**
16 **end**
17 *return* $\mathcal{S}^{\mathcal{A}}$

---

**Theorem 3.** *For a given $k \in \mathbb{Z}^+$, let $\mathcal{S}^{\mathcal{A}}$ is the set obtained by Algorithm 2 and $\mathcal{S}^{OPT}$ denotes the optimal set of size $k$. Then $\mathcal{D}(\mathcal{S}^{\mathcal{A}}) \geq (1 - \frac{1}{e}) \cdot \mathcal{D}(\mathcal{S}^{OPT})$.*

It can be observed that in the worst case, the time requirement of Algorithm 2 is the same as Algorithm 1. However, it can be observed that to choose any iteration we need to store the information of the marginal gains of the nodes in the previous iteration. For this, we need to consume $\mathcal{O}(n)$ space. Hence, Theorem 4 holds.

**Theorem 4.** *The running time and space requirement of Algorithm 2 is of $\mathcal{O}(n^3)$ and $\mathcal{O}(n)$, respectively.*

In our experiments, we observe that Algorithm 2 takes much less time compared to Algorithm 1. However, we can improve this further by using the notion of Pruned Submodularity Graph, which is stated in Definition 9.

**Definition 9 (Pruned Submodularity Graph).** *Given a submodular function $\mathcal{D}$ defined with its ground set $V(\mathcal{G})$, we construct the pruned submodularity graph $G(V, E, w)$ as follows. For every vertex $v \in V(\mathcal{G})$, we create one vertex for the graph $G$ and between every pair of vertices $u$ and $v$ of $G$ there will be an edge. The weight $w_{u \to v}$ is defined using Eq. No. 4.*

$$w_{u \to v} = \mathcal{D}(v|u) - \mathcal{D}(u|V(\mathcal{G}) \setminus \{u\}) \tag{4}$$

The significance of this weight is as follows. $w_{u \to v}$ measures the worst case loss in maximizing $\mathcal{D}(.)$ on the reduced ground set $V'$ which contains $u$ but does not contain $v$. Here, $\mathcal{D}(v|u)$ denotes the maximum possible gain that $v$ can offer when $u$ is already contained in it. $\mathcal{D}(u|V(\mathcal{G}) \setminus \{u\})$ signifies the minimal possible contribution that the node $u$ can make. Now, we define the notion of the divergence of a subset of the nodes.

**Definition 10 (Divergence of a node).** *Given the pruned submodularity graph $G(V, E, w)$, the divergence of a node $v \in V(G)$ for the set of nodes $V' \subseteq V(G)$ is denoted by $w_{V'v}$ and defined using Eq. No. 5.*

$$w_{V'v} = \min_{x \in V'} w_{xv} \tag{5}$$

The overall idea of this algorithm is as follows. Given an uncertain graph, first, we create its corresponding pruned submodularity graph as mentioned in Definition 9. Subsequently, we apply the algorithm proposed by Zhou et al. [15] to reduce the ground set. Successively, we apply the incremental greedy approach to find out the required number of nodes. Algorithm 3 describes the procedure in the form of pseudocode.

Now, we present a few lemmas that will together imply the running time and space requirement of Algorithm 3. Lemma 2 describes the running time and space requirement for the construction of the pruned submodularity graph. Lemma 3 talks about the time and space requirement for reducing the ground set. Finally Lemma 4 talks about the time and space requirement for selecting $k$ sized set from the reduced ground set.

**Lemma 2.** *Given an uncertain graph $\mathcal{G}(V, \mathcal{E}, \mathcal{P})$ its corresponding pruned submodularity graph $G(V, E, w)$ can be constructed in $\mathcal{O}(n^3)$ time using $\mathcal{O}(n^2)$ space.*

*Proof.* It is easy to observe that this is the time and space requirement of Algorithm 3 from Line No. 1 to 8. It is easy to observe that to store the pruned submodularity graph $\mathcal{O}(n^2)$ space is required where $n$ is the number of nodes of the input uncertain graph. Consider any two vertices $v_u$ and $v_w$. As mentioned in Eq. No. 1, the weight of the edge $(v_u v_w)$ can be expressed as follows:

$$w_{v_u \to v_w} = \mathcal{D}(v_u \cup v_w) - \mathcal{D}(v_w) - \mathcal{D}(V(G)) + \mathcal{D}(V(G) \setminus \{v_w\}) \tag{6}$$

Equation No. 6 can be obtained from Eq. No. 4 by expanding each term in terms of marginal gain in dominance. It is easy to observe that the first, second and fourth quantities can be computed in $\mathcal{O}(n)$ time using $\mathcal{O}(n)$ space. The value of the third quantity is $n$ and can be computed in $\mathcal{O}(1)$ time. So, for a vertex pair their edge weight can be computed in $\mathcal{O}(n)$ time using $\mathcal{O}(n)$ space. This $\mathcal{O}(n)$ space can be reused every time for computing edge weights for different vertex pairs. There are $\mathcal{O}(n^2)$ many vertex pairs. Hence, the total time and space requirement for constructing the pruned submodularity graph will be of $\mathcal{O}(n^3)$ time and $\mathcal{O}(n^2)$ space.

**Lemma 3.** *Reduction of the ground set element from $\mathcal{V}$ to $\mathcal{V}'$ will take $\mathcal{O}(n^2 \log n)$ time and $\mathcal{O}(n)$ space.*

---

**Algorithm 3:** Algorithm for Solving the Dominance Maximization Problem using Pruned Submodularity Graph

---

**Input** : The Uncertain Graph $\mathcal{G}(\mathcal{V}, \mathcal{E}, \mathcal{P})$, Cost Function $C : \mathcal{V}(\mathcal{G}) \longrightarrow \mathbb{Z}^+$,
Budget $B$.
**Output**: The Set $\mathcal{S}$

1   $\mathcal{V}(\mathcal{G}) \longleftarrow \emptyset, \mathcal{E}(\mathcal{G}) \longleftarrow \emptyset$;
2   **for** *All $u \in V(G)$* **do**
3     |   $\mathcal{V}(\mathcal{G}) \longleftarrow \mathcal{V}(\mathcal{G}) \cup \{v_u\}$;
4   **end**
5   **for** *Every Pair $v_u, v_w \in \mathcal{V}(\mathcal{G})$* **do**
6     |   $\mathcal{E}(\mathcal{G}) \longleftarrow \mathcal{E}(\mathcal{G}) \cup (v_u, v_w)$;
7     |   Calculate $w_{v_u \to v_w}$ using Equation No. 4;
8   **end**
9   $\mathcal{V}' \longleftarrow \emptyset; \mathcal{U} \longleftarrow \emptyset, n \longleftarrow |\mathcal{V}|$;
10 **while** $|\mathcal{V}| > r \log n$ **do**
11   |   Sample $r \log n$ items uniformly at random from $\mathcal{V}$ and place them in $\mathcal{U}$;
12   |   $\mathcal{V} \longleftarrow \mathcal{V} \setminus \mathcal{U}, \mathcal{V}' \longleftarrow \mathcal{V}' \cup \mathcal{U}$;
13   |   **for** *All $v \in \mathcal{V}$* **do**
14   |     |   $w_{\mathcal{U}v} \longleftarrow \underset{u \in \mathcal{U}}{min} [\mathcal{D}(v|u) - \mathcal{D}(u|\mathcal{V} \setminus \{u\})]$;
15   |   **end**
16   |   Remove $(1 - \frac{1}{\sqrt{c}}) \cdot |\mathcal{V}|$ many elements from $\mathcal{V}$ having the smallest $w_{\mathcal{U}v}$ value;
17 **end**
18 $\mathcal{S}^A \longleftarrow \emptyset$;
19 **for** $i = 1$ *to $k$* **do**
20   |   $u^* \longleftarrow \underset{u \in \mathcal{V}' \setminus \mathcal{S}^A}{argmax} \mathcal{D}(\mathcal{S}^A \cup \{u\}) - \mathcal{D}(\mathcal{S}^A)$;
21   |   $\mathcal{S}^A \longleftarrow \mathcal{S}^A \cup \{u^*\}$;
22 **end**
23 *return $\mathcal{S}$*;

---

*Proof.* It is easy to observe that this is the time and space requirement from Line No. 9 to 17 of Algorithm 3. The running time is dependent on the number of times the while loop will be executed. It can be observed that in each iteration

the ground set is reduced by $(1 - \frac{1}{\sqrt{n}})$ fraction, and hence, the number of iterations required to exhaust the ground set is of $\mathcal{O}(\log_{\sqrt{c}} n)$. The time requirement
for each iteration will be as follows. Sampling $r \log n$ many elements uniformly
at random from the ground set will take $\mathcal{O}(\log n)$ time. Also, reducing these elements from the ground set $\mathcal{V}$ and putting these elements in the set $\mathcal{V}'$ will take
$\mathcal{O}(\log n)$ time. The for loop from Line No. 13 to 15 will run at most $\mathcal{O}(n)$ times.
The number of elements in $\mathcal{U}$ can be at most $\mathcal{O}(n)$. For any $v \in \mathcal{V}$ computing
$w_{\mathcal{U}v}$ will take $\mathcal{O}(n)$ time. So the time requirement for executing the for loop will
be of $\mathcal{O}(n^2)$. Now, the nodes in $\mathcal{V}$ are sorted based on their $w_{\mathcal{U}v}$ values and this
step takes $\mathcal{O}(n \log n)$ time. Then removing the required set of items will take
$\mathcal{O}(n)$ time. So, in each iteration of the while loop, the time requirement will be
$\mathcal{O}(\log n + n^2 + n \log n) = \mathcal{O}(n^2)$. So the time requirement for the construction
of the reduced ground set (i.e., the execution time of the while loop) will take
$\mathcal{O}(n^2 \cdot \log_{\sqrt{c}} n) = \mathcal{O}(n^2 \cdot \log n)$ time. In this part, only two lists $\mathcal{V}'$ and $\mathcal{U}$ are
there and both of them will take $\mathcal{O}(n)$ space. Hence, the statement is proved.

It is important to observe that the final selection of nodes will be done from the
reduced ground set. Zhou et al. [15] gave a probabilistic estimation about the
size of the reduced ground set which is stated in Theorem 5.

**Theorem 5.** *The size of* $\mathcal{V}'$ *will be* $\frac{cp}{\log \sqrt{c}} K \log^2 n$ *with high probability, where* $p$
*and* $K$ *is related as* $r = p.c.K$.

As mentioned in Theorem 5, there are $\mathcal{O}(\log^2 n)$ many elements in the reduced
ground set. From this reduced ground set we select $k$ many nodes. Now, if $k \geq$
$\frac{cp}{\log \sqrt{c}} K \log^2 n$ then all the elements of the reduced ground set can be returned.
Otherwise, $k$ of them has to be chosen. So the Lines from 18 to 22 as well as the
statement of Lemma 4 only makes sense only when the number of elements in
the reduced ground set is much more than the number of nodes that we want to
choose (i.e., the value of $k$).

**Lemma 4.** *In Algorithm 3, from the reduced ground set finding out the set* $\mathcal{S}^A$
*will take* $\mathcal{O}(k \cdot n \cdot \log^2 n)$ *time and* $\mathcal{O}(\log^2 n)$ *space.*

*Proof.* From Line No. 18 to 23, Algorithm 3 will choose $k$ many nodes from
$\mathcal{O}(\log^2 n)$ many nodes. The number of times the for loop of Line No. 19 will
be executed for $\mathcal{O}(k)$ times. In each iteration, in the worst case, the number of
marginal gain computation will be of $\mathcal{O}(\log^2 n)$. In each marginal gain computations, the time requirement is of $\mathcal{O}(n)$. Hence the time requirement will be of
$\mathcal{O}(k \cdot n \cdot \log^2 n)$. In the worst case, the value of $k$ will be of $\mathcal{O}(\log^2 n)$, and hence,
$\mathcal{O}(\log^2 n)$ space is required.

Lemma 2, 3, and 4 together imply Theorem 6.

**Theorem 6.** *Computational time and space requirement of Algorithm 3 is of*
$\mathcal{O}(n^3 + k \cdot n \cdot \log^2 n)$ *and* $\mathcal{O}(n^2)$, *respectively.*

Now, we state the quality of solution returned by Algorithm 3 and this follows
from [15].

**Theorem 7.** *Let $\mathcal{S}^{OPT}$ denotes an optimal set of size $k$ that maximizes the dominance function. $\mathcal{S}^A$ denotes the $k$ sized set returned by Algorithm 3. Then the following relation will always hold:*

$$\mathcal{D}(\mathcal{S}^A) \geq (1 - \frac{1}{e}) \cdot (\mathcal{D}(\mathcal{S}^{OPT}) - 2 \cdot k \cdot \epsilon) \tag{7}$$

## 4 Experimental Evaluation

In this section we describe the experimental evaluation of the proposed solution approaches. Initially, we start by describing the datasets.

*Datasets.* We have used the following three datasets in our experiments: Zachary Karate Club [7], Email-Eu-Core Network [14], Gnutella Peer-to-Peer Network [8]. In Figs. 1, 2, and 3, we abbreviate these datasets as Zachary, Email, and P2P, respectively. All these datasets have been used extensively in network science literature. The mentioned links can be used to explore more about these datasets.

*Experimental Setup.* In our experiments the following two parameters need to be set: Edge Probability and the value of $k$. The following three edge probability setting has been adopted in the literature, namely *Uniform*, *Trivalancy*, and *Weighted Cascade (WC)*. In the uniform setting, all the edges of the graph have the same probability value, i.e., $(uv) \in E(\mathcal{G})$, $\mathcal{P}(u, v) = p_c$ where $p_c \in (0, 1]$. In this study, we consider three different values of $p_c$ as 0.1, 0.01, and 0.001. Also, $p_c$ does not include the value 0 because it signifies the corresponding edge does not exist. In the trivalancy setting, the existence probability of every edge is assigned uniformly at random from the set $\{0.1, 0.01, 0.001\}$. In the weighted cascade model, any edge $(uv) \in E(\mathcal{G})$ will have the existence probability of $\frac{1}{deg(v)}$. We use the following values of $k$: 10, 15, 20, 25, and 30.

*Baseline Methods.* We use the following two baseline methods to compare the performance of the proposed solution approaches, namely, *Random* and *Top-k*. In the Random method for a given value of $k$, we randomly pick up $k$ nodes. This method will take $\mathcal{O}(k)$ time. In the Top-$k$ method, we compute the individual dominance of every node and sort these nodes based on this value. From the sorted list of nodes, we pick Top-$k$ nodes and return them. This method will take $\mathcal{O}(n^2)$ time.

*Experimental Results and Discussions.* In this section, we describe the experimental results along with the discussions. Figure 1 shows the budget vs. dominance plots for all the datasets where the edge probability follows the uniform setting with the values 0.1, 0.01, and 0.001. From these figures, we observe that for a fixed probability and for a fixed budget value the proposed solution approaches lead to more dominance value compared to the baseline methods. As an example, for the P2P Network dataset with the probability value 0.1

and $k = 50$, the dominance value due to the set selected by Random, Top-$k$, Prunned Submodularity+Incremental Greedy, Incremental Greedy, Incremental Greedy Optimized are 122.4, 309.48, 322.01, 322.01, and 322.01, respectively. Also, for any fixed algorithm and probability value if we increase the budget then quite naturally the dominance value is also increasing. As an example for the Zachary Karate Club Dataset and probability value 0.1, when the budget is increasing from 5 to 20 to 50, the dominance value is increased from 10.02 to 25.57 to 55.57. Also, we observe that for a fixed budget and fixed algorithm, if we decrease the probability value then the dominance value is also decreasing. As an example, for the Email dataset and $k = 5$ when the probability value is decreased from 0.1 to 0.01 to 0.001 the dominance value is decreased from 114.7 to 17.11 to 6.22.

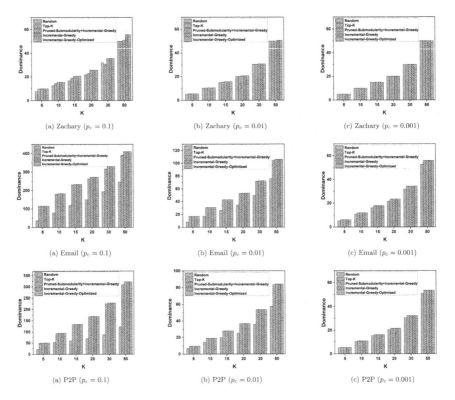

**Fig. 1.** Budget Vs. Dominance Plots for the Uniform Probability Setting (on three different probability values 0.1, 0.01, and 0.001) on Different Datasets

Figure 2 shows the budget vs. dominance plots for the datasets for weighted cascade and trivalancy probability setting. As an example for the P2P Network dataset for the weighted cascade probability setting when the budget value is increased from 20 to 50, the dominance value is increased from 202 to 352.

However, for the trivalancy setting when the budget value is increased from 20 to 50, the dominance value is increased from 81.84 to 168.33. The observations made earlier for the uniform probability setting, continues to hold in these two probability setting as well.

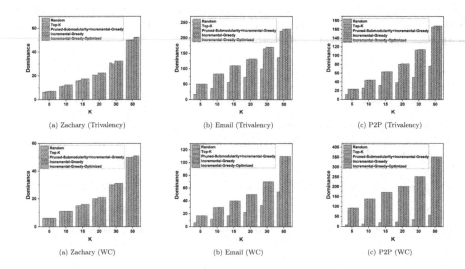

**Fig. 2.** Budget Vs. Dominance Plots for Trivalancy and Weighted Cascade Setting on Different Datasets

Figure 3 shows the budget vs. computational time plots for different datasets and different probability settings. Due to space limitation, in case of uniform probability setting we only consider $p_c = 0.1$. From the figures, we observe that the computational time requirement of the baseline methods is very less compared to the proposed solution approaches. As an example for the E-mail dataset with trivalancy setting and $k = 20$ the execution time due to the set selected by Random, Top-$k$, Prunned Submodularity+Incremental Greedy, Incremental Greedy, Incremental Greedy Optimized are 0.0004 s, 0.425 s, 101.32 s, 46.4 s, and 1.82 s, respectively. However, the same for $k = 50$ are 0.001 s, 0.428 s, 110.66 Secs, 244.38 s, and 4.05 s, respectively. From these values, we also observe that as the value of $k$ increases, the execution time increases. This is obvious because each algorithm has to do additional processing to select more elements in the set. In addition, for smaller values of $k$ the Prunned Submodularity+Incremental Greedy takes more time than the normal Incremental Greedy. But as $k$ increases, the benefit of using Prunned Submodularity Graph technique becomes evident. This is due to the following reason. To use the Prunned Submodularity+Incremental Greedy Approach, we have to construct the combinatorial object called pruned submodularity graph. As the value of $k$ increases, if we do not have this object, then we have to do a lot of marginal gain computations. However, if we prune the ground set and then we choose $k$ nodes then we can reduce the number of marginal gain computations. Hence, Prunned Sub-modularity+Incremental Greedy takes lesser time compared to the Incremental Greedy Approach. The Incremental Greedy Optimized Method takes the least time, and much faster than the Incremental Greedy Approach while giving the same dominance values. This is due to the exploitation of the submodularity of the dominance function. Our final conclusion is that the proposed solution approaches leads to more dominance value compared to the baseline methods with reasonable computational overhead.

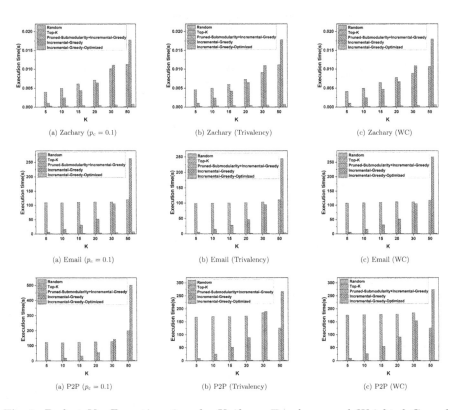

**Fig. 3.** Budget Vs. Execution time for Uniform, Trivalancy and Weighted Cascade Setting on Different Datasets

## 5   Conclusion and Future Directions

In this paper, we have introduced and studied the problem of dominance maximization in the context of uncertain graphs. We have shown that this problem exactly resembles the problem of finding the minimum size dominating set for a deterministic graph which is a classic and well-studied problem in algorithmic graph theory. We have shown that the dominance function is nonnegative, monotone, and sub-modular. An incremental greedy approach based on the marginal gain in dominance leads to an approximation algorithm with the performance guarantee of $(1 - \frac{1}{e})$. This algorithm has been improved by exploiting the sub-modularity property of the dominance function. We have also proposed a solution approach that uses a combinatorial object called Pruned Submodularity Graph. All the proposed methodologies have been analyzed to understand their time and space requirements, as well as the performance guarantees. A number of experiments have been carried out with real-world datasets to demonstrate the performance of the proposed solution approaches. Our future study on this problem will remain concentrated on developing efficient local search techniques.

**Acknowledgements.** The work of Dr. Suman Banerjee is supported with the Seed Grant sponsored by the Indian Institute of Technology Jammu (Grant No.: SG100047).

# References

1. Aggarwal, C.C., Li, Y., Wang, J., Wang, J.: Frequent pattern mining with uncertain data. In: Proceedings of the 15th ACM SIGKDD International Conference on Knowledge Discovery and Data Mining, pp. 29–38 (2009)
2. Aggarwal, C.C., Philip, S.Y.: A survey of uncertain data algorithms and applications. IEEE Trans. Knowl. Data Eng. **21**(5), 609–623 (2008)
3. Baïou, M., Barahona, F.: The dominating set polytope via facility location. In: Fouilhoux, P., Gouveia, L.E.N., Mahjoub, A.R., Paschos, V.T. (eds.) ISCO 2014. LNCS, vol. 8596, pp. 38–49. Springer, Cham (2014). https://doi.org/10.1007/978-3-319-09174-7_4
4. Banerjee, S.: A survey on mining and analysis of uncertain graphs. Knowl. Inf. Syst. **64**(7), 1653–1689 (2022)
5. Chen, G., Zhang, X., Wang, Z.J., Li, F.: Robust support vector data description for outlier detection with noise or uncertain data. Knowl.-Based Syst. **90**, 129–137 (2015)
6. Dai, F., Wu, J.: On constructing K-connected K-dominating set in wireless networks. In: 19th IEEE International Parallel and Distributed Processing Symposium, p. 10. IEEE (2005)
7. Kunegis, J.: KONECT: the Koblenz network collection. In: Proceedings of the 22nd International Conference on World Wide Web, pp. 1343–1350 (2013)
8. Leskovec, J., Kleinberg, J., Faloutsos, C.: Graph evolution: densification and shrinking diameters. ACM Trans. Knowl. Discovery Data (TKDD) **1**(1), 2-es (2007)
9. Leskovec, J., Krause, A., Guestrin, C., Faloutsos, C., VanBriesen, J., Glance, N.: Cost-effective outbreak detection in networks. In: Proceedings of the 13th ACM SIGKDD International Conference on Knowledge Discovery and Data Mining, pp. 420–429 (2007)
10. Nemhauser, G.L., Wolsey, L.A., Fisher, M.L.: An analysis of approximations for maximizing submodular set functions-i. Math. Program. **14**, 265–294 (1978)
11. Raghavan, S., Zhang, R.: Rapid influence maximization on social networks: the positive influence dominating set problem. INFORMS J. Comput. **34**(3), 1345–1365 (2022)

12. Tomar, D., Sathappan, S.: A method for handling clustering of uncertain data. In: 2016 International Conference on Advances in Human Machine Interaction (HMI), pp. 1–5. IEEE (2016)
13. Xie, Z., Xu, Y., Hu, Q.: Uncertain data classification with additive kernel support vector machine. Data Knowl. Eng. **117**, 87–97 (2018)
14. Yin, H., Benson, A.R., Leskovec, J., Gleich, D.F.: Local higher-order graph clustering. In: Proceedings of the 23rd ACM SIGKDD International Conference on Knowledge Discovery and Data Mining, pp. 555–564 (2017)
15. Zhou, T., Ouyang, H., Bilmes, J., Chang, Y., Guestrin, C.: Scaling submodular maximization via pruned submodularity graphs. In: Artificial Intelligence and Statistics, pp. 316–324. PMLR (2017)

# LAGCL: Towards Stable and Automated Graph Contrastive Learning

Hengrui Gu[1], Ying Wang[2], and Xin Wang[1(✉)]

[1] School of Artificial Intelligence, Jilin University, Changchun 130012, China
guhr22@mails.jlu.edu.cn
[2] College of Computer Science and Technology, Jilin University, Changchun 130012, China
{wangying2010,xinwang}@jlu.edu.cn

**Abstract.** Graph Neural Networks (GNNs) have emerged as essential deep learning methods for modeling and analyzing graph structured data. Meanwhile, due to scaling and label sparsity issues, there is an increasing need to develop graph self-supervised representation learning, especially graph contrastive learning. However, existing contrastive method are either based on time-consuming trial-and-error selection of augmentation or employing intense adversarial training to automatically choose augmentation while easily leading to unstable training process and model degradation. In this work, we address above issues by proposing **L**aplace-smoothing based **A**utomated **G**raph**CL** (**LAGCL**), an automated version of GraphCL. LAGCL leverages laplace-smoothing to smooth the intensity of adversarial training and enriches the diversity of augmentations when performing contrastive learning, further improving downstream task performance over other counterparts. Extensive experiments on various datasets show that LAGCL outperforms most baselines on unsupervised learning task and consistently provide robust representations over different backbones.

**Keywords:** Graph contrastive learning · Adversarial training · Laplace smoothing

## 1 Introduction

Graph Neural Networks (GNNs) have become increasingly promising since many real-world data can be represented as graphs, such as social networks, molecules and financial data. Numerous variants of GNNs have been proposed to achieve state-of-the-art performances in graph-based tasks. It's worth mentioning that self-supervised learning on graph-structured data has raised significant interests recently. Over the last years, many works propose methods to learn high-quality frozen graph representations, as it allows for efficient training on downstream tasks with less label data [12–14]. Among many others, graph contrastive learning methods currently achieve state-of-the-art performance in self-supervised

© The Author(s), under exclusive license to Springer Nature Switzerland AG 2023
X. Yang et al. (Eds.): ADMA 2023, LNAI 14178, pp. 247–260, 2023.
https://doi.org/10.1007/978-3-031-46671-7_17

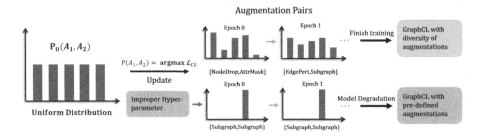

**Fig. 1.** An illustration of model degradation problem that frequently occurs during JOAO training.

representation learning. Specifically, the **GraphCL** [31] framework is the original work of augmentation-based graph contrastive learning methods. However, the effectiveness of GraphCL hinges on ad-hoc data augmentation due to the various nature of graph data, which have to be manually picked per dataset, by either rules of thumb or trial-and-errors. However, this method relies on repeated experiments, and researchers need to investigate the best performance on a validation set for a specific dataset, which is not always available [6]. Based on the observation, we argue that leveraging only two types augmentations to generate contrastive views of raw graph data largely restrict the amount of semantic information encoder can learn. That significantly limits the usability and applicability of GraphCL in various domains.

Holding the same view with us, to address the drawback and improve capability of GraphCL, You et al. (2021) propose **JOAO** [30], an adversarial-training-based bi-level optimization framework, which automates the training of GraphCL by selecting the data augmentation automatically according to a trainable joint probability distribution.

**Motivating Example.** Although JOAO frees the training process of GraphCL from manually picking fixed augmentations, but in practical training, we find that JOAO is highly sensitive to hyperparameters. As shown in Fig. 1, under most hyperparameter settings, the trainable augmentation distribution in JOAO quickly converges to a Dirac distribution and remains relatively fixed during the entire training process. This means that the encoder can be considered as using only two fixed augmentations throughout the entire training process. Additionally, due to the special nature of adversarial maximization optimization for updating the augmentation distribution, the loss values fluctuate throughout the entire training process, as shown in Fig. 2, under different batchsize settings. Therefore, it is difficult for researchers to judge the training status of the model and adjust hyperparameters such as batchsize and epochs based on the loss curve.

Thus, the existing model exhibits several issues that need to be addressed. Firstly, the convergence process of loss is always accompanied by fluctuation, which may mislead researchers to misjudge the convergence condition of encoder parameters and negatively impact the quality of learned graph representation.

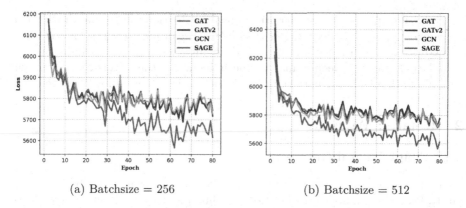

(a) Batchsize = 256          (b) Batchsize = 512

**Fig. 2.** An example of loss fluctuation problem that frequently occurs during JOAO training. The above two subfigures denotes loss curve of JOAO training with altering batchsize across different backbones

Additionally, the trainable joint probability distribution is sensitive to the choice of hyperparameters, and improper selection can cause the distribution to excessively prefer specific augmentations(e.g. `Subgraph`), leading to degraded performance as a general instantiation of GraphCL. We attribute this problem to the min-max optimization, which introduces excessive adversarial perturbation. Based on an intuitive hypothesis, we suggest that only a proper intensity of adversarial perturbation can benefit the performance of contrastive learning models.

**Contributions.** This paper aims at solving the crucial and inherent issue appearing in JOAO. That aforementioned example suggests that we ought to smooth the extent of adversarial perturbation to reduce the training difficulty of GraphCL and meanwhile enrich the semantic information encoder can learn. Hence, we propose a novel version of automatic and adaptive GraphCL framework, dubbed **L**aplace-smoothing based **A**utomated **G**raph**CL** (**LAGCL**). LAGCL outperforms the original GraphCL and JOAO on most datasets by providing a moderate learning process for unsupervised representation learning. In summary, our paper's contributions are as follows:

- Employing GraphCL as basic model, we propose LAGCL framework to address the inherent drawbacks of JOAO. LAGCL liberates GraphCL from the need for trial-and-error or empirical rules, and effectively overcomes the frequently-occurring unstable training issues present in JOAO.
- Laplace-smoothing, the inspiration for LAGCL, also called additive smoothing, is a technique used to smooth categorical data. LAGCL leverages this technique to incorporate adversarial factor into the expected joint augmentation distribution, realizing smooth and controllable adversarial perturbations that is not available previously.
- Extensive experiments on 5 real-world datasets demonstrate that LAGCL outperforms JOAO on all datasets and outperforms GraphCL on 4 datasets

under unsupervised representation linear evaluation. This quantitative evidence supports our hypothesis that LAGCL can enrich the semantic information learned by the encoder. Additionally, we conduct an indispensable ablation study to deeply analyze the influence of the adversarial coefficient and backbone on our performance, further verifying our above hypothesis.

## 2    Related Works

### 2.1    Graph Neural Networks

In recent years, graph neural networks (GNNs) have shown great ability in various tasks, including node classification [15], link prediction [32], and graph classification [5]. A graph can be denoted as $G = (V, E)$ with the matrix $X$ collecting all node features. Generally, GNNs adopt the message-passing mechanism to aggregate information along edge in the input graph according to graph connectivity so as to iteratively learn node representations. So far, numerous graph neural networks with different structures have been proposed by researchers like GCN [15], GraphSAGE [9], GAT [24], GIN [29] and so on. Here we leverage the Graph Convolutional Network (GCN) to illustrate such basic message-passing procedures:

$$X^{(k+1)} = \sigma(\tilde{D}^{-\frac{1}{2}}\tilde{A}\tilde{D}^{-\frac{1}{2}}X^{(k)}W^{(k)}) \tag{1}$$

where $X^{(k)}$ is the representation matrix at the $k$-layer and $X^{(0)} = X$. $\tilde{A} = A + I$ is the adjacency matrix of the graph $G$ with self connections added and $\tilde{D}$ is its diagonal degree matrix. $\sigma(\cdot)$ is the activation function and $W^{(K)}$ is the trainable linear-transformation weight matrix of the $k$-th layer. Graph representations are usually gained by applying aggregation (**sum** or **max** or **mean**) operation on all node representations. In addition, in order to alleviate the shortage of labels existing in a few downstream tasks, self-supervised GNNs are investigated to learn better representations [4].

### 2.2    Graph Contrastive Learning

Contrastive learning leverages self-supervision signal constructed by proper transformations to learn generalizable representations and has achieved great success in visual domain. Graph contrast learning follows the basic idea of contrast learning and has recently emerged as a promising direction. Unlike traditional methods for graph representation learning that attempt to reconstruct adjacency matrices [16] or adopt random-walk to generate structural encoding [8,21], graph contrastive learning enhances the similarity of representations of positive samples aggregated from GNNs to better capture structural information and currently achieves state-of-the-art performance in self-supervised learning.

Conventionally, data augmentation is necessary for constructing positive and negative samples required for contrastive learning. Thus, existing work has proposed many data augmentation techniques for general graph-structured data [10,20,31,34], e.g., node dropping, edge perturbation, graph diffusion, attribute masking, subgraph. Specifically, GraphCL [31] investigates multiple augmentations for graph-level representation learning and analyses their inherent suitability. MVGRL [10] contrast views generated from graph diffusion to encode global information into representations. GRACE [33] constructs node-node pairs by using edge removing and feature masking. In addition to above, it's worth mentioning that some of the most recent work conduct graph contrast learning in augmentation-free way and has also achieved competitive performance, such as AF-GCL [25] and SimGRACE [28].

Contrastive methods avoid a time-consuming generation step in data space [7]. In practice, we usually leverage data augmentations to generate multiple views for each instance. From the perspective of perturbation invariance, two views generated from the same instance are deemed as a positive pair, while two views generated from different instances are regarded as a negative pair [27]. The fundamental idea of contrastive learning is to maximize the agreement of views in positive pairs while minimize the agreement of views in negative pairs so that GNNs can extract high-quality representations that are tolerant to data transformation [28].

## 3   Methodology

In this section, we provide a detailed description of LAGCL. Unlike JOAO [30], LAGCL employs a unique approach to incorporate adversarial training. We begin with an overview of the GraphCL framework [31] and then proceed to elaborate on the LAGCL framework, culminating in an exposition of its behavioral intuitions.

### 3.1   Notations and Preliminaries

Let $G$ represent the input graph variable. We denote $G$ sampled from sample distribution $P_G$ and the augmented graph $\widetilde{G}$ satisfying: $\widetilde{G} \sim P_{(\widetilde{G}|G)}$, where $P_{(\widetilde{G}|G)}$ is the augmentation distribution conditioned on the original graph. In practice, guided by human prior knowledge or trial-and-errors, we select two random augmentation transformation $A_1, A_2$ from an augmentation pool as $\mathcal{A} = \{\texttt{NodeDrop}, \texttt{EdgePert}, \texttt{AttrMask}, \texttt{Subgraph}, \texttt{Identical}\}$. To obtain contrastive views, two augmentation transformations are applied to the input graph $G$. Finally, an embedding model is utilized to extract graph-level representations in a latent space for both views and maximize agreement between them by optimizing the following contrastive loss function [31] (Fig. 3):

$$\mathcal{L}_{CL} = \mathbb{E}_{P_{\widetilde{G}_1}} \left\{ -\mathbb{E}_{P_{(\widetilde{G}_2|\widetilde{G}_1)}} T(E(\widetilde{G}_1), E(\widetilde{G}_2)) + \log(\mathbb{E}_{P_{\widetilde{G}_2}} e^{T(E(\widetilde{G}_1), E(\widetilde{G}_2))}) \right\}, \quad (2)$$

where $\widetilde{G}_1 = A_1(G), \widetilde{G}_2 = A_2(G)$ are both contrastive views, $E(\cdot)$ is embedding model and consisted of $f(\cdot)$ and $g(\cdot)$, the GNN-based encoder combined with subsequent two-layer perceptron (MLP). $T(u,v) = \frac{\text{sim}(u,v)=u^T v/\|u\|\|v\|}{\tau}$ means

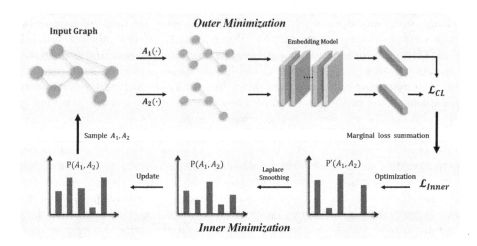

**Fig. 3.** An illustration of the proposed LAGCL framework is presented, wherein two augmentations are sampled based on a joint augmentation distribution $P(A_1, A_2)$ and applied to the input graph. In practice, we instantiate the embedding model as GNN-based encoder and MLP to generate representations of contrastive views.

a similarity measure function and $\tau$ denotes the temperature parameter. The above contrastive loss $\mathcal{L}_{CL}$ essentially fits the formulation of the InfoNCE loss [19,23] such that minimizing Eq. 2 is equivalent to maximizing a lower bound of the mutual information between the latent representations of two contrastive views. However, in GraphCL, $A_1, A_2$ is pre-defined by manual selection, which means the augmentation distribution $P_{(\tilde{G}|G)}$ is a constant Dirac distribution. Simply relying on GraphCL presents an inevitable challenge of selecting the most appropriate augmentation pair to achieve better performance. Additionally, using unalterable augmentation types throughout the entire training process may limit the structural semantic information that can be learned by GNN-based encoders, ultimately impacting the quality of frozen graph representations.

### 3.2    Description of LAGCL

Motivated from [30], in order to consider the variation of augmentation during training, the Eq. 2 can be rewritten into the following form:

$$
\begin{aligned}
\min_\theta \; & \mathcal{L}_{CL}(G, A_1, A_2, \theta) \\
= & \left( -\mathbb{E}_{P_G \times P_{(A_1, A_2)}} \frac{\mathrm{sim}(E_\theta(A_1(G)), E_\theta(A_2(G)))}{\tau} \right) \\
& + \mathbb{E}_{P_G \times P_{A_1}} \log\left( \mathbb{E}_{P_{G'} \times P_{A_2}} \exp\left( \frac{\mathrm{sim}(E_\theta(A_1(G)), E_\theta(A_2(G')))}{\tau} \right) \right),
\end{aligned}
\tag{3}
$$

where $A_1, A_2$ are the augmentation operators defined before, $P_{(A_1, A_2)}$ is the joint augmentation distribution, $(A_1, A_2) \sim P_{(A_1, A_2)}$, $P_{A_1}$ and $P_{A_2}$ are the marginal distributions, $P_G$ is sample distribution as aforementioned, and $P_{G'} = P_G$ acts as the negative sampling distribution for the rest part of graph samples in a batch. The same as Sect. 3.1, $E_\theta(\cdot) = f(\cdot) \circ g(\cdot)$ is parameterized by $\theta$. This rewritten form Eq. 3 introduces augmentation operator into loss function $\mathcal{L}_{CL}$ and makes it possible to update the probability distribution during training process.

Unlike JOAO, we take into account an obvious intuition: the most challenging data augmentation for our contrastive loss may undermine the perturbation-invariance assumption, which is the fundamental assumption of augmentation-based contrastive learning, and then prevent the model from learning enough similarity semantic information in the self-supervised scenario. Therefore, we designed a two-layer optimization framework that enables automatic and adaptive selection of the easiest data augmentation w.r.t. the current loss when performing GraphCL:

$$
\begin{aligned}
&\min_\theta \mathcal{L}_{CL}(G, A_1, A_2, \theta) \\
&\text{s.t. } P'_{(A_1,A_2)} \in \arg\min_{P_{(A'_1,A'_2)}} \left\{ \mathcal{L}_{CL}(G, A'_1, A'_2, \theta) - \gamma \cdot H(A'_1, A'_2) \right\},
\end{aligned}
\tag{4}
$$

where $P'_{(A_1,A_2)}$ is the raw augmentation distribution and required to be further processed, $H(\cdot, \cdot)$ is the joint entropy regularization term that facilitates variety of augmentations, $\gamma \in \mathbb{R} \geq 0$ is a hyper-parameter controlling the weight of the regularization. Specifically, $H(A_1, A_2) = \sum_{i=1}^{|\mathcal{A}|} \sum_{j=1}^{|\mathcal{A}|} -p_{ij} \cdot \log p_{ij}$ where $p_{ij} = P(A_1 = A^i, A_2 = A^j)$, i.e. the probability of selecting the $i$-th and $j$-th data augmentation jointly.

Aiming to improve the generalization, robustness and transferability [22,26] of our framework, we employ Laplace-smoothing to post-process $P'_{(A_1,A_2)}$, which could be viewed as controllable adversarial training, as presented in Eq. 5.

$$
\boldsymbol{p} = \frac{\boldsymbol{p}' + \beta}{1 + |\mathcal{A}|^2 \cdot \beta}.
\tag{5}
$$

where $\boldsymbol{p}'$ denotes raw probability vector w.r.t. $P'_{(A_1,A_2)}$, $\boldsymbol{p}' = [p'_{ij}]$, $i, j = 1, \cdots, |\mathcal{A}|$, while $\boldsymbol{p}$ denotes refined probability vector w.r.t. $P_{(A_1,A_2)}$, $\boldsymbol{p} = [p_{ij}]$, $i, j = 1, \cdots, |\mathcal{A}|$. $\beta$ is the hyper-parameter weighting the intensity of adversarial perturbation, known as **adversarial coefficient**. We use Laplace-smoothing technique to refine the raw augmentations distribution in order to incorporate adversarial factor into the joint augmentation distribution. The intuitive insight of this operation is to make the challenging augmentation could be a little more likely to be involved in training, aligning with the idea of adversarial training.

By leveraging alternating gradient descent algorithm, we instantiate LAGCL as the form of two-layer optimization, alternating between outer minimization and inner minimization.

**Outer Minimization.** To update parameters of the embedding model $E_\theta(\cdot)$, LAGCL performs outer minimization. The minimization optimizes the loss $\mathcal{L}_{CL}$ using gradient descent [17] and updates the parameters $\theta$ of the embedding model.

**Inner Minimization.** To update the raw joint augmentation distribution $P'_{(A_1,A_2)}$, i.e. $p'$, LAGCL performs inner minimization. The objective function in Inner Minimization is the following rewritten form of Eq. 4:

$$\mathcal{L}_{inner} = \mathcal{L}_{CL}(G, A'_1, A'_2, \theta) - \gamma \cdot H(A'_1, A'_2), \tag{6}$$

where $\mathcal{L}_{inner}$ denotes the inner minimization loss. Since it is intractable to directly calculate the gradient of Eq. 6 w.r.t. $p'$, we rewrite Eq. 6 as the following expectation approximation:

$$\mathcal{L}_{inner} \approx \sum_{i=1}^{|\mathcal{A}|}\sum_{j=1}^{|\mathcal{A}|} p_{ij} l_{CL}(G, A_1^i, A_2^j, \theta) - \sum_{i=1}^{|\mathcal{A}|}\sum_{j=1}^{|\mathcal{A}|} -p_{ij}\log p_{ij}, \tag{7}$$

where $p_{ij} = P(A_1 = A^i, A_2 = A^j)$. In the Eq. 7, we approximate the original loss $\mathcal{L}_{CL}$ with the summation of marginal loss related to $p_{ij}$. To sum up, the objective function can be written in the following form:

$$P'_{(A_1,A_2)} \in \arg\min_p \left\{ \sum_{i=1}^{|\mathcal{A}|}\sum_{j=1}^{|\mathcal{A}|} p_{ij} l_{CL}(G, A_1^i, A_2^j, \theta) - \sum_{i=1}^{|\mathcal{A}|}\sum_{j=1}^{|\mathcal{A}|} -p_{ij}\log p_{ij} \right\}. \tag{8}$$

where $\mathcal{L}_{inner}$ is a strongly-concave function w.r.t. $p$. Hence, we can use the value of outer minimization loss $\mathcal{L}_{CL}$ under augmentation pair $(A_1 = A^i, A_2 = A^j)$ to approximate $l_{CL}(G, A^i, A^j, \theta)$. Projected gradient descent [3] is a useful optimization technique that allows us to update $p'$ while ensuring that it satisfies the probability vector constraint. By projecting the updated $p'$ onto the probability simplex, we can ensure that it remains a valid probability distribution. The particular optimization process is as follows:

$$
\begin{aligned}
b &= p^{(n-1)} - \eta \nabla_p \mathcal{L}_{inner}(G, p^{(n-1)}, \theta), \\
p'^{(n)} &= (b - \mu\mathbf{1})_+.
\end{aligned}
\tag{9}
$$

where $\eta \in \mathbb{R} > 0$ is the learning rate of inner minimization, $(\cdot)_+$ is the element-wise non-negative operator, and $\mu$ is the root of the equation $\mathbf{1}^T(b - \mu\mathbf{1})_+ = 1$. $\mu$ can be efficiently found via the bi-section method [1,30].

## 4   Experiments

In this section, we conduct empirical evaluations of our proposed method, LAGCL, against other state-of-the-art self-supervised methods, including GraphCL and JOAO, to answer the following research questions:

- **RQ1:** How effective is LAGCL on unsupervised representation learning?
- **RQ2:** Whether our hypothesis about the intensity of the adversarial perturbation holds?
- **RQ3:** Can LAGCL's performance remain robust against different GNN backbone?

### 4.1 Datasets

**Datasets:** To demonstrate the effectiveness of LAGCL on unsupervised representation learning, we have selected a set of datasets from the TUDataset benchmark [18], which covers a diverse range of fields such as small molecules, bioinformatics, and relational networks. Here we list them as below:

- MUTAG and NCI1 are molecules datasets. In a molecular graph, nodes represent atoms and edges linking corresponding nodes represent chemical bonds. The labels of molecular graphs are determined by the molecular compositions.
- PROTEINS is a bioinformatic dataset including 1113 graphs. In each graph, the nodes represent amino acids, and two nodes are connected by an edge if their distance is less than 6 Angstroms. The labels of these graphs indicate whether they are enzymes or not.
- IMDB-BINARY and REDDIT-BINARY are both relation networks, respectively including movie collaboration graph in IMDB and online discussions graph on Reddit. We use them to evaluate performance of LAGCL on dense and large-scale graph.

### 4.2 Experiment Settings

We conduct a group of quantitative experimental comparisons to evaluate the proposed LAGCL. We choose several methods including graph2vec, InfoGraph, GraphCL and JOAO as baselines. These results of baselines for experimental comparison refer to the published papers [30].

In our evaluation, we adopt GIN with 3 layers and 32 hidden dimensions as backbone, following the same setting in previous work [30]. We train all models on NVIDIA TITAN RTX with early stopping strategy based on accuracy on validation set. We use the ADAM optimizer with a cosine annealing learning rate schedule, i.e., we set learning rate $\alpha_{outer} = \frac{1}{2} \cdot \alpha_{base} \left( 1 + \cos(\frac{t}{T_{max}}\pi) \right)$ with $t$ the current epoch and $T_{max}$ the maximum number of training epochs.

Three hyper-parameters in LAGCL are fine-tuned to improve the average performance. We tune the learning rate $\alpha_{base}$ measuring the update step in **Outer minimization** in the range of $\{0.1, 0.01, 0.005\}$. We tune the learning rate $\eta$ measuring the update step in **Inner minimization** in the range of $\{0.1, 0.05, 0.01\}$. We tune the adversarial coefficient $\beta$ weighting the intensity of Laplace-smoothing in the range of $\{0.01, 0.005, 0\}$.

**Table 1.** The classification accuracies of unsupervised representation learning task. We apply 5 independent runs on each dataset and report the means and standard deviations. **Bold** numbers indicate the top-2 accuracy.

| Method | NCI1 | PROTEINS | MUTAG | RDT-B | IMDB-B |
|---|---|---|---|---|---|
| graph2vec | 73.22±1.81 | 73.30 ± 2.05 | 83.15 ± 9.25 | 75.78 ± 1.03 | 71.10 ± 0.54 |
| InfoGraph | 76.20 ± 1.06 | 74.44 ± 0.31 | **89.01 ± 1.13** | 82.50 ± 1.42 | **73.03 ± 0.87** |
| GraphCL | 77.87 ± 0.41 | 74.39 ± 0.45 | 86.80 ± 1.34 | **89.53 ± 0.84** | 71.14 ± 0.44 |
| JOAO | **78.07 ± 0.47** | **74.55 ± 0.41** | 87.35 ± 1.02 | 85.29 ± 1.35 | 70.21 ± 3.08 |
| LAGCL(Ours) | **79.05 ± 0.57** | **74.69 ± 0.64** | **89.34 ± 1.06** | **88.50 ± 0.44** | **71.55 ± 0.45** |

### 4.3   Evaluations on LAGCL

We compare the classification accuracy of LAGCL with other baselines. Results shown in Table 1 indicate that LAGCL achieves top-2 accuracy on all 5 datasets. Moreover, LAGCL outperforms than JOAO on all datasets. The effectiveness of the proposed LAGCL has been verified.

It is noteworthy that GraphCL, which selects augmentations manually through trial-and-error, only achieves the best performance on RDT-B. This result further supports our hypothesis that different augmentations inject various types of semantic knowledge that benefit the encoder. It is important to note that RDT-B is relatively large-scale in our benchmark, which raises the question of the special properties of large-scale graph data in graph contrastive learning. However, due to computational limitations, we leave this for future research.

### 4.4   Ablation Study

**Ablation on Adversarial Coefficient $\beta$.** To empirically validate our hypothesis that the intensity of adversarial perturbation may impact the training stability of the model and limit the semantic information that the encoder can learn, we perform an ablation study to investigate the effect of the adversarial coefficient $\beta$. This study further confirms that LAGCL's introduction of controllable adversarial perturbation is an improvement over JOAO. We follow the fine-tuned hyper-parameters setting and only alter the value of $\beta$ as $\{0.1, 0.01, 0.005, 0\}$ to analyse the influence of different adversarial coefficient values to the model.

As shown in Fig. 4, the loss curve of LAGCL becomes more unstable with the increasing value of $\beta$. Especially when $\beta = 0.1$ (means high-intensity adversarial-perturbation), the curve fluctuates wildly throughout the whole training process, which probably cause researchers to misjudge the training progress and let model train more epochs, leading to usual overfitting problem [11].

Moreover, the results from Table. 2 verifies our another hypothesis that smooth adversarial-perturbation injects more similarity semantic information created by pretext task into model and then improves the downstream task performance. LAGCL achieves the best accuracy with moderate $\beta = 0.01, 0.005$ and the worst accuracy with $\beta = 0.1$. It should be noted that even when

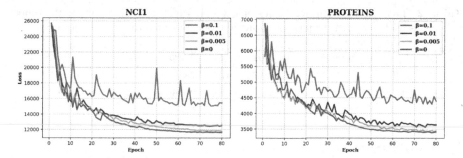

**Fig. 4.** Loss curves of Outer Minimization in LAGCL on datasets NCI1 and PRO-TEINS with different $\beta$ values.

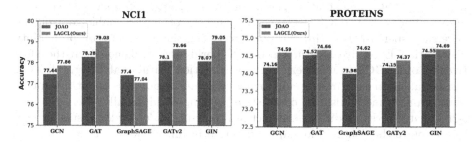

**Fig. 5.** Performance comparison of LAGCL and JOAO w.r.t. different encoder back-bones on datasets NCI1 and PROTEINS.

discarding adversarial-perturbation (means $\beta = 0$), LAGCL outperforms than high-intensity situation (means $\beta = 0.1$), i.e., the similarity semantic informa-tion is undermined by unrestricted adversarial-perturbation. These observations tells us to weigh carefully how to balance the adversarial-perturbations with the perturbation-invariance assumption.

**Ablation on Backbone.** Furthermore, we conduct ablation experiments to study the impact of using different backbones and evaluate the robustness of our proposed model.

Figure 5 compares the performance of JOAO and LAGCL on NCI1 and PRO-TEINS using different backbones including GCN [15], GAT [24], GraphSAGE [9], GATv2 [2], GIN [29] with same setting as in Sect. 4.2 (GAT and GATv2 acti-vate multi-heads mechanism with 3 heads). We find that LAGCL consistently outperforms JOAO almost on all backbones. Besides, the performance of the two methods over different backbone is relatively stable. These results demonstrate that LAGCL consistently provides robust representations across different back-bones and represents an improvement over JOAO. This enhances the practical efficacy and applicability of our model.

**Table 2.** The classification accuracies of LAGCL with different $\beta$ values. We report the means and standard deviations. **Bold** number indicates the highest accuracy.

| $\beta$ | 0.1 | 0.01 | 0.005 | 0 |
|---|---|---|---|---|
| NCI1 | $78.24 \pm 0.35$ | $\mathbf{79.05 \pm 0.57}$ | $78.43 \pm 0.41$ | $78.39 \pm 0.73$ |
| PROTEINS | $74.18 \pm 0.46$ | $74.33 \pm 0.56$ | $\mathbf{74.69 \pm 0.64}$ | $74.59 \pm 0.89$ |

## 5   Conclusion

This paper focuses on addressing the challenge of unstable training and model degradation in existing methods by proposing an automatic, adaptive, and controllable augmentation-based contrastive learning framework. The proposed framework, LAGCL, leverages Laplace-smoothing to enable controllable adversarial-perturbation through post-processing the augmentation probability distribution. Extensive experiments on real-world datasets demonstrate the effectiveness of LAGCL in unsupervised representation learning at the graph-level. Further ablation experiments empirically verify our hypotheses about the underlying training mechanism of contrastive learning. Additionally, there are also some related research directions which still remains under-explored such as how to accelerate the marginal loss approximation in Inner Minimization.

**Acknowledgment.** This work was supported by a grant from the National Natural Science Foundation of China under grants (No. 62272191), and the Foundation of the National Key Research and Development of China (No. 2021ZD0112500), and the International Science and Technology Cooperation Program of Jilin Province (No. 20230402076GH), and the Science and Technology Development Program of Jilin Province (No. 20220201153GX).

## References

1. Boyd, S., Boyd, S.P., Vandenberghe, L.: Convex Optimization. Cambridge University Press (2004)
2. Brody, S., Alon, U., Yahav, E.: How attentive are graph attention networks? arXiv preprint arXiv:2105.14491 (2021)
3. Chen, Y., Wainwright, M.J.: Fast low-rank estimation by projected gradient descent: general statistical and algorithmic guarantees. arXiv preprint arXiv:1509.03025 (2015)
4. Dai, E., Wang, S.: Towards prototype-based self-explainable graph neural network. arXiv preprint arXiv:2210.01974 (2022)
5. Defferrard, M., Bresson, X., Vandergheynst, P.: Convolutional neural networks on graphs with fast localized spectral filtering. In: Advances in Neural Information Processing Systems, vol. 29 (2016)
6. Dwivedi, V.P., Joshi, C.K., Laurent, T., Bengio, Y., Bresson, X.: Benchmarking graph neural networks (2020)
7. Grill, J.B., et al.: Bootstrap your own latent-a new approach to self-supervised learning. Adv. Neural. Inf. Process. Syst. **33**, 21271–21284 (2020)

8. Grover, A., Leskovec, J.: node2vec: scalable feature learning for networks. In: Proceedings of the 22nd ACM SIGKDD International Conference on Knowledge Discovery and Data Mining, pp. 855–864 (2016)
9. Hamilton, W., Ying, Z., Leskovec, J.: Inductive representation learning on large graphs. In: Advances in Neural Information Processing Systems, vol. 30 (2017)
10. Hassani, K., Khasahmadi, A.H.: Contrastive multi-view representation learning on graphs. In: International Conference on Machine Learning, pp. 4116–4126. PMLR (2020)
11. Hawkins, D.M.: The problem of overfitting. J. Chem. Inf. Comput. Sci. **44**(1), 1–12 (2004)
12. Hu, W., et al.: Strategies for pre-training graph neural networks. arXiv preprint arXiv:1905.12265 (2019)
13. Huang, T., Pei, Y., Menkovski, V., Pechenizkiy, M.: Hop-count based self-supervised anomaly detection on attributed networks. arXiv preprint arXiv:2104.07917 (2021)
14. Jin, W., et al.: Self-supervised learning on graphs: deep insights and new direction. arXiv preprint arXiv:2006.10141 (2020)
15. Kipf, T.N., Welling, M.: Semi-supervised classification with graph convolutional networks. arXiv preprint arXiv:1609.02907 (2016)
16. Kipf, T.N., Welling, M.: Variational graph auto-encoders. arXiv preprint arXiv:1611.07308 (2016)
17. Lemaréchal, C.: Cauchy and the gradient method. Doc. Math. Extra **251**(254), 10 (2012)
18. Morris, C., Kriege, N.M., Bause, F., Kersting, K., Mutzel, P., Neumann, M.: TUDataset: a collection of benchmark datasets for learning with graphs. arXiv preprint arXiv:2007.08663 (2020)
19. Oord, A.V.D., Li, Y., Vinyals, O.: Representation learning with contrastive predictive coding. arXiv preprint arXiv:1807.03748 (2018)
20. Peng, Z., et al.: Graph representation learning via graphical mutual information maximization. In: Proceedings of the Web Conference 2020, pp. 259–270 (2020)
21. Perozzi, B., Al-Rfou, R., Skiena, S.: DeepWalk: online learning of social representations. In: Proceedings of the 20th ACM SIGKDD International Conference on Knowledge Discovery and Data Mining, pp. 701–710 (2014)
22. Robey, A., Hassani, H., Pappas, G.J.: Model-based robust deep learning: generalizing to natural, out-of-distribution data. arXiv preprint arXiv:2005.10247 (2020)
23. Tschannen, M., Djolonga, J., Rubenstein, P.K., Gelly, S., Lucic, M.: On mutual information maximization for representation learning. arXiv preprint arXiv:1907.13625 (2019)
24. Veličković, P., Cucurull, G., Casanova, A., Romero, A., Lio, P., Bengio, Y.: Graph attention networks. arXiv preprint arXiv:1710.10903 (2017)
25. Wang, H., Zhang, J., Zhu, Q., Huang, W.: Augmentation-free graph contrastive learning. arXiv preprint arXiv:2204.04874 (2022)
26. Wang, J., et al.: Towards a unified min-max framework for adversarial exploration and robustness. arXiv: Learning (2019)
27. Wu, L., Lin, H., Tan, C., Gao, Z., Li, S.Z.: Self-supervised learning on graphs: contrastive, generative, or predictive. IEEE Trans. Knowl. Data Eng. (2021)
28. Xia, J., Wu, L., Chen, J., Hu, B., Li, S.Z.: SimGRACE: a simple framework for graph contrastive learning without data augmentation. In: Proceedings of the ACM Web Conference 2022, pp. 1070–1079 (2022)
29. Xu, K., Hu, W., Leskovec, J., Jegelka, S.: How powerful are graph neural networks? arXiv preprint arXiv:1810.00826 (2018)

30. You, Y., Chen, T., Shen, Y., Wang, Z.: Graph contrastive learning automated. In: International Conference on Machine Learning, pp. 12121–12132. PMLR (2021)
31. You, Y., Chen, T., Sui, Y., Chen, T., Wang, Z., Shen, Y.: Graph contrastive learning with augmentations. Adv. Neural. Inf. Process. Syst. **33**, 5812–5823 (2020)
32. Zhang, M., Chen, Y.: Link prediction based on graph neural networks. In: Advances in Neural Information Processing Systems, vol. 31 (2018)
33. Zhu, Y., Xu, Y., Yu, F., Liu, Q., Wu, S., Wang, L.: Deep graph contrastive representation learning. arXiv preprint arXiv:2006.04131 (2020)
34. Zhu, Y., Xu, Y., Yu, F., Liu, Q., Wu, S., Wang, L.: Graph contrastive learning with adaptive augmentation. In: Proceedings of the Web Conference 2021, pp. 2069–2080 (2021)

# Discriminative Graph-Level Anomaly Detection via Dual-Students-Teacher Model

Fu Lin[1,2]([✉]), Xuexiong Luo[2], Jia Wu[2], Jian Yang[2], Shan Xue[2], Zitong Wang[1], and Haonan Gong[1]

[1] Wuhan University, Wuhan, Hubei, China
{linfu,zitongwang,gonghaonan}@whu.edu.cn
[2] Macquarie University, Sydney, NSW, Australia
xuexiong.luo@hdr.mq.edu.au, {jia.wu,jian.yang,emma.xue}@mq.edu.au

**Abstract.** Different from the current node-level anomaly detection task, the goal of graph-level anomaly detection is to find abnormal graphs that significantly differ from others in a graph set. Due to the scarcity of research on the work of graph-level anomaly detection, the detailed description of graph-level anomaly is insufficient. Furthermore, existing works focus on capturing anomalous graph information to learn better graph representations, but they ignore the importance of an effective anomaly score function for evaluating abnormal graphs. Thus, in this work, we first define anomalous graph information including node and graph property anomalies in a graph set and adopt node-level and graph-level information differences to identify them, respectively. Then, we introduce a discriminative graph-level anomaly detection framework with dual-students-teacher model, where the teacher model with a heuristic loss are trained to make graph representations more divergent. Then, two competing student models trained by normal and abnormal graphs respectively fit graph representations of the teacher model in terms of node-level and graph-level representation perspectives. Finally, we combine representation errors between two student models to discriminatively distinguish anomalous graphs. Extensive experiment analysis demonstrates that our method is effective for the graph-level anomaly detection task on graph datasets in the real world.(The source code is at https://github.com/whb605/GLADST.git).

**Keywords:** graph anomaly detection · graph neural networks · dual-students-teacher model

## 1 Introduction

Graph anomaly detection investigation has already become a hot topic in academic and industry communities in the past few years. Researchers aim to design a more effective anomaly detection method to detect existing anomalous information on graph datasets [19,25]. Besides, they also actively explore practical application scenarios based on graph anomaly detection task, such as abnormal account detection on financial transaction platforms [26], fake information monitor on social websites [8,22] and intrusion detection in cyber security [30]. However, most of the current research pays more attention to analyzing abnormal nodes from a graph, i.e., node-level anomaly detection. For example, DOMINANT [3], ComGA [16] and DAGAD [12] models utilize deep graph neural networks (GNNs) [24] to capture various node anomalies including local, global and community structure anomalies in the single graph. Even though these methods have achieved great success, a new problem involving how to detect existing anomalous graphs within a set of graphs is worth further exploration, and it also has the huge practical value, such as distinguishing abnormal molecule graphs for molecule property prediction.

Due to the obvious difference between node-level anomaly detection and graph-level anomaly detection, the previous approaches are not appropriate for graph-level anomaly work. Thus, we initially need to explore the key problem, this being which form does an abnormal graph take compared with other normal graphs. According to the intuitive analysis, an abnormal graph will represent a significant difference in node and graph properties. Specifically, a certain node may contain anomalous attribute information and have an abnormal connection with neighbors. For example, when we monitor bank account transactions in a certain region, abnormal accounts will show the account identity information anomaly and many abnormal transaction connections with others. Furthermore, the graph property anomaly more shows the difference of the whole graph structure information. For example, the molecule graph with two benzene rings is abnormal compared with other molecule graphs that only have a benzene ring on a molecule graph set. Thus, two key anomaly definitions above are conducive to find out anomalous graphs within a graph set. Another key problem is to design an effective anomaly score function to judge which graph is abnormal. Besides, the score function has the power to discriminatively distinguish normal and abnormal graphs without the influence of graph data types. It is worth mentioning that there is a limited quantity of research about the graph-level anomaly detection problem, such as GLocalKD [18], GLADC [17], and iGAD [27] methods. But these methods either ignore two types of graph anomaly form mentioned or they do not consider a discriminative anomaly score function for anomalous graph detection.

Based on the aforementioned discussion, in this article, we design a discriminative **Graph Level Anomaly Detection** framework by building a competitive dual-**Students-Teacher** model named **GLADST**. The proposed GLADST framework consists of one teacher model, two student models and an anomaly score function, where the backbone of these models are GNNs. Specifically, we first train the teacher model with a heuristic loss to make learned graph representations more divergent, which can help better capture complex graph information pattern. Then we train student model A with normal graphs to fit the graph representation distribution of the teacher model from node-level and graph-level representation perspectives. Similarly, student model B is trained by abnormal graphs according to the way above. The key idea of the design is that two competing student models can effectively learn normal and abnormal graph representation patterns, respectively and node-level and graph-level representation errors achieve node and graph properties anomaly detection, respectively. Thus, given a test graph, if the graph is normal, graph representations of student model A will better match graph representations of the teacher model, and student model B will keep away. In other words, node-level and graph-level representation errors of student model A and the teacher model will be smaller, but for student model B, these will be larger. Besides, to discriminatively distinguish abnormal graphs and normal graphs, we design a competitive anomaly score function based on the representation error value of two student models, which makes the anomaly score of the real abnormal graph larger than the normal graph. Thus, our work has the following three key contributions:

- We explore the graph-level anomaly detection problem and define existing anomalous graph information including node and graph property anomalies. Furthermore, we jointly utilize node-level representation and graph-level representation errors to detect these anomalies.
- We introduce a discriminative graph-level anomaly detection framework relying on dual-students-teacher model. Specifically, the special training way is beneficial to better learning normal and abnormal graph representation patterns. And the anomaly score function relies on the representation error value of two student models, it can optimize the effectiveness of abnormal graph detection.
- We conduct performance comparison experiments with baselines and model analysis experiments to illustrate the efficiency of GLADST.

## 2   Related Work

Benefiting from advanced deep learning techniques, graph anomaly detection research based on GNNs has attracted considerable interest recently. In light of the difference between anomalous objects, graph anomaly detection can be categorized into the two types below:

**Node-Level Anomaly Detection** (NLAD) is to discover abnormal nodes which are different from other nodes in structure and attribute information. And NLAD is used to identify abnormal nodes by inputting a graph. DOMINANT [3] first employed GNNs for the NLAD task. It utilizes GNNs to learn graph representations and then constructs the reconstruction errors of graph structure and node attribute to capture abnormal nodes. Afterwards, many methods based on GNNs [4,10,12,16] focus on analyzing different types of node anomalies, such as local, global, and community structure anomalies in graphs. In addition, some NLAD methods based on graph contrastive learning are proposed [2,7,15,29] and they built different contrast pairs of node and subgraph to better exploit rich graph information for anomalous node detection.

**Graph-Level Anomaly Detection** (GLAD) detects abnormal graphs that have the obvious difference with other graphs in a graph set. Besides, GLAD is clearly different from NLAD and the aforementioned methods are unsuitable for the GLAD task. Thus, several research works have explored this issue. For example, GLocalKD [18] utilized the predictor network to learn normal graph representations of the random network by global and local graph representation distillation and abnormal graphs will show obvious graph representation errors in the framework. But the method easily fails these graph data that abnormal graph representation pattern is not very obvious within the graph set. GLADC [17] used disturbed features to construct contrastive instances to improve the performance of GLAD. iGAD [27] designed a new graph neural network to investigate anomalous attributes and substructures to learn graph representations. Other two methods [11,14] focus more on the out-of-distribution problem of graph data. Although these methods achieve great performance in GLAD task, they lack an effective anomaly score function to keep competitive performance on different types of abnormal graph data. Thus, we propose a discriminative GLAD framework, where two competing student models learn normal and abnormal graph representation patterns respectively by a trained teacher model. Then the representation error value between two student models can be significantly distinguished abnormal and normal graphs.

In addition to the advancements in graph neural networks for anomaly detection, there are several notable developments in graph processing techniques that contribute to the field. For example, Hooi, B. et al. proposed a method [5] to analyze the real-world graphs on fraud attacks. DenseAlert and DenseStream [21] can focus on detecting dense subtensors to discover the anomalies. Spade [6] is another method that proposed three fundamental peeling sequence reordering techniques. It can effectively detect fraudulent communities.

## 3    Definition and Problem Statement

*Definition 1 (Graph)* $G = (\mathcal{V}_G, \mathcal{X}_G, \mathcal{E}_G, \mathcal{A}_G)$ represents a graph, where $\mathcal{V}_G = \{v_1, v_2, ..., v_n\}$ denotes the node set and $x_i \in \mathcal{X}_G$ is the attribute feature of node $v_i \in \mathcal{V}_G$. $\mathcal{X}_G$ is the attribute feature matrix. We call the graph $G$ *plain graph* if it doesn't have the attribute information, otherwise, it is called *attributed graph*. $e_{ij} \in \mathcal{E}_G$ is the edge between $v_i$ and $v_j$. $\mathcal{A}_G$ is the adjacency matrix, $\mathcal{A}_G(i,j) = 1$ denotes that nodes $v_i$ and $v_j$ have an edge between them; and $\mathcal{A}_G(i,j) = 0$ otherwise.

*Definition 2 (Graph-level Anomaly)* Given a graph dataset $\mathcal{G} = \{G_1, G_2, ..., G_m\}$ with each graph $G \in \mathcal{G}$ denoted by $G = (\mathcal{V}_G, \mathcal{X}_G, \mathcal{E}_G, \mathcal{A}_G)$. **Node property anomaly** is where the node of given graph $G$ has anomalous attributes and abnormal connections with neighbors compared with normal graphs. **Graph property anomaly** is when the structure construction of graph $G$ is inconsistent with others in $\mathcal{G}$ from the global view.

We aim to learn an anomaly evaluation function $f : \mathcal{G} \rightarrow \mathbb{R}$ with parameter $\Theta$ on the graph set $\mathcal{G}$, and the return value of function $f(\hat{G}_i; \Theta) > f(\hat{G}_j; \Theta)$ when the input graph $\hat{G}_i$ is more like an anomaly graph than $\hat{G}_j$.

# 4    Framework of GLADST

To capture normal and abnormal graph representation patterns respectively and learn an effective anomaly score function, we propose a discriminative graph-level anomaly detection framework. As shown in Fig. 1, the framework is composed of a dual-students-teacher model and a discriminative anomaly score function, and the detailed operation is introduced as follows:

## 4.1    Dual-Students-Teacher Model

As GLocalKD framework [18] used a predictor network to capture normal graph representation pattern of the random network by the knowledge distillation method, but this way easily gets suboptimal performance when abnormal graph pattern is difficult to be distinguished. We consider to design a dual-students-teacher model to overcome the above problem.

**Trained Teacher Model.** The teacher model is a graph convolutional network (GCN) [9] that aggregates node's neighbors feature information to update itself feature, to learn graph representations. The teacher model takes matrix $\mathcal{A}_G$ and $\mathcal{X}_G$ as input and then uses GCN to map each node $v_i \in \mathcal{V}_G$ into the representation space. We define $h_i^l$ as the hidden representation of node $v_i$ at the $l^{th}$ layer :

$$h_i^l = ReLU\left(\tilde{\mathcal{D}}_G^{-1/2}\tilde{\mathcal{A}}_G\tilde{\mathcal{D}}_G^{-1/2}h_i^{l-1}\Theta^{l-1}\right), \tag{1}$$

where the $(l-1)^{th}$ layer's weight parameters are $\Theta^{l-1}$ and the node representation is $h_i^{l-1}$. $\tilde{\mathcal{A}}_G = \mathcal{A}_G + \mathcal{I}_G$ and $\mathcal{I}_G$ denotes the identity matrix. $|G|$ is the

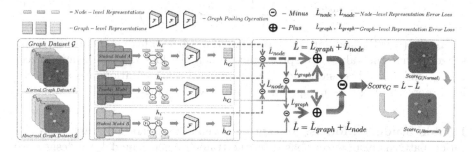

**Fig. 1.** The framework of the proposed GLADST. We first train the teacher model with a heuristic loss to learn node-level and graph-level representations on a given graph dataset. Then, we train the student model A to obtain normal node-level and graph-level representation patterns respectively by the teacher model under node-level and graph-level representation error loss optimization. Similarly, the student model B is trained to obtain abnormal node-level and graph-level representation patterns. And the backbone of these models are GNNs. Finally, the value of representation error between student A and student B is as the anomaly score to identify anomalous graphs.

number of nodes and $\tilde{\mathcal{D}}_G$ is the corresponding diagonal degree matrix:

$$\tilde{\mathcal{D}}_G\left(i,i\right) = \sum_{j=1}^{|G|} \tilde{\mathcal{A}}_G\left(i,j\right), \tag{2}$$

where the feature vector in $\mathcal{X}_G$ is used as the initialized input of node representations, $i.e$, the $0^{th}$ layer's $h_i^0 = \mathcal{X}_G(i,:)$. The plain graph $G$ does not have the parameter $\mathcal{X}_G$, following [18,28], so we construct a simple $\mathcal{X}_G$ by using the node degree information.

The $h_i$ of the last layer is the model's output of node-level representations. We use the max-pooling operation for all node representations on $d$ dimension space, and learn graph-level representations $h_G$ :

$$h_G = [\max_{i=1}^{n} h_{i,1}, \max_{i=1}^{n} h_{i,2}, ..., \max_{i=1}^{n} h_{i,d}]. \tag{3}$$

We utilize a dataset comprising graphs $\mathcal{G}$ to train the teacher model and initialize the model weights $\Theta$ randomly. Specifically, we develop a heuristic loss to form the teacher model and the training purpose is to make learned graph representations more divergent, which can help student models better capture normal and abnormal graph patterns. The training losses are as:

$$L_{teacher} = \frac{1}{\left(L_{graph} + L_{node}\right)}, \tag{4}$$

$$L_{graph} = \frac{1}{|\mathcal{G}|} \sum_{G \in \mathcal{G}} Std\left(h_G\right), \tag{5}$$

$$L_{node} = \frac{1}{|\mathcal{G}|} \sum_{G \in \mathcal{G}} \left(\frac{1}{|G|} \sum_{v_i \in \mathcal{V}_G} Std\left(h_i\right)\right), \tag{6}$$

where $Std\left(.\right)$ is the standard deviation function where a high standard deviation suggests that the values are spread out. When the loss $L_{teacher}$ is minimized, $L_{graph}$ and $L_{node}$ become larger, $i.e$, graph-level representations $h_G$ and node-level representations $h_i$ are spread out over a wider range.

**Double Student Models.** To capture normal and abnormal graph representation patterns respectively based on the trained teacher model, we design two competing student models that are indispensable in the training process and their backbones are GCN model [9] with exactly the same structure as the teacher model. Then, we will describe the work flow of two student models.

1. We initially input the normal graph dataset $\hat{\mathcal{G}} \in \mathcal{G}$ into the trained teacher model above to acquire node-level representations $h_i$ and graph-level representations $h_G$. Then, we also train the student model A based on graph set $\hat{\mathcal{G}}$ to learn node-level representations $\hat{h}_i$ and graph-level representations $\hat{h}_G$.

2. We construct node-level representation error loss $\hat{L}_{node}$ and graph-level representation error loss $\hat{L}_{graph}$ with the trained teacher model, which aims to catch normal node-level and graph-level representation patterns. The two losses as:

$$\hat{L}_{graph} = \frac{1}{|\mathcal{G}|} \sum_{G \in \mathcal{G}} f_d \left( h_G, \hat{h}_G \right), \tag{7}$$

$$\hat{L}_{node} = \frac{1}{|\mathcal{G}|} \sum_{G \in \mathcal{G}} \left( \frac{1}{|G|} \sum_{v_i \in \mathcal{V}_G} f_d \left( h_i, \hat{h}_i \right) \right), \tag{8}$$

where $f_d(\cdot, \cdot)$ is the function to calculate the difference between two graph representations. Here, we can choose *mean square error (MSE)* function.

3. We can learn a final normal graph representation pattern by the following loss as:

$$\hat{L} = \hat{L}_{graph} + \hat{L}_{node}. \tag{9}$$

4. Finally, we use abnormal graph dataset $\check{\mathcal{G}} \in \mathcal{G}$ to train the student model B, and the training process of student B is the same as that of model A as described above. And we also can learn a final abnormal graph representation pattern $\check{L}$.

## 4.2   Discriminative Anomaly Score Function

In our framework, we consider node-level representation error and graph-level representation error to detect two categories of graph anomaly respectively: *node property anomaly* and *graph property anomaly*. Specifically, we propose a dual-students-teacher model above to capture normal and abnormal graph representation patterns. Thus, when we input a test graph sample $G$, the discriminative anomaly score function is designed as follows:

$$Score_G = \left( \left\| h_G - \hat{h}_G \right\|^2 + \frac{1}{|G|} \sum_{v_i \in \mathcal{V}_G} \left\| h_i - \hat{h}_i \right\|^2 \right)$$
$$- \left( \left\| h_G - \check{h}_G \right\|^2 + \frac{1}{|G|} \sum_{v_i \in \mathcal{V}_G} \left\| h_i - \check{h}_i \right\|^2 \right). \tag{10}$$

If the value of $Score_G$ is larger, the probability that graph $G$ is an abnormal graph is greater.

## 4.3   Theoretical Analysis

We use model $\phi$ to represent the teacher model, model $\hat{\phi}$ to represent student model A and model $\check{\phi}$ to represent student model B. Given a test graph sample $G \in \mathcal{G}$, $\phi_G^*$ denotes the representations outputs of the teacher model, $\hat{\phi}_G^*$ and

$\check{\phi}_G^*$ denote the representation outputs of these two student models, respectively. The score of anomaly is simplified as follows:

$$Score_G = \hat{S}_G - \check{S}_G = \left\| \phi_G^* - \hat{\phi}_G^* \right\|^2 - \left\| \phi_G^* - \check{\phi}_G^* \right\|^2 . \tag{11}$$

At the training stage, we first use $\mathcal{G}$ to train the teacher model $\phi$ with a heuristic loss. Then we use normal graphs $\hat{\mathcal{G}}$ to train student model $\hat{\phi}$ and use abnormal graphs $\check{\mathcal{G}}$ to train student model $\check{\phi}$. We want node-level representations and graph-level representations of two student models on each training sample to be as close as possible to the corresponding representations of the teacher model, respectively. Thus, other training graphs with similar patterns will have small prediction errors between them in the student model A. The situation is similar for the student model B. Specifically, given a normal graph sample $G$, its patterns are similar to many other training graphs of normal graph set $\hat{\mathcal{G}}$, and the loss error $\hat{S}_G$ is small, by contrast, $\check{S}_G$ is large because its patterns are dissimilar to many other training graphs of abnormal graph set $\check{\mathcal{G}}$. Thus, if $G$ is normal, $\hat{S}_G$ is small and $\check{S}_G$ is large, after normalization, $Score_G = \hat{S}_G - \check{S}_G < 0$ under ideal conditions. Otherwise, if $G$ is abnormal, $\hat{S}_G$ is large and $\check{S}_G$ is small, after normalization, the anomaly score $Score_G = \hat{S}_G - \check{S}_G > 0$ under ideal conditions. Obviously, the anomaly score above can be significantly distinguished normal and abnormal graphs compared with current baselines whose anomaly scores only rely on simple graph representation errors.

## 5    Experiments

### 5.1    Datasets

**Table 1.** The information of experimental datasets.

| Datasets | Graphs | Avg-nodes | Avg-edges |
|----------|--------|-----------|-----------|
| HSE | 8,417 | 16.89 | 17.23 |
| MMP | 7,558 | 17.62 | 17.98 |
| P53 | 8,903 | 17.92 | 18.34 |
| PPAR | 8,451 | 17.38 | 17.72 |
| AIDS | 2,000 | 15.69 | 16.20 |
| BZR | 405 | 35.75 | 38.36 |
| COX2 | 467 | 41.22 | 43.45 |
| DHFR | 756 | 42.43 | 44.54 |
| NCI1 | 4,110 | 29.87 | 32.30 |
| ENZYMES | 600 | 32.63 | 62.14 |
| PROTEINS | 1,113 | 39.06 | 72.82 |
| COLLAB | 5,000 | 74.49 | 2,457.78 |

**Table 2.** Anomaly detection performance measured mean value of AUC (%) and standard deviation (%) when graph data of label 0 is graph anomaly.

| Datasets | FGSD-IF | FGSD-LOF | FGSD-OCSVM | GLocalKD | GOOD-D | GLADST |
|---|---|---|---|---|---|---|
| HSE | $39.38 \pm 1.35$ | $43.44 \pm 2.49$ | $42.24 \pm 4.43$ | $59.25 \pm 1.09$ | $\mathbf{69.39} \pm 1.05$ | $54.76 \pm 2.12$ |
| MMP | $67.78 \pm 0.90$ | $57.00 \pm 1.98$ | $52.14 \pm 2.97$ | $32.43 \pm 0.81$ | $\mathbf{69.76} \pm 8.10$ | $68.50 \pm 0.72$ |
| P53 | $66.94 \pm 3.67$ | $56.55 \pm 3.55$ | $48.63 \pm 2.76$ | $33.35 \pm 3.34$ | $62.51 \pm 1.85$ | $\mathbf{68.86} \pm 3.51$ |
| PPAR | $34.49 \pm 4.03$ | $46.41 \pm 4.81$ | $50.45 \pm 3.86$ | $65.46 \pm 4.05$ | $\mathbf{66.65} \pm 1.47$ | $61.75 \pm 3.12$ |
| AIDS | $\mathbf{99.38} \pm 0.89$ | $87.73 \pm 4.89$ | $86.20 \pm 4.22$ | $96.61 \pm 0.53$ | $92.58 \pm 1.36$ | $97.65 \pm 0.98$ |
| BZR | $44.51 \pm 6.06$ | $49.56 \pm 8.73$ | $41.15 \pm 6.44$ | $67.12 \pm 8.71$ | $74.84 \pm 5.40$ | $\mathbf{81.60} \pm 2.80$ |
| COX2 | $56.49 \pm 3.45$ | $56.71 \pm 4.85$ | $54.23 \pm 5.81$ | $52.13 \pm 7.24$ | $61.17 \pm 7.49$ | $\mathbf{63.35} \pm 7.44$ |
| DHFR | $51.62 \pm 5.25$ | $49.20 \pm 5.94$ | $55.89 \pm 4.45$ | $63.11 \pm 3.38$ | $61.17 \pm 4.82$ | $\mathbf{76.67} \pm 2.63$ |
| NCI1 | $33.19 \pm 1.59$ | $53.93 \pm 2.18$ | $50.18 \pm 2.58$ | $68.32 \pm 1.47$ | $60.32 \pm 2.39$ | $\mathbf{68.44} \pm 0.81$ |
| ENZYMES | $48.51 \pm 5.96$ | $38.98 \pm 6.57$ | $42.80 \pm 8.79$ | $55.27 \pm 1.40$ | $63.10 \pm 4.29$ | $\mathbf{71.77} \pm 5.84$ |
| PROTEINS | $75.40 \pm 2.79$ | $59.79 \pm 3.64$ | $33.63 \pm 1.64$ | $68.55 \pm 5.31$ | $72.18 \pm 3.96$ | $\mathbf{79.60} \pm 3.93$ |
| COLLAB | $45.42 \pm 1.49$ | $61.47 \pm 1.27$ | $37.55 \pm 1.26$ | $51.95 \pm 1.36$ | $\mathbf{70.55} \pm 2.15$ | $52.76 \pm 1.52$ |

We perform experiments to showcase the efficiency and adaptability of the model we proposed on diverse datasets. Therefore, we choose twelve public and available real-world datasets and their statistics are given in Table 1. HSE, MMP, p53, PPAR are real graph anomalies. They are chemical compounds with complex and different structures in toxicology studies and the unique structure may make the compound activity different in certain conditions. Furthermore, these datasets have been categorized into test and training sets in original setting and here we mix them up and redivide them in our experiment. AIDS, BZR, COX2, DHFR and NCI1 are molecule datasets, where every node symbolizes an atom in the molecule, and every edge symbolizes a chemical bond. ENZYMES and PROTEINS are protein datasets. The difference is that nodes here mean amino acids, and edges indicate that the connected nodes are relatively close. COLLAB is a social network dataset. The nodes are individuals, and the connections are edges. Thus, the coverage of experimental datasets is wide enough to examine the capability of our model. Besides, the degree information of the node is chosen as the node attribute feature for these plain graph data, according to [18,28]. It is worth noting that all these datasets including real graph anomaly and classification graph data are bifurcated into two categories and the label setting is 0 and 1. Thus, we select label 0 and 1 as graph anomaly label respectively to evaluate the performance of GLADST.

## 5.2  Baselines

In the field of graph-level anomaly detection, few effective methods are put into use. Therefore, we perform the experiment with representatives from both the recent methods and the traditional methods. Firstly, we choose GLocalKD [18] as one of the baselines, which is a new deep learning method to detect graph-level anomaly. GLocalKD is capable of devising graph representations

**Table 3.** Anomaly detection performance measured mean value of AUC (%) and standard deviation (%) when graph data of label 1 is graph anomaly.

| Datasets | FGSD-IF | FGSD-LOF | FGSD-OCSVM | GLocalKD | GOOD-D | GLADST |
|---|---|---|---|---|---|---|
| HSE | **60.62** ± 1.35 | 56.56 ± 2.49 | 57.76 ± 4.43 | 40.92 ± 0.98 | 54.83 ± 3.32 | 55.47 ± 3.23 |
| MMP | 32.22 ± 0.90 | 43.00 ± 1.94 | 47.86 ± 2.97 | 68.11 ± 0.80 | 52.38 ± 4.72 | **68.55** ± 1.80 |
| P53 | 33.06 ± 3.67 | 43.45 ± 3.55 | 51.37 ± 2.76 | 66.98 ± 3.32 | 59.13 ± 4.82 | **69.61** ± 3.61 |
| PPAR | **65.51** ± 4.03 | 53.59 ± 4.81 | 49.55 ± 3.86 | 34.69 ± 4.06 | 57.03 ± 2.87 | 61.32 ± 2.88 |
| AIDS | 0.62 ± 0.89 | 12.27 ± 4.89 | 13.80 ± 4.22 | 95.10 ± 1.87 | 14.28 ± 7.77 | **97.67** ± 0.81 |
| BZR | 55.49 ± 6.06 | 50.44 ± 8.73 | 58.85 ± 6.44 | 62.57 ± 7.32 | 29.92 ± 9.58 | **81.02** ± 3.00 |
| COX2 | 43.51 ± 3.45 | 43.29 ± 4.85 | 45.77 ± 5.81 | 62.21 ± 5.35 | 42.13 ± 1.45 | **63.05** ± 9.59 |
| DHFR | 48.38 ± 5.25 | 50.80 ± 5.94 | 44.11 ± 4.45 | 55.05 ± 3.58 | 61.61 ± 4.84 | **77.36** ± 3.49 |
| NCI1 | 66.81 ± 1.59 | 46.07 ± 2.18 | 49.81 ± 2.58 | 31.77 ± 1.53 | 34.26 ± 2.36 | **68.12** ± 1.60 |
| ENZYMES | 47.98 ± 3.20 | 45.21 ± 5.50 | 56.05 ± 7.08 | 47.91 ± 6.17 | 54.22 ± 4.96 | **69.43** ± 9.14 |
| PROTEINS | 24.60 ± 2.79 | 40.21 ± 3.64 | 66.36 ± 1.64 | 56.17 ± 3.46 | 72.35 ± 3.34 | **78.91** ± 3.28 |
| COLLAB | 65.28 ± 1.50 | 45.17 ± 1.68 | 75.51 ± 1.48 | 67.42 ± 2.06 | 50.46 ± 2.84 | **77.65** ± 6.33 |

and is able to detect both local-anomaly and global-anomaly graphs better, owing to the usage of joint random distillation. In addition, some traditional methods are chosen for comparison. We select FGSD [23] as the model for graph representation learning. Then, it is used to drive the certain anomaly detection algorithm for GLAD, including isolation forest (IF) [13], local outlier factor (LOF) [1] and one-class support vector machine (OCSVM) [20], hence the FGSD-IF, FGSD-LOF, and FGSD-OCSVM are included in our baselines. Furthermore, we choose a unique and recently published model, GOOD-D [14], which is an unsupervised graph out-of-distribution detection method based on contrastive learning.

### 5.3 Parameter Settings

Three models are used in the GLADST experiments, one teacher model and two student models. The identical graph encoder is applied, which is made up of double GCN layers, whose dimensions are $d$-512-256, where $d$ denotes the attribute features' dimension size in the datasets for training. For GLocalKD, we choose the recommended default parameters. We use different algorithms (IF, LOF and OCSVM) to drive FGSD while choosing the same default parameters. The paper which proposes GOOD-D provides unique parameters for each dataset, and we use them in the experiment.

### 5.4 Anomaly Detection Performance

To prove that our model performs well in many cases, we evaluate the performance of our model through comparing it with the baselines on all the aforementioned datasets. We employ 5-fold cross-validation to train these approaches and record the average AUC results along with their standard deviation. The evaluation metric then becomes the criteria by which we judge the models' effectiveness according to the previous graph anomaly detection works [16,18]. Furthermore, to determine the influence of the selection of abnormal labels, we use different signs of graph anomaly to examine the models. In a word, we set graph data of label 0 as abnormal graphs, and graph data of label 1 is normal graphs, otherwise.

The AUC scores of GLADST and baselines are shown in Table 2 and Table 3, respectively. We use label 1 or 0 as a sign of graph anomaly to observe the degree of its influence on all models. Based on experimental results presented in Table 2, it is obvious that the AUC results of GLADST are much better than those of the baselines most of the time, except for several datasets. Our model only obtains lower scores on HSE, MMP, PPAR, AIDS and COLLAB, and the gap between our model and the highest-scoring model is small. From the Table 3, GLADST outperforms all baselines apart from HSE and PPAR. Besides, the improvement of anomaly detection performance is obvious on p53, BZR, DHFR, ENZYMES and COLLAB.

In addition, it is obvious that our model is less susceptible to interference from the selection of graph anomaly label. This is due to the symmetry of our model, which means that we have one student model trained with normal graphs and another trained with abnormal graphs. In contrast, the influence is much greater for the baselines, especially FGSD.

## 5.5  Ablation Study

We also perform an ablation experiment on GLADST, focusing on the importance of the teacher model, node-level representation error loss ($L_{node}$), and graph-level representation error loss ($L_{graph}$). Therefore, we remove each part separately to observe their effect. Firstly, we train the student models with an untrained teacher model; then, we remove the node-level loss and graph-level loss of GLADST respectively. During the experiment, we record all the average scores and standard deviations of these models. To arrive at a high-confidence conclusion, we choose six datasets (BZE, COX2, DHFR, PROTEINS, MMP, and p53) for examination, and the sign of graph anomaly is label 0. To show our results more clearly, we present a series of graphs in Fig. 2 showing the rating scores and standard deviations for reference.

From Fig. 2, the GLADST model shows a significant improvement when compared to the model with an untrained teacher on most of the datasets.Furthermore, when we remove the graph-level loss in the model, the scores on most datasets decrease dramatically. But the model without node-level loss seems to be only a little affected. Overall, our model makes progress in terms of performance. The results demonstrate that the design of the node-level and graph-level losses are effective to achieve node property and graph property anomaly detection. The results also show that significant improvement is made

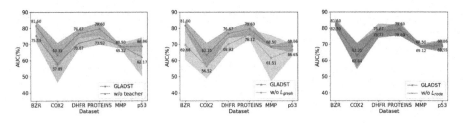

**Fig. 2.** The AUC results of model variants. The lines represent the AUC values of each model, and the shadows show the standard division.

due to the trained teacher which contributes to making a more obvious distinction between the feature from the teacher and the feature from the student model without training.

### 5.6   Efficiency Analysis

In this section, we mainly explore the impact of varying the number of abnormal samples in both the test and training sets. To begin with, we choose label 0 as the sign of graph anomaly and the same datasets in our ablation study. We next divide each dataset into a test set and a training set using the ratio of four to one. Then, we separate the anomalies from both sets. After this, we add 10% of the anomalies into the training set each time to train the model and do the same thing with the test set. Our results are shown in Fig. 3.

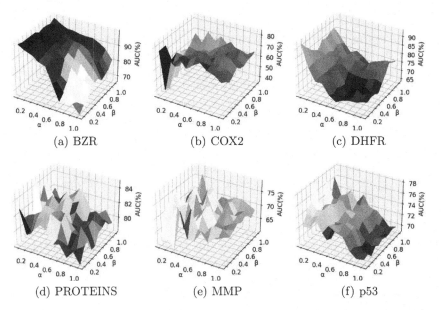

**Fig. 3.** The AUC results of GLADST under a different number of anomalies in the test set and training set. $\alpha$ represents the proportion of the anomalies to the original total anomalies of the test set. Similarly, $\beta$ represents this in the training set.

Figure 3 illustrates that with an increase in anomalies in the training set, the AUC results fluctuate and improve to a certain extent when the anomalies in the testing set stay the same. With an increase in abnormal samples in the test set, the results of AUC decrease sharply when the training set is invariant. The overall AUC results change to be within a stable and acceptable range. Furthermore, it is obvious that even when the quantity of anomalies in the training set or test set is not large, GLADST remains valid.

## 5.7   Visualization Analysis

For an anomalous graph, we consider node property and graph property anomalies in the graph and apply node-level representation and graph-level representation error losses to achieve anomaly detection in the GLAD task. To intuitively represent the effect of the proposed GLADST for anomalous graph detection, we first train our method based on the dual-students-teacher model on the training set for the DHFR dataset and then give the test graphs to evaluate its efficiency. We visualize the experiment results in Fig. 4. Thus, we can

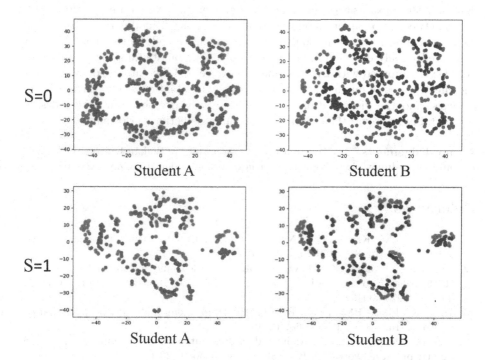

**Fig. 4.** A visualization of GLADST performance on the DHFR dataset, where red denotes the graph feature representations of the trained teacher model, and green and blue denote the graph feature representations for student A with normal graphs training and student B with abnormal graphs training, respectively. $S = 0$ and $S = 1$ denote the inputs of models with normal graphs and abnormal graphs, respectively. (Color figure online)

see that when we input the normal graphs $S = 0$ into student A and student B, respectively, the feature representation of student A is closer to the feature representation of the teacher model than student B. That is to say, the joint error loss $\hat{S}_G$ of student A is smaller and the $\check{S}_G$ of student B is larger, so the difference between them as an anomaly score will be less than 0, which is judged as a normal graph. Similarly, for the abnormal graphs $S = 1$, the anomaly score will be greater than 0, which is judged as an abnormal graph. The practical analysis above demonstrates the effectiveness of utilizing node-level and graph-level representation error losses to perform anomalous graph detection and designing a dual-students-teacher model to train the GLAD framework.

## 6    Conclusion

We explore the key problem that node property and graph property anomalies are very important to anomalous graph detection. To design a powerful evaluation mechanism to distinguish anomalous graphs, we introduce a discriminative graph-level anomaly detection framework via dual-students-teacher model. Through the optimization of node-level and graph-level representation error losses between two student models and a trained teacher model, respectively, the value of representation error between two student models as the score function can be effectively detected anomalous graphs. The outstanding performance of GLADST compared with five baselines on twelve real-life datasets demonstrates the effectiveness of our method. Furthermore, the ablation study, the efficiency analysis, and the visualization experiments also verify that our model design considerably improves the graph-level anomaly detection performance.

**Acknowledgements.** The research work is supported by Wuhan University People's Hospital Cross-Innovation Talent Project Foundation under JCRCZN-2022-008.

## References

1. Breunig, M.M., Kriegel, H.P., Ng, R.T., Sander, J.: LOF: identifying density-based local outliers. In: SIGMOD, pp. 93–104 (2000)
2. Chen, B., et al.: GCCAD: graph contrastive learning for anomaly detection. IEEE Trans. Knowl. Data Eng. **35**(8), 8037–8051 (2022). https://doi.org/10.1109/TKDE.2022.3200459
3. Ding, K., Li, J., Bhanushali, R., Liu, H.: Deep anomaly detection on attributed networks. In: SDM, pp. 594–602 (2019)
4. Fan, H., Zhang, F., Li, Z.: Anomalydae: dual autoencoder for anomaly detection on attributed networks. In: ICASSP, pp. 5685–5689 (2020)
5. Hooi, B., Song, H.A., Beutel, A., Shah, N., Shin, K., Faloutsos, C.: Fraudar: bounding graph fraud in the face of camouflage. In: KDD, pp. 895–904 (2016)
6. Jiang, J., Li, Y., He, B., Hooi, B., Chen, J., Kang, J.K.Z.: Spade: a real-time fraud detection framework on evolving graphs. In: VLDB, vol. 16, pp. 461–469 (2022)
7. Jin, M., Liu, Y., Zheng, Y., Chi, L., Li, Y.F., Pan, S.: Anemone: graph anomaly detection with multi-scale contrastive learning. In: CIKM, pp. 3122–3126 (2021)

8. Khattar, D., Goud, J.S., Gupta, M., Varma, V.: Mvae: Multimodal variational autoencoder for fake news detection. In: WWW, pp. 2915–2921 (2019)
9. Kipf, T.N., Welling, M.: Semi-supervised classification with graph convolutional networks. arXiv preprint arXiv:1609.02907 (2016)
10. Li, Y., Huang, X., Li, J., Du, M., Zou, N.: SpecAE: spectral autoencoder for anomaly detection in attributed networks. In: CIKM, pp. 2233–2236 (2019)
11. Li, Z., Wu, Q., Nie, F., Yan, J.: GraphDE: a generative framework for debiased learning and out-of-distribution detection on graphs. In: NeurIPS, vol. 35, pp. 30277–30290 (2022)
12. Liu, F., et al.: DAGAD: data augmentation for graph anomaly detection. arXiv preprint arXiv:2210.09766 (2022)
13. Liu, F.T., Ting, K.M., Zhou, Z.H.: Isolation forest. In: ICDM, pp. 413–422 (2008)
14. Liu, Y., Ding, K., Liu, H., Pan, S.: Good-d: on unsupervised graph out-of-distribution detection. In: WSDM, pp. 339–347 (2023)
15. Liu, Y., Li, Z., Pan, S., Gong, C., Zhou, C., Karypis, G.: Anomaly detection on attributed networks via contrastive self-supervised learning. IEEE Trans. Neural Netw. Learn. Syst. **33**(6), 2378–2392 (2021)
16. Luo, X., et al.: Comga: community-aware attributed graph anomaly detection. In: WSDM, pp. 657–665 (2022)
17. Luo, X., et al.: Deep graph level anomaly detection with contrastive learning. Sci. Rep. **12**(1), 1–11 (2022)
18. Ma, R., Pang, G., Chen, L., van den Hengel, A.: Deep graph-level anomaly detection by Glocal knowledge distillation. In: WSDM, pp. 704–714 (2022)
19. Ma, X., et al.: A comprehensive survey on graph anomaly detection with deep learning. IEEE Trans. Knowl. Data Eng. (2021). https://doi.org/10.1109/TKDE.2021.3118815
20. Schölkopf, B., Williamson, R.C., Smola, A., Shawe-Taylor, J., Platt, J.: Support vector method for novelty detection. In: NeurIPS, vol. 12 (1999)
21. Shin, K., Hooi, B., Kim, J., Faloutsos, C.: Densealert: incremental dense-subtensor detection in tensor streams. In: KDD, pp. 1057–1066 (2017)
22. Shu, K., Sliva, A.,Wang, S., Tang, J., Liu, H.: Fake news detection on social media: a data mining perspective. In: KDD, vol. 19, pp. 22–36 (2017)
23. Verma, S., Zhang, Z.L.: Hunt for the unique, stable, sparse and fast feature learning on graphs. In: NeurIPS. vol. 30 (2017)
24. Wu, Z., Pan, S., Chen, F., Long, G., Zhang, C., Philip, S.Y.: A comprehensive survey on graph neural networks. IEEE Trans. Neural Netw. Learn. Syst. **32**(1), 4–24 (2020)
25. Yang, Z., et al.: A comprehensive survey of graph-level learning. arXiv preprint arXiv:2301.05860 (2023)
26. Zhang, G., Li, Z., Huang, J., Wu, J., Zhou, C., Yang, J., Gao, J.: efraudcom: an e-commerce fraud detection system via competitive graph neural networks. ACM Trans. Inf. Syst. **40**(3), 1–29 (2022)
27. Zhang, G., et al.: Dual-discriminative graph neural network for imbalanced graph-level anomaly detection. In: NeurIPS, vol. 35, pp. 24144–24157 (2022)
28. Zhang, M., Cui, Z., Neumann, M., Chen, Y.: An end-to-end deep learning architecture for graph classification. In: AAAI, p. 4438–4445 (2018)

29. Zheng, Y., Jin, M., Liu, Y., Chi, L., Phan, K.T., Chen, Y.P.P.: Generative and con-trastive self-supervised learning for graph anomaly detection. IEEE Trans. Knowl. Data Eng. (2021). doi:https://doi.org/10.1109/TKDE.2021.3119326
30. Zhou, X., et al.: Hierarchical adversarial attacks against graph-neural-network-based IoT network intrusion detection system. IEEE Internet Things J. **9**(12), 9310–9319 (2021)

# Common-Truss-Based Community Search on Multilayer Graphs

Xudong Liu and Zhaonian Zou[✉]

Harbin Institute of Technology, Harbin 150001, Heilongjiang, China
{xdliu,znzou}@hit.edu.cn

**Abstract.** Multilayer graphs have a multitude of applications in diverse fields, ranging from biology to social networks. In the context of the community search in multilayer graphs, the goal is to identify a target multilayer graph that contains query vertices. However, existing algorithms do not succeed in finding a common graph in subgraphs, while also ensuring that the multilayer graph is sufficiently dense. To address this limitation, we propose the CTruss model, a novel structure for multilayer graphs. This structure aims to find a common graph that exists on each layer in the subgraph. Next, we present a Baseline algorithm to find a multilayer graph containing the query vertices. We also propose Top-Down and Bottom-Up algorithms that can compute the CTruss quickly under different conditions. Finally, we conduct experiments to demonstrate that our proposed model can achieve a higher density in some cases compared to the state of art models such as FirmTruss and $K$-core.

**Keywords:** Multilayer Graph · Community Search · Truss

## 1 Introduction

The problem of graphs has attracted a lot of attention in the past decades, and many realistic data have been abstracted as graphs and analyzed accordingly. However, the single-layer graph has many limitations when dealing with a greater number of edge types. Therefore, the multilayer graph is proposed, which can solve the problem accurately when there are more edge types. Here we use multiplex in multilayer graph [11], which means the vertices of the multilayer graph are replicated over layers, and each layer represents a particular aspect of the connection between the vertices. For example, Fig. 1 is a multilayer graph that has three layers.

Communities are parts of the graph with a few ties with the rest of the system [6], which can be used in social networks. But in many problems such as recommender systems, people only aim to find the community containing the query vertices. So there are many works about community search. Community search is to find communities containing given query vertices.

Supported by the National Natural Science Foundation of China (No. 62072138).

X. Yang et al. (Eds.): ADMA 2023, LNAI 14178, pp. 277–291, 2023.
https://doi.org/10.1007/978-3-031-46671-7_19

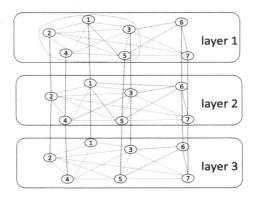

**Fig. 1.** A multilayer graph with three layers

To solve the problem, many kinds of models have been proposed on single-layer graphs, $k$-core [20], $k$-truss [2], $k$-plex [23], $k$-clique [17] and so on. All of them can find a dense subgraph efficiently on a single-layer graph. Meanwhile, there are also some models on multilayer graphs, such as FirmTruss [1], FirmCore [8], d-CC [25], multilayer $k$-truss [9]. However, they can't find a common graph on multilayer graphs, just like finding a pattern in the frequent subgraph. Therefore, we propose the CTruss. Unlike other models, our model focuses on the following three points: (1) We find a common graph that exists on each layer. (2) The common graph is a connected $k$-truss graph. (3) The density of the subgraph is high. Using our model, if we consider each layer of multilayer graphs as a relation of vertices in a time period, we can find a dense graph of multilayer graphs like frequent patterns, which has never been proposed.

Then, we propose a Common Truss community search problem that aims to find a CTruss in a multilayer graph that contains the query vertices and has the maximum density. To address this problem, we first design a Baseline algorithm to compute the CTruss in multilayer graphs directly and explicitly. Next, we design a Bottom-Up algorithm and a Top-Down algorithm to compute the CTruss in multilayer graphs, respectively, which perform differently in cases. The Bottom-Up algorithm gradually adds layers and removes some edges until the set of edges is empty or until the layer is empty. The Top-Down algorithm finds the CTruss containing the query vertices on all layers by initially finding the set of edges on all layers, noting that this set of edges may be empty. Then gradually decreasing the set of layers, each decreases a layer, which may result in adding some edges until the number of layers is less than the minimum number of layers limit.

**Contributions.** Our contributions are mainly as follows:

(1) We propose a novel CTruss model to find a dense common graph on multi-layer graphs.

(2) We define the problem of Common Truss community search (CTCS).

(3) We introduce three algorithms, namely the Baseline algorithm, Bottom-Up algorithm, and Top-Down algorithm.

(4) We design experiments and demonstrate that our algorithm can efficiently compute the CTruss of multilayer graphs and that our model can get a high density.

The paper is organized as follows. Section 2 presents the work related to CTCS, including community search, frequent subgraph pattering, and cohesive subgraph mining. Section 3 presents our CTruss model and the problem of CTCS. Sections 4-6 present three different algorithms to solve CTCS. The performance of algorithms and our model is evaluated with various real datasets in Sect. 7. Lastly, the conclusion is presented in Sect. 8.

## 2   Related Work

**Community Search.** The problem of community search was first proposed by Sozio et al. [20], then the problem of community search in $k$-core and the solution were proposed. After that, Wang et al. [22] proposed the problem of finding a subgraph that both satisfies the conditions of $k$-truss and ensures the minimum distance from the query vertices. After that, [15, 17, 20, 23] studied the solution to the single-layer graph community discovery problem around the degree of vertices, and [2, 10] gave different ways to define the truss around the edges. At the same time, various graph structures were proposed, such as directed graphs [16], attributed graphs [5], weighted graphs [21], uncertain graphs [14], dynamic graphs [18]. and so on.

**Frequent Subgraph Pattern Mining.** Given a graph database $S$, frequent subgraph pattern mining is to find a pattern $D$ such that $D$ is subgraph isomorphic to a minimum threshold proportion of graphs in $S$ [24]. Our work is different from this because in our work, graphs don't have labels, and the number of vertices on each layer is the same.

**Cohesive Subgraph Mining.** There are many models to mine a cohesive subgraph in graphs. On single-layer graphs, lots of work have been done [13]. On multilayer graphs, there are also many structures. For example, Zhu et al. [25] defined $d$-CC, to find a subset of multilayer graphs such that the vertex degree on each layer is greater than or equal to $k$, and the set covers the maximum number of vertices. Behrouz et al. [1] proposed a firm truss model to find the multilayer graph where for any edge, there exists some layer on which the truss number of the edge is greater than or equal to $k$. Huang et al. [9] studied the problem of making the truss number of edges on a specified number of layers in a multilayer graph greater than or equal to $k$. But in our model, the CTruss should exist on each layer.

# 3   Problem Definition

In this section, we introduce some basic concepts and formally define the problem studied in this paper.

## 3.1   Multilayer Graphs

A multilayer graph is a triple $G = (V, E, L)$, where $V$ is a set of vertices, $L$ is a set of layer numbers, and $E \subseteq V \times V \times L$ is a set of edges. Any edge $(u, v, l) \in E$ indicates that the vertices $u$ and $v$ are adjacent on layer $l$. For $l \in L$, let $E_l = \{(u, v) | (u, v, l) \in E\}$ be the set of edges on layer $l$, and the $l$-th layer is the single-layer graph $G_l = (V, E_l)$. The set of neighbors of a vertex $v$ on layer $l$ is denoted as $N_l(v)$, and the degree of $v$ on layer $l$ is denoted as $d_l(v)$. A multilayer graph $G' = (V', E', L')$ is a subgraph of another multilayer graph $G = (V, E, L)$ if $V' \subseteq V$, $L' \subseteq L$, and $E' \subseteq E$. For two vertex $u, v \in V$ and a layer $l \in L$, if there is a path $v_1 \to v_2 \to \cdots \to v_n$ such that $v_1 = u$, $v_n = v$, and $(v_i, v_{i+1}, l) \in E$ for $1 \leq i \leq n - 1$, $u$ is said to be connected to $v$ on layer $l$.

## 3.2   Truss

In a single-layer graph $G = (V, E)$, a cycle of three vertices $u, v, w \in V$ is called a triangle, denoted as $\triangle_{u,v,w}$. The support of an edge $e \in E$ in $G$, denoted as $\sup(e)$, is the number of triangles that contain $e$. A subgraph $H$ of $G$ is a $k$-truss if each edge in $H$ has a support greater than or equal to $k - 2$ in $H$. The truss number of an edge $e$ in $G$, denoted as $\text{truss}(e)$ is the largest number $k$ such that $e$ is contained in a $k$-truss in $G$.

## 3.3   Common Truss Based Community Search

In this section, we propose the common truss model and give a definition of the common truss community search problem.

There have been various ways to extend the definition of $k$-truss from single-layer graphs to multilayer graphs. For example, in the model proposed by Huang et al. [9] ensures that for any layer $l$, $G_l$ is a $k$-truss. However, all of them are not suitable to find common subgraphs that co-occur in multiple layers of a multilayer graph. Therefore, we extend the definition of $k$-truss to the following definition of common truss.

**Definition 1 (Common Truss).** *Given a multilayer graph* $G = (V, E, L)$, $L' \subseteq L$ *and an integer* $k$, *a graph* $H = (V_H, E_H)$ *is called a common* $k$-truss *($k$-CTruss for short) on the layers in* $L'$ *if* $H$ *is a* $k$-truss *and is a subgraph of* $G_i$ *for all* $i \in L'$.

In Fig. 1, the subgraph formed by the orange edges is a 4-CTruss on layers 1, 2 and 3.

**Definition 2 (Maximal Common Truss).** *Given a multilayer graph* $G = (V, E, L)$, $L' \subseteq L$ *and an integer* $k$, *a* $k$-*CTruss* $H$ *on* $L'$ *is maximal if there is no other* $k$-*CTruss* $H'$ *on* $L'$ *such that* $H$ *is a subgraph of* $H'$.

According to the property of $k$-truss, we easily have the following property.

*Property 1 (Hierarchy).* Given a multilayer graph $G = (V, E, L)$ and $L' \subseteq L$, if $H$ is a $k$-CTruss on $L'$, $H$ must be a $(k-1)$-CTruss on $L'$.

In community search problem settings, given a set of query vertices, there can be many maximal connected $k$-CTrusses that contain all the query vertices. To differentiate these $k$-CTrusses, we compare their density.

**Definition 3 (Density).** *Given a multilayer graph* $G = (V, E, L)$, $L' \subseteq L$ *and an integer* $k$, *the density of a* $k$-*CTruss* $H$ *on* $L'$ *is given by*

$$\frac{|E_H|}{|V_H|} \cdot |L'|^{\beta}, \tag{1}$$

*where* $\beta$ *is a parameter.*

Based on the CTruss model and the density measure, we give a formal definition of the <u>C</u>ommon <u>T</u>russ <u>C</u>ommunity <u>S</u>earch (CTCS) problem: Given a multilayer graph $G = (V, E, L)$, a set of query vertices $Q \subseteq V$, two positive integers $s$ and $k$, the CTCS problem is to find the maximal connected $k$-CTruss $H$ and the associated set of layers $L_H \subseteq L$ such that

(1) $|L_H| \geq s$, that is, $H$ occurs in at least $s$ layers;
(2) $Q \subseteq V_H$, that is, $H$ connects all the query vertices in $Q$;
(3) $H$ has the largest density $\rho(H, L_H)$ among all $k$-CTrusses that satisfy both conditions (1) and (2).

**Lemma 1 (Uniqueness).** *Given a multilayer graph* $G = (V, E, L)$, $L' \subseteq L$ *and an integer* $k$, *the maximal connected* $k$-*CTruss* $H$ *on* $L'$ *such that* $Q \subseteq V_H$ *is unique.*

*Proof.* Suppose there are two different maximal connected $k$-CTrusses $H_1 = (V_1, E_1)$ and $H_2 = (V_2, E_2)$ on $L'$, where $Q \subseteq V_1$ and $Q \subseteq V_2$. The vertices in $V_1$ and $V_2$ must be connected to each other via the vertices in $Q$, and each edge $e \in E_1 \cup E_2$ still has truss$(e) \geq k$. Therefore, the graph $H' = (V_1 \cup V_2, E_1 \cup E_2)$ is a $k$-CTruss on $L'$, which contradicts with the maximality of $H_1$ and $H_2$.

## 4  Baseline Alogrithm

In this section, we propose a baseline algorithm to solve the CTCS problem. Algorithm 1 presents the pseudocode of the algorithm. Given a multilayer graph $G = (V, E, L)$, two positive integers $k$, $s$, and a set of query vertices $Q$ as input, the algorithm runs as follows: First, we enumerate all the subsets of $L$ with sizes

**Algorithm 1.** CTCS-Baseline

---

**Input:** A multilayer graph $G = (V, E, L)$, two positive integers $k$, $s$, and a set $Q$ of query vertices

**Output:** the $k$-CTruss $H$ attaining the maximum density $\rho_{max}$ and the associated subset $L_H$ of layers

1: $\mathcal{L} \leftarrow \{L'|L' \subseteq L, |L'| \geq s\}$
2: $\rho_{max} \leftarrow 0, H_{max} \leftarrow \emptyset, L_{max} \leftarrow \emptyset$
3: **for** $L' \in \mathcal{L}$ **do**
4:     $E' \leftarrow \bigcap_{i \in L'} E_i$
5:     $V' \leftarrow \{v|\exists u \in V, \text{s.t.}(u,v) \in E'\}$
6:     $H \leftarrow \mathsf{MaximalConnectedTruss}((V', E'), Q, k)$
7:     **if** $E_H \neq \emptyset$ **then**
8:         $\rho \leftarrow \frac{|E_H|}{|V_H|}|L'|^\beta$
9:         **if** $\rho > \rho_{max}$ **then**
10:            $\rho_{max} \leftarrow \rho, H_{max} \leftarrow H, L_{max} \leftarrow L'$
11: **return** $H_{max}, L_{max}$

---

no less than $s$ (line 1). Then, $H_{max}, \rho_{max}, L_{max}$ are initialized as $\emptyset, 0, \emptyset$ (line 2). For all $L' \in \mathcal{L}$, we compute the set $E'$ of common edges on all layers in $L'$ (line 3–4) and use the algorithm proposed by Wang et al. [22] to compute the maximal connected $k$-truss on the single-layer graph induced by all edges in $E'$ that connects all query vertices in $Q$ (line 6). If the edge set of $H$ is not empty (line 7), we compute the density of $H$ by Eq. (1) (line 8) and update the $k$-CTruss with the highest density discovered so far and its associated layer subset (line 10). At the end of the algorithm, the CTruss with the highest density and the associated set of layers are returned.

Since the baseline algorithm iterates over all subsets in $\mathcal{L}$, it incurs a large number of repeated operations at line 6 to compute $k$-trusses. Particularly, for two subsets $L_1, L_2 \in \mathcal{L}$, if $L_1 \cap L_2 \neq \emptyset$, the set $E'_1$ of common edges on $L_1$ and the set $E'_2$ of common edges on $L_2$ may have a significant overlap. On the graphs induced by $E'_1$ and $E'_2$, Procedure $\mathsf{MaximalConnectedTruss}$ at line 6 will execute many repeated operations. To reduce repeated operations, we propose two algorithms in the rest of the paper.

**Theorem 1 (Baseline Complexity).** *Algorithm 1 takes $O(2^{|L|} * |E|^{1.5})$ time, and $O(E)$ space.*

*Proof.* There are $2^{|L|}$ subsets of $L$ in total. We need $O(|E|)$ to execute the intersection of the edge set, and $O(|E|^{1.5})$ to execute the function MaximalConnectedTruss, So Algorithm 1 takes $O(2^{|L|} * |E|^{1.5})$ time.

## 5   Bottom-Up Search Algorithm

In this section, we propose a bottom-up search algorithm to solve the CTCS problem. Distinguished from the baseline algorithm which treats the subsets of $L$ independently, the bottom-up algorithm organizes all subsets of $L$ into a search

---

**Algorithm 2.** CTCS-BU

---

**Input:** A multilayer graph $G = (V, E, L)$, two positive integers $k$, $s$, and a set $Q$ of query vertices

**Output:** the $k$-CTruss $H$ attaining the maximum density $\rho_{max}$ and the associated subset $L_H$ of layers

1: $\rho_{max} \leftarrow 0$, $H_{max} \leftarrow \emptyset$, $L_{max} \leftarrow \emptyset$
2: **for** $l \in L$ **do**
3:     $\phi_l \leftarrow$ Decompose$(G_l)$
4: $\phi \leftarrow \{\phi_l | l \in L\}$
5: **for all** $l \in L$ **do**
6:     $T \leftarrow$ MaximalConnectedTruss$(G_l, Q, k)$
7:     $H, L_H, \rho \leftarrow$ BU$(G, d, s, Q, T, \{l\}, \phi)$
8:     **if** $\rho > \rho_{max}$ **then**
9:         $\rho_{max} \leftarrow \rho$, $H_{max} \leftarrow H$, $L_{max} \leftarrow L$
10: **return** $H_{max}, \rho_{max}, L_{max}$

---

tree as illustrated in Fig. 2a. In the search tree, a subset $L' \subseteq L$ is the parent of another subset $L'' \subseteq L$ if $|L''| = |L'| + 1$, and all the layer numbers in $L'$ are less than the layer number of $L'' \setminus L'$. The root of the search tree is $\emptyset$. The algorithm traverses the search tree. For each subset $L'$ of layers, the algorithm finds the maximal connected $k$-CTruss on $L'$ that contains the given set $Q$ of query vertices. The bottom-up algorithm makes use of the following property to reduce repeated operations.

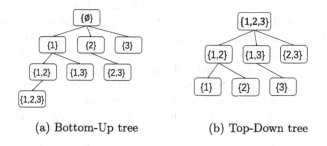

(a) Bottom-Up tree        (b) Top-Down tree

**Fig. 2.** Illustration of search trees.

**Lemma 2.** *For two subsets $L_1, L_2 \subseteq L$, let $H_1$ and $H_2$ be the maximal connected $k$-CTruss on $L_1$ and $L_2$ that contain the query vertices in $Q$, respectively. If $L_1 \subseteq L_2$, $H_2$ is a subgraph of $H_1$.*

*Proof.* We let $H_1 = (V_1, E_1), H_2 = (V_2, E_2)$. For any edge $e \in E_2$, any $l \in L_2$, $e \in E_l, truss(e) \geq k$. Because $L_1 \subseteq L_2$, for any $l' \in L_1$, $e \in E_{l'}$. So $e \in E_2$, which means $E_2 \subseteq E_1$, $H_2$ is a subgraph of $H_1$.

According to this lemma, if the result obtained on a subset $L'$ of layers does not contain all the query vertices in $Q$, any superset of $L'$ needs not to be considered

---

**Algorithm 3.** BU

---

**Input:** A multilayer graph $G = (V, E, L)$, two positive integers $k$, $s$, the set of query
vertices $Q$, $H = (V_H, E_H)$, $L_H$, truss number $\phi$ of edges in $G$
**Output:** $k$-CTruss $H_{max}$, its density $\rho_{max}$, the associated subset of layers $L_{max}$
1: $\rho_{max} \leftarrow 0, H_{max} \leftarrow \emptyset, L_{max} \leftarrow \emptyset$
2: **if** $Q \subsetneq V_H$ **then**
3:    **return** $H_{max}, \rho_{max}, L_{max}$
4: **if** $|L_H| \geq s$ **then**
5:    $\rho \leftarrow \frac{|E_H|}{|V_H|}|L'|^\beta$
6:    $\rho_{max} \leftarrow \rho, H_{max} \leftarrow H, L_{max} \leftarrow L_H$
7: $\tau \leftarrow \{l|max(L_H) < l \leq max(L)\}$
8: **for all** $l \in \tau$ **do**
9:    $E_{del} \leftarrow \{e \in E_H | e \notin E_l \text{ or } \phi_l(e) < k\}$
10:    $E_H \leftarrow E_H - E_{del}, V_H \leftarrow \{u|\exists v, (u, v) \in E_H\}$
11:    $T \leftarrow \text{MaximalConnectedTruss}((V_H, E_H), Q, k)$
12:    **if** $E_T \neq \emptyset$ **then**
13:       $H', \rho, L' \leftarrow \text{BU}(G, k, s, Q, T, L_H \cup \{l\}, \phi)$
14:       **if** $\rho > \rho_{max}$ **then**
15:          $\rho_{max} \leftarrow \rho, H_{max} \leftarrow H', L_{max} \leftarrow L'$
16: **return** $H_{max}, \rho_{max}, L_{max}$

---

and can be pruned from the search space. Algorithm 2 presents the pseudocode
of the bottom-up algorithm.

Given a multilayer graph $G = (V, E, L)$, two positive integers $k$, $s$, and a
set of query vertices $Q$ as input, the CTCS-BU algorithm calculates the CTruss
$H$ with the maximal density $\rho_{max}$ and the associated subset $L_H$ of layers. It
runs as follows: We use $H_{max}$, $\rho_{max}$, $L_{max}$ to keep the currently computed $k$-
CTruss with maximal density, its density, and the associated subset of layers.
In line 1, $H_{max}$, $\rho_{max}$, $L_{max}$ are initialized. Then, for each layer $l \in L$, we
call the function Decompose to compuse the truss number of each edge on $G_l$
(lines 2–4). As it makes no sense to compute the maximal connected $k$-CTruss
on $L'$, where $L' = \emptyset$, i.e., the root of the search tree, the algorithm explores the
search tree from each set of single layer $l$, $l \in L$, using the **for** loop in lines 5–
10. Specifically, it computes the maximal connected CTruss $T$ that contains $Q$
on layer $l$ by calling function MaximalConnectedTruss. After that, the algorithm
invokes Procedure BU (Algorithm 3) to perform the depth-first search(DFS) on
the search tree from the layer set $\{l\}$. The procedure returns the $k$-CTruss $H$
on layers $L_H$ of the maximum density $\rho$ met during the DFS. If $\rho$ is larger
than $\rho_{max}$, $H_{max}$, $\rho_{max}$, $L_{max}$ are accordingly updated to $H$, $\rho$, $L$, respectively
(lines 9–10).

Procedure BU (Algorithm 3) works as follows. First, we initialize $H_{max}$, $\rho_{max}$,
$L_{max}$ to $\emptyset, 0, \emptyset$ (line 1). If $Q \subsetneq V_H$, there is no need to explore the children of
$L_H$, so the procedure terminates (lines 2–3). If the size of $L_H$ is no less than $s$,
the algorithm updates the maximum density $\rho_{max}$, $H_{max}$, and $L_{max}$ (lines 4–6).
We set $\tau$ as the set of all the possible $l$ such that $L_H \cup \{l\}$ is a child of $L_H$ in the
Bottom Up search tree (line 7). For each $l$ in $\tau$, we add it into $L_H$ to construct

---

**Algorithm 4.** CTCS-TD

---

**Input:** A multilayer graph $G = (V, E, L)$, two positive integers $k$, $s$, a set of query vertices $Q$

**Output:** the $k$-CTruss $H$ attaining the maximum density $\rho_{max}$ and the associated subset $L_H$ of layers

1: $E_H \leftarrow \bigcap_{l \in L} E_l, V_H \leftarrow \{u | \exists v, (u, v) \in E_H\}$
2: $T \leftarrow$ MaximalConnectedTruss$((V_H, E_H), Q, k)$
3: **if** $V_T \neq \emptyset$ **then**
4:     **return** TD$(G, d, s, Q, T, L, True)$
5: **else**
6:     **return** TD$(G, d, s, Q, T, L, False)$

---

a child of $L_H$. Then, we delete the edges in $E_H$ but not in $E_l$ and the edges with truss numbers less than $k$ on $E_l$ (lines 8–10). Next, the algorithm calls the function MaximalConnectedTruss to compute the maximal connected $k$-truss subgraph $T$ such that $Q \subseteq V_T$ (line 11). If $T$ doesn't exist, there is no need to explore the children of $L_H \cup \{l\}$ according to Lemma 2. Otherwise, we continue to execute the procedure BU on $T$ to find the $k$-CTruss $H'$ on layers $L_H \cup \{l\}$ with the maximum density $\rho$ within $T$. If $\rho > \rho_{max}$, the $\rho_{max}$, $H_{max}$, and $L_{max}$ are updated to $\rho$, $H'$, and $L'$ (lines 12–15). Finally, the algorithm returns $H_{max}$, $L_{max}$, and $\rho_{max}$.

## 6  Top-Down Search Algorithm

The bottom-up algorithm has an intrinsic limitation: it requires traversing at least $s$ levels in the search tree to reach the subsets of no less than $s$ layers. When $s$ is large, particularly, $s > |L|/2$, the performance of the bottom-up algorithm is poor. To overcome this limitation, we propose a top-down search algorithm to solve the CTCS problem. In the top-down algorithm, the subsets of $L$ are organized into a search tree as illustrated in Fig. 2b. Given a set of layer $L' \subseteq L$, the minimum delete layer of $L'$ is $min(\{l \in L | l \notin L'\})$. In the search tree, a subset $L' \subseteq L$ is the parent of another subset $L'' \subseteq L$ if $L'' \subseteq L'$, $|L'| = |L''| + 1$, all the layer numbers in $L'$ are less than the layer number of $L'' \setminus L'$, and $L'' \setminus L'$ is smaller than the minimum delete number of $L'$. The root of the search tree is $L$. Algorithm 4 presents the pseudocode of the top-down algorithm.

Given a multilayer graph $G = (V, E, L)$, two positive integers $k$, $s$, and a set of query vertices $Q$ as input, the CTCS-TD algorithm calculates the CTruss $H$ with the maximal density $\rho_{max}$ and the associated subset $L_H$ of layers. It runs as follows. First, we compute the edges shared among all layers (line 1). Then, the algorithm calls the function MaximalConnectedTruss to compute the connected $k$-truss $T$ containing $Q$ (line 2). In the procedure TD (Algorithm 5), a flag is used to indicate whether $T$ is a maximal connected $k$-CTruss and $V_T$ containing $Q$. If $T$ exists, we call the procedure TD with flag $= True$ (lines 3–4). Otherwise, the procedure TD is called with flag $= False$ (lines 5–6).

---

**Algorithm 5.** TD

---

**Input:** A multilayer graph $G = (V, E, L)$, two positive integers $k$, $s$, the set of query vertices $Q$, $H = (V_H, E_H)$, $L_H$, a flag indicating whether $H$ is a maximal connected $k$-CTruss with $Q \subseteq V_H$
**Output:** $k$-CTruss $H_{max}$, its density $\rho_{max}$, the associated subset of layers $L_{max}$
1: $H_{max} \leftarrow \emptyset$, $\rho_{max} \leftarrow 0$, $L_{max} \leftarrow \emptyset$
2: **if** $|L_H| < s$ **then**
3:     **return** $H_{max}, \rho_{max}, L_{max}$
4: **if** $flag$ **then**
5:     $\rho \leftarrow \frac{|E_H|}{|V_H|} |L'|^\beta$
6:     $H_{max} \leftarrow (V_H, E_H), \rho_{max} \leftarrow \rho, L_{max} \leftarrow L_H$
7:     $l' \leftarrow min(\{l \in L | l \notin L_H\})$
8:     $\tau \leftarrow \{l | min(L') \leq l < l'\}$
9: **for all** $l \in \tau$ **do**
10:     $\psi \leftarrow \bigcap_{i \in L_H \setminus \{l\}} E_i - E_l, \gamma \leftarrow \emptyset$
11:     $E_H \leftarrow E_H \cup \phi, V_H \leftarrow \{u | \exists v, (u, v) \in E_H\}$
12:     **if** $not\ flag$ **and** $Q \subseteq V_H$ **then**
13:         $T \leftarrow$ MaximalConnectedTruss$(H, Q, k)$
14:         **if** $T \neq \emptyset$ **then**
15:             $E_H \leftarrow T, V_H \leftarrow \{u | \exists v, (u, v) \in E_H\}, flag \leftarrow True$
16:     **else if** $flag$ **then**
17:         $\psi \leftarrow \{e | e \in \psi$ and $sup(e) < k - 2\}$
18:         **while** $\psi \neq \emptyset$ **do**
19:             $E_H \leftarrow E_H - \psi$
20:             $\psi \leftarrow \{e | e \in \psi$ and $sup(e) < k - 2\}$
21:         $V_H \leftarrow \{v | v \in V$ and $deg(v) \neq 0$ and $v$ is connected to $Q\}$
22:         $E_H \leftarrow \{e(u,v) | e \in E_H$ and $u \in V_H$ and $v \in V_H\}$
23:     $H', \rho', L' \leftarrow$ TD$(G, d, s, Q, H, L_H - \{l\}, flag)$
24:     **if** $\rho' > \rho_{max}$ **then**
25:         $\rho_{max} \leftarrow \rho', H_{max} \leftarrow H', L_{max} \leftarrow L'$
26: **return** $H_{max}, \rho_{max}, L_{max}$

---

The TD function works as follows. First, we initialize $H_{max}$, $\rho_{max}$, $L_{max}$ to $\emptyset, 0, \emptyset$ (line 1). Then, the algorithm checks whether the size of $L_H$ is no less than $s$. If it is already less than $s$, there is no need to explore the children of $L_H$, so the procedure terminates (lines 2–3). If flag is $True$, we calculate the density $\rho(H, L_H)$ and update $H_{max}$, $\rho_{max}$, and $L_{max}$ to $H, \rho, L_H$ (lines 4–6). After that, we compute the minimum delete layer of $L_H$ and the set of $\tau$ of all the possible layer $l$ such that $L_H - \{l\}$ is a child of $L_H$ in the Top Down search tree (lines 7–8). For each $l$ in $\tau$, we delete it from $L_H$ to construct a child of $L_H$. Then, we compute the set $\psi$ of edges on $L_H$ but not on layer $l$. We add the edges in $\psi$ to $E_H$, and update $V_H$ accordingly (lines 8–9). There are three different cases that need to be handled separately before exploring the child of $L_H - \{l\}$:

**Case1:** When flag is $False$ and $Q \subsetneq V_H$, it means that there is no $k$-CTruss connected to $Q$ in $H$. Therefore, we continue executing the procedure TD on $H$.

**Case2:** When flag is $False$ and $Q \subseteq V_H$, the algorithm calls the function MaximalConnectedTruss to compute the maximal connected $k$-truss subgraph $T$ such that $Q \subseteq V_T$. If $T$ exists, we explore the children of $L_H - \{l\}$ on $T$ with flag $= True$. Otherwise, we continue executing the procedure TD on $H$ (lines 10–13).

**Case3:** When flag is $True$, it indicates that $H$ is already a $k$-CTruss on $L_H$. According to Lemma 2, we only need to check whether the support of each edge in $\psi$ is no smaller than $k - 2$. If not, they will be deleted from $\psi$ until the supports of all edge in $\psi$ are no less than $k - 2$ (lines 16–20). Then, we update $V_H$ and $E_H$ accordingly (lines 21–22)

Next, we invoke the procedure TD to find the $k$-CTruss on layers $L_H - \{l\}$ with the maximum density $\rho$. $\rho_{max}$, $H_{max}$, and $L_{max}$ are accordingly updated if a denser $k$-CTruss is found (lines 23–25). Finally, the algorithm returns $H_{max}$, the maximum density $\rho_{max}$, and the associated layers $L_{max}$ (line 26).

# 7 Experiments

In this section, we present the experimental evaluation of our model and the proposed algorithms.

All algorithm implementations and codes for our experiments are available at https://github.com/lxdgogogo/CTruss.

## 7.1 Experimental Setup

The proposed algorithms are implemented in Python, and the experiments were performed on a server with Intel Xeon E5-2620 v4 CPU and 220GB of RAM, running Ubuntu 18.04. The experiments were performed on several multilayer graph datasets. The sizes of the datasets are shown in Table 1. The datasets cover a wide range of domains from social networks to protein composition [3,4,12,19]. In the experiments, we select the vertices with the largest total degree in all layers as the query vertices, and the parameters $k$ and $s$ are set randomly. For the density measure, we set $\beta$ to 2.

## 7.2 Efficiency

In this experiment, we compare the execution time of three algorithms proposed in this paper. Figure 3 shows the experimental results. We can see that, in most cases, the top-down algorithm outperforms the bottom-up algorithm, and the bottom-up algorithm outperforms the baseline algorithm. The reasons are explained as follows:

**Table 1.** Statistics of Datasets.

| Dataset | $|V|$ | $|E|$ | $|L|$ |
|---------|-------|-------|-------|
| RM | 91 | 14K | 10 |
| Terrorist | 79 | 2.2K | 14 |
| Homo | 18K | 153K | 7 |
| DBLP | 513 | 1.0M | 10 |

(1) The time of Bottom-Up is shorter than that of Baseline in most cases, which is because BU reduces many unnecessary computations during the search. If there are no query vertices in a layer, there is no need to continue the search, so the search space is significantly reduced.

(2) Top-Down is faster than Bottom-Up in more cases because in the top-down search process, if there is already a connected truss containing the query vertices on some layers, it is only necessary in the next recursion to ensure that the support of the new added edges is not less than $k - 2$, and there is no need to check the existing trusses. At the same time, if $s$ is large, the Top-Down algorithm will stop earlier.

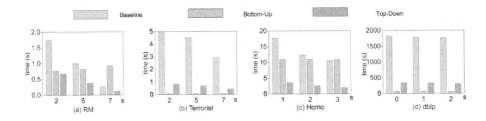

**Fig. 3.** Execution time of three different algorithms w.r.t. $s$

## 7.3   Density

In this experiment, we compare the density of the subgraphs obtained by different models. Figure 4 shows that our model has a high density in certain cases when compared to the FirmTruss [1] and $K$-core models [7]. The reason is that our model ensures that each edge exists in all layers. Therefore, the community found on this basis is not sparse on each layer. In contrast, FirmTruss only guarantees the edge exists on some layers, which may result in a density of 0 on other layers. Because $K$-core only guarantees that the degree of each vertex in the graph is no less than $k$, while the $k$-truss guarantees that each edge in the graph can form at least $k - 2$ triangles, so our model yields a higher density than $K$-core in some cases.

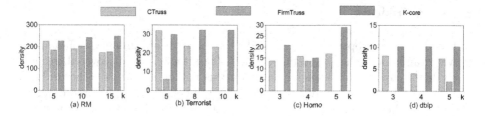

**Fig. 4.** The density of three different models and different $k$

## 7.4   Case Study: Terrorist

We conduct a case study to assess the quality of the CTruss model on the Terrorist dataset and show the practical implications of the model. In this case, a vertex represents a terrorist, an edge indicates a relationship between two terrorists, and a layer means a relationship type. Using Abdul Rohim as a query vertex, we aim to find a community that collaborates with at least 2 relationship types with him. Figure 5 shows this community, which consists of 20 terrorists. In this community, an edge denotes that two terrorists are friends and communicate with each other. Finding the largest density of CTruss with a density of 23.4, we calculate the average number of the criminal group in this community as 11.7.

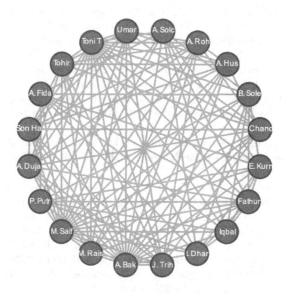

**Fig. 5.** The terrorist community

# 8    Conclusions

In this paper, we propose a CTruss model to search for a common graph on multilayer graphs, and the properties of the CTruss model are discovered, respectively hierarchy and uniqueness. Based on this model, we propose the problem of Common Truss community search (CTCS). Then, we design three different algorithms, Baseline, Bottom-Up, and Top-Down algorithms to solve the problem, and the algorithm complexity is analyzed. Finally, we design experiments to demonstrate that our model yields higher community density in many cases and that the Top-Down algorithm requires lower cost. Finally, we conduct a case study on the Terrorist dataset to access the practical utility of the model of the CTruss model.

# References

1. Behrouz, A., Hashemi, F., Lakshmanan, L.V.S.: Firmtruss community search in multilayer networks. arXiv preprint arXiv:2205.00742 (2022)
2. Cohen, J.: Trusses: cohesive subgraphs for social network analysis. National security agency Technical Report **16**(3.1) (2008)
3. dblp (2011). http://en.wikipedia.org/wiki/DBLP
4. De Domenico, M., Porter, M.A., Arenas, A.: Muxviz: a tool for multilayer analysis and visualization of networks. J. Complex Netw. **3**(2), 159–176 (2015)
5. Fang, Y., Cheng, R., Chen, Y., Luo, S., Jiafeng, H.: Effective and efficient attributed community search. VLDB J. **26**, 803–828 (2017)
6. Fortunato, S.: Community detection in graphs. Phys. Rep. **486**(3–5), 75–174 (2010)
7. Galimberti, E., Bonchi, F., Gullo, F.: Core decomposition and densest subgraph in multilayer networks. In: Proceedings of the 2017 ACM on Conference on Information and Knowledge Management, pp. 1807–1816 (2017)
8. Hashemi, F., Behrouz, A., Lakshmanan, L.V.S.: Firmcore decomposition of multilayer networks. In: Proceedings of the ACM Web Conference 2022, pp. 1589–1600 (2022)
9. Huang, H., Linghu, Q., Zhang, F., Ouyang, D., Yang, S.: Truss decomposition on multilayer graphs. In: 2021 IEEE International Conference on Big Data (Big Data), pp. 5912–5915. IEEE (2021)
10. Huang, X., Cheng, H., Qin, L., Tian, W., Yu, J.X.: Querying k-truss community in large and dynamic graphs. In: Proceedings of the 2014 ACM SIGMOD International Conference on Management of Data, pp. 1311–1322 (2014)
11. Kivelä, M., et al.: Multilayer networks. J. Complex Netw. **2**(3), 203–271 (2014)
12. Kunegis, J.: Konect: the koblenz network collection. In: Proceedings of the 22nd International Conference on World Wide Web, pp. 1343–1350 (2013)
13. Victor E., et al.: A survey of algorithms for dense subgraph discovery. In: Managing and Mining Graph Data, pp. 303–336 (2010)
14. Li, R.-H., Su, J., Qin, L., Yu, J.X., Dai, Q.: Persistent community search in temporal networks. In: 2018 IEEE 34th International Conference on Data Engineering (ICDE), pp. 797–808. IEEE (2018)
15. Liu, D., Zou, Z.: On flexible cohesive subgraph mining. World Wide Web **25**(2), 535–567 (2022)

16. Liu, Q., Zhao, M., Huang, X., Xu, J., Gao, Y.: Truss-based community search over large directed graphs. In: Proceedings of the 2020 ACM SIGMOD International Conference on Management of Data, pp. 2183–2197 (2020)
17. Lu, C., Yu, J.X., Wei, H., Zhang, Y.: Finding the maximum clique in massive graphs. Proc. VLDB Endowment 10(11), 1538–1549 (2017)
18. Lu, Z., Zhu, Y., Zhong, M., Yu, J.X.: On time-optimal (k, p)-core community search in dynamic graphs. In: 2022 IEEE 38th International Conference on Data Engineering (ICDE), pp. 1396–1407. IEEE (2022)
19. Roberts, N., Everton, S.: The noordin top terrorist network data (2011). http://www.thearda.com/Archive/Files/Descriptions/
20. Sozio, M., Gionis, A.: The community-search problem and how to plan a successful cocktail party. In: Proceedings of the 16th ACM SIGKDD International Conference on Knowledge Discovery and Data Mining, pp. 939–948 (2010)
21. Sun, L., Huang, X., Li, R.-H., Choi, B., Jianliang, X.: Index-based intimate-core community search in large weighted graphs. IEEE Trans. Knowl. Data Eng. 34(9), 4313–4327 (2020)
22. Wang, J., Cheng, J.: Truss decomposition in massive networks. arXiv preprint arXiv:1205.6693 (2012)
23. Wang, Y., Jian, X., Yang, Z., Li, J.: Query optimal k-plex based community in graphs. Data Sci. Eng. 2, 257–273 (2017)
24. Yan, X., Han, J.: gspan: graph-based substructure pattern mining. In: Proceedings of 2002 IEEE International Conference on Data Mining, 2002, pp. 721–724. IEEE (2002)
25. Zhu, R., Zou, Z., Li, J.: Diversified coherent core search on multi-layer graphs. In: 2018 IEEE 34th International Conference on Data Engineering (ICDE), pp. 701–712. IEEE (2018)

# Learning to Predict Shortest Path Distance

Zhixin Qu[1][✉], Zixiao Zong[2], and Jianfang Zhang[3]

[1] Shenyang Aerospace University, Shenyang 110136, Liaoning, China
331318641@qq.com
[2] University of California, Irvine, CA 92697, USA
[3] China National Institute of Standardization, Beijing 450047, China

**Abstract.** As graph data emerging in various application e.g., biology, social network, scalable graph methods are required to analyze such data. However, traditional graph algorithms cannot meet the need due to their high complexity of both time and space. In this paper, a machine learning based method is proposed to predict shortest path under the case where complete accuracy is not required. A feed-forward neural network classification model and a regression model are constructed to deal with graphs with discrete path distance and continuous path distance, respectively. In addition, we further improve the above models to adapt the dynamic changes of graph data in the real world. The results on real-world datasets show that the proposed method can approach the shortest distance with lower error rate than the comparison methods. We also evaluated different graph embedding methods and training set construction methods on the experimental results.

**Keywords:** graph embedding · machine learning · shortest path distance

## 1 Introduction

The computation of the shortest path distance has always been a classic research topic in graph theory. Traditional graph algorithms such as Dijkstra or A* search algorithm can return accurate results, but due to its high time complexity, it cannot scale to large networks. Therefore, landmark-based distance oracle is widely used to optimize the query algorithm of the shortest path distance [1–3]. However, traversal-based distance calculation from landmark vertices to non-landmark vertices is still unavoidable. Though embedding techniques [4,5] for mapping graph vertices to specific low-dimension graph coordinate system, e.g., Euclidean spaces or hyperbolic space, are computational efficient, the accuracy of distance calculation is strictly limited by the representational capacity of the coordinate space. Utilizing the capabilities of machine learning can provide a solution to this problem, particularly in scenarios where complete accuracy is not essential and some degree of error tolerance is acceptable, such as in predicting URL distances for search purposes.

X. Yang et al. (Eds.): ADMA 2023, LNAI 14178, pp. 292–307, 2023.
https://doi.org/10.1007/978-3-031-46671-7_20

This paper proposes a method of predicting the shortest path distance on a large graph using graph embedding and machine learning. Firstly, graph embedding methods, Node2vec [6] and LINE [7], are considered to obtain the embedding vector of graph vertices. Secondly, using the idea of landmark [4], a small number of vertices on the graph are selected as the landmarks. Next, for each landmark vertice and every vertice from a sampled non-landmarks vertice set, a vertices pair set is constructed. Then, the training set $S_{train}$ and the validating set $S_{validate}$ are generated from it where the feature vector of each pair of vertices is the fusion of embeddings of both vertices and the label of each pair of vertices is the value of the shortest path distance between the two vertices. Thus, supervised learning methods can be used to train a model for predicting the shortest path distance between any two vertices in the graph efficiently and effectively. Particularly, a feed-forward neural network model is constructed [8] to solve the graph with small average distance shortest path and integral edge weight. While, for the graph with non-integral edge weights, a regression model is constructed by modeling the prediction of shortest path distance as a function fitting problem. For the dynamic graph, a method is proposed to determine the necessity of recalculating the shortest path distance between two vertices by judging whether the distance may be affected by the change of the graph.

Our experimental results on real data set show that the proposed machine learning based shortest path distance prediction method is able to achieve high accuracy and low prediction error.

The main contributions of this paper are as follows:

(1) The idea of using graph embedding, landmark, and machine learning models to solve the shortest path distance prediction problem on large-scale graphs is proposed.
(2) The proposed method can be adapted to used in scenarios of unweighted graphs, weighted graphs, and dynamic graphs.
(3) We present the influence of different embedding methods and different methods of constructing training sets according to the experimental results, and make theoretical analysis.

## 2   Related Work

Finding the shortest path between two vertices on a graph is called the single-source shortest-path (SSSP) problem. Cormen et al. [9] proposed a method based on breadth-first search (BFS) for the SSSP problem whose time complexity is $O(n + m)$ for graphs with $n$ vertices and $m$ edges. The Dijkstra algorithm [10] solves the SSSP from a single source to all other vertices on directed graphs without negative weights, and the time complexity is $O(n^2)$. This algorithm is especially effective in calculating some graphs with larger edge weights, but can only deal with graphs without negative weights and static graphs. Fredman and Tarjan [11] improved the Dijkstra algorithm using Fibonacci heap, and reached a time complexity of $O(n\log n + m)$. Fredman and Willard [12] proposed another method to improve the time complexity to $O(m + n\log n/ \log\log n)$ by AF heap.

Since the above methods usually have poor time overhead when querying the shortest path distance on large-scale graphs, researchers propose various methods based on the *distance oracle* technique. A distance oracle based methods has two stages, i.e., preprocessing and query. In the preprocessing stage, in order to calculate the distance, most methods [1,2] have adopted the idea of landmarks to construct the distance oracle index. Here, landmarks are a small number of vertices selected from the graph with certain strategy. Distance from landmarks to all other vertices are pre-calculated. Thus, in the query stage, the index can return the distance to be queried in a constant time without runtime calculation. This paper uses landmarks method to construct the training set for the learner. The typical representative of this kind of predictive query method is the algorithm proposed by Thorup et al. [1], which can preprocess the graph data under the time complexity of $O(kmn^{1/k})$, construct the index with the space complexity of $O(kn^{1+1/k})$, and complete the query with about $O(k)$ time complexity, where $k$ represents the parameter of the upper bound of the prediction error.

The shortest path calculation on dynamic graphs is a practical problem. Most of the existing work [13–15] are based on the shortest path tree. When the graph changes, the update range is limited to the vertices affected by the topology change, so as to achieve the purpose of controlling redundant updates, and then minimize the topology of the existing shortest path tree and maintain the shortest path property. This paper attempts to extend this traditional method to the learning model.

Orion, Rigel and other algorithms [4,5,16] have adopted graph embedding and landmark-based methods to preprocess the graph, which has some similarities with this work. Orion and Rigel embed the graph into the Euclidean coordinate system and hyperbolic coordinate system respectively. The shortest path distance between two vertices can be quickly calculated by using the embedded vertice coordinates. However, these embedding methods rely on selecting vertices with certain centrality, which still has the potential to lead to an efficiency bottleneck. Moreover, the accuracy of their distance calculation is strictly limited by the representational capacity of the specific coordinate space. Such algorithms still cannot meet the stringent requirements of some application scenarios, so the shortest path distance prediction on large-scale graphs still has improvement room.

## 3    Preliminaries

In order to facilitate the representation of the method, let $G = (V, E, w)$ be a directed weighted graph, where $V$ is the set of all vertices on the graph, $E$ is the set of all edges on the graph, and $w(u, v)$ is the weight of the directed edge $(u, v)$. Let the function $\oslash(v)$ be the function of obtaining the graph embedding vector by graph embedding technique, where $v$ is the vertice index, and the function result is the embedding vector corresponding to the vertice $v$. In addition, $l(u, v)$ is the distance calculation function between two vertices on the graph, where $u$

and $v$ are two vertices on the graph. The function result is the shortest distance between $u$ and $v$ vertices calculated by traditional graph algorithms such as Dijkstra or BFS algorithm.

**Feature Vector Fusion.** This paper uses the above mentioned Node2vec and LINE and other graph embeddings. By the edge list $E$ of the graph data and the label value of the vertice $u$, the embedding vector $\oslash(u)$ of graph data is obtained. After the embedding vector is obtained, the shortest distance between any two vertices is the label. The feature vector can be obtained by fusing the embedding vectors of the two vertices in some way. This paper attempts to obtain the following three feature vectors, as shown in Table 1.

**Table 1.** Methods to obtain feature vectors.

| Operation name | Hadamard | Average | Subtraction |
|---|---|---|---|
| Definition | $\oslash(u) \cdot \oslash(v)$ | $[\oslash(u) + \oslash(v)]/2$ | $\oslash(u) - \oslash(v)$ |

**Landmark Vertices Selection.** Landmark-based distance oracle is used to accelerate the distance query between vertices in the graph by calculating and storing the distance information between vertice pairs in advance. Both the efficiency and accuracy of the shortest path distance calculation are related to the selection of landmarks. A more complicated landmark selection method typically results in lower efficiency but better expressiveness, hence better accuracy for calculating the shortest path distance, and vice versa. Two basic landmark selection strategies are considered in the following sections.

(1) *Random*: Randomly select a number of vertices from all vertices as landmarks.
(2) *High Degree*: Real-world graphs such as social networks typically demonstrate a power-law distribution of vertices [17].

That means that vertices with high degrees are at the center of such graphs, also known as central vertices. The selection of these vertices may result in better accuracy, according to the research conducted by Potamias et al. [18].

## 4    Method

### 4.1    Algorithmic Framework

The overall framework of the algorithm is shown in Fig. 1.

In the construction of training set and test set, it is necessary to select the appropriate graph embedding method, feature vector fusion method and landmark selection method. Choose a selection strategy either from Random or High

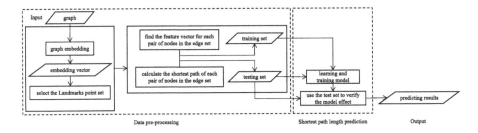

**Fig. 1.** Overall Algorithmic Framework

Degree to select a set of landmarks points, and find all edges $(u, v)$ constitutes a set $S_{train}$, where $u \in S_{landmarks}$ and $v \in (G.V - S_{landmarks})$. The construction process of the test set repeats the construction process of the training set: after using the Random strategy to select vertices, the set $S_{validate}$ is obtained by using the same edge selection method as the construction training set.

Then, the traditional graph algorithm (such as BFS algorithm) is used to calculate the shortest distance between each pair of vertices in the $S_{train}$ set and process it into the label form required in the algorithm. Next, the graph embedding vector is calculated for each pair of vertices in the $S_{train}$ set, and an operation strategy is determined in Table 1 to obtain the feature vector. This allows each pair of vertices in the $S_{train}$ to have a feature vector obtained by a graph embedding vector, and a label representing the distance, so that the training set is constructed. Input the training set into feedforward The neural network classification model or polynomial regression model can be trained to obtain the model. In this paper, a representative three-layer feedforward neural network is selected. The dimension of the input layer depends on the dimension of the vector obtained after graph embedding. The dimension of the output layer depends on the number of classifications in the classification model. The activation function uses ReLU, the loss function is mean square error, and the stochastic gradient descent algorithm is used for training. The polynomial regression model is fitted by the least square method.

## 4.2   Shortest Path Distance Prediction Based on Classification

For unweighted graphs, the distance between two vertices is an integer. As there are only reachable and unreachable cases between any two vertices on the graph, the distance must be an integer. The values of all distances can form a finite set when the distance between any two vertices on the graph is confirmed. The prediction of the shortest path distance between any two vertices must belong to this set. Obviously, this can be solved with classifiers by taking it as a typical classification problem.

As we know, in social networks, according to the six-degree separation theory and practical experience, if the distance between two vertices is greater than or equal to 6, the two vertices can be considered to be disconnected. Therefore, to balance efficiency and accuracy for the majority of graphs, the shortest path distance calculation threshold is set to 5 in this paper. That is, when the distance between two vertices exceeds 5, two vertices are considered to be disconnected (the distance here is 5 refers to the number of edges on the shortest path between two vertices). Of course, in practice, larger threshold can be used in our algorithm framework to obtain better accuracy.

We use an $n$-dimensional vector to represent the label. In initialization, each dimension in the vector is set to 0. To label the distance of a path, the $m$-th position in the vector is set to 1 ($m \in [0, n-2]$), which means that the distance is $m$. Note that if the n-1 position is 1, it predicts that two vertices are unreachable. In some scenarios, the weighted graph with weight 1 is equivalent to the unweighted graph.

### 4.3 Shortest Path Distance Prediction Based on Regression

For some non-integral weighted graphs, the shortest path distance between two vertices is non-discrete. Therefore, it is not feasible to model the results of each distance as nominal class labels as in the form of classifiers. A feasible scheme to solve the problem is to use regression model. Regression is a supervised learning technique that models the relationship between input and output variables. Given a set of training examples, regression learns a function that maps input features to output values Training a regression model is equivalent to performing function fitting. In order to prevent the different dimensions of the input feature vector from eliminating the influence of some dimensions on the results due to the different magnitudes, normalization is performed on feature vector as preprocessing. The labels in this case are floating-point values, rather than bit vectors in the above classification model.

### 4.4 Shortest Path Distance Prediction on Dynamic Graph

It is of practical significance to study the algorithm on dynamic graphs, because in the real world, graph data is often dynamic: new vertices are inserted or existing vertices are deleted, new edges are added or existing edges are deleted, and the weights on the original edges are changed. Traditional algorithms, such as Dijkstra and Floyd-Warshall, have their application scenarios on static graphs. When the structure of the graph changes, these algorithms must recalculate the entire graph, even if the changed structure has no effect on the properties between the two vertices being queried. This makes them extremely expensive on dynamic graphs and does not meet performance requirements. The distance changes in the dynamic graph changes are limited to a few fixed paths, and their number is relatively small compared to the total number of graph paths. Considering that the change on the dynamic graph in reality is often limited, it can be regarded as a dynamic graph with limited change. It can be seen that when the graph changes, the space-time complexity of the query is only related to the size of the graph structure change.

---

**Algorithm 1.** Shortest path distance query algorithm on dynamic graph

---

**Input:** Graph G
    Reachable threshold THRESHOLD
    Changed edge list *changed_edges_list*
    The threshold *sp_threshold* for the shortest path to change.
    Vertice pairs $(u_t, v_t)$ waiting for prediction.
**Output:** *Predict_distance* of vertice pairs to be predicted
1: *embedding_vector* $= \oslash(G)$
2: $S_{landmarks} = $ pick_landmarks$(G.V)$
3: initial *labels, features*
4: **for** $(u,v)$ where $u \in S_{landmarks} \wedge v \in (G.V\text{-}S_{landmarks}) \wedge v$ is reachable for $u$: **do**
5:     get label$(u,v)$ from calculating distance$(u, v, THRESHOLD)$
6:     *embed_u_scale* $=$ scale$(\oslash(u))$; *embed_v_scale* $=$ scale$(\oslash(v))$;
7:     get feature$(u,v)$ from opi(*embed_u_scale, embed_v_scale*), label$(l(u,v))$
8: **end for**
9: reg $=$ regress(*labels, features*)
10: $G = $ dynamic_change$(G, changed\_edges\_list)$
11: predict_distance $=$ reg.predict($op_i(\oslash(u_t), \oslash(v_t))$)
12: initialize *shortest_path* $= +\infty$
13: **for** each $i \in changed\_edges\_list$: **do**
14:     *predict_distance1* $=$ reg.predict($op_i(\oslash(u_t), \oslash(i[0]))$)
15:     *predict_distance2* $=$ reg.predict($op_i(\oslash(i[0]), \oslash(i[1]))$)
16:     *predict_distance3* $=$ reg.predict($op_i(\oslash(i[1]), \oslash(v_t))$)
17:     sum $=$ *predict_distance1* $+$ *predict_distance2* $+$ *predict_label3*
18:     **if** *sum* $<$ *shortest_path* : **then**
19:         shortest_path $=$ sum
20:     **end if**
21: **end for**
22: **if** abs(*shortest_path* - *predict_distance*)$<=$ *sp_threshold*: **then**
23:     return re-calculate$(u_t, v_t)$
24: **else**
25:     return *predict_distance*
26: **end if**

---

Algorithm 1 shows the shortest path distance query algorithm on the dynamic graph. Among them, regress $(x,y)$ is a regression function, and the feature vector is normalized, that is, the scale() function in line 6. For each pair of vertices $(u,v)$, where $u \in$ Slandmarks, $v \in (V\text{-}S_{landmarks})$, the $op_i$ $(\oslash(u),\oslash(v))$ of each pair of vertices is used as the feature vector, and $l(u,v)$ is transformed into the label vector label as the parameter of the regression function. The 11th to 18th lines of the algorithm are used to learn which edge changes the weight to make the distance between the two vertices to be predicted the shortest ; lines 19 to 20 are used to calculate the difference between the above shortest distance and the predicted shortest distance on the original graph, so as to judge whether the edge with changed weight is in the shortest path on the original graph. If the change range is large, it indicates that the predicted shortest path on the original graph has not been changed. If it is small, it indicates that the original shortest path is changed by the edge with changed weight. At this time, the shortest path distance between the two vertices to be predicted needs to be recalculated. In addition, a good threshold value can make the algorithm achieve a good balance between making full use of existing learning models and recalculating to maintain accuracy.

# 5    Result

This section mainly evaluates the accuracy and efficiency of the above proposed algorithms. By evaluating different graph embedding algorithms and landmark selection algorithms on different test data sets, the experimental results are analyzed and summarized.

## 5.1    Experimental Configuration

The experiment is carried out under the Microsoft Window10 system with an Intel®$Core^{TM}$ i5-6200U 2.30GHZ CPU and 8GB DDR3 running main memory.

The datasets are from Stanford Network Analysis Project [19], which aims general network analysis. In order to test the proposed algorithms, five datasets, i.e., wiki [20], ca-GrQc [21], p2p-Gnutella04 [21], Slashdot0811 [22], and youtube [2] were selected from the data set disclosed by the project, which provided different types of network data for the experiment in this paper. The attributes of the five data sets are shown in Table 2.

**Table 2.** Data Set Attributes.

| Dataset | wiki | Ca-GrQc | p2p-Gnutella04 | Slashdot0811 | youtube |
|---|---|---|---|---|---|
| # vertices | 2405 | 5242 | 10876 | 77360 | 15088 |
| # edges | 15358 | 28968 | 39994 | 905468 | 19923067 |
| Is directed | True | True | True | True | True |

## 5.2    Shortest Path Distance Prediction Based on Classification

The classifier used in the experiment is a single hidden layer forward feedback neural network. The vector dimension obtained by the input layer and the graph embedding method is related to the acquisition method of the feature vector, which is 128 in the experiment. The hidden layer has 500 neurons, and the output layer is 6 dimensions. The first five dimensions represent distance, and the sixth dimension represents unreachable.

The selection strategy of different modeling methods is used to evaluate the accuracy on different data sets. The distance relationship only lists the experimental data on two representative data sets as shown in Table 3, the experimental results of other data sets with similar data performance are omitted.

**Table 3.** Prediction accuracy of different modeling strategies.

| Graph embedding | Feature vector | Dataset | | | |
|---|---|---|---|---|---|
| | | ca-GrQc | | p2p-Gnutella04 | |
| | | Random | High degree | Random | High degree |
| LINE | Hadamard | 0.9026 | **0.9518** | **1.0** | 0.8703 |
| | Subtraction | 0.888 | 0.7613 | 1.0 | 0.0567 |
| | Average | 0.8062 | 0.7945 | 0.998 | 0.981 |
| Node2vec | Hadamar | 0.8736 | 0.6168 | 1.0 | 0.8736 |
| | Subtraction | 0.6608 | 0.1232 | 0.9273 | 0.9919 |
| | Average | 0.9263 | 0.8062 | 1.0 | 0.996 |

It can be found that the most accurate method combination for prediction results in ca-GrQc is: LINE + High degree + Hadamard ; considering the average result and the variance of the result, the best way to obtain the feature vector is Average. However, when comparing the selection of different landmarks on the p2p-Gnutella04 dataset, it is found that the accuracy of Random strategy is close to 100%. The reason may be that because of the random method, after the selection of landmarks reaches a certain number, the samples of the classifier can be selected to any point on the graph, and the proportion of various types of samples is approximately the same as the complete graph data, that is, it can be considered to cover the entire graph. Therefore, if the graph data is sparser, the trained classifier is more inclined to classify the category into the most likely case, that is, unreachable; the test samples are also randomly selected, and the sample distribution is uneven. When predicting, the model tends to think that the test sample is unreachable, and the test sample is very likely to be unreachable, resulting in a very high accuracy rate when adopting the Random strategy.

In contrast, the accuracy of the landmarks selection strategy using High degree on the p2p-Gnutella04 dataset is relatively low. This paper believes that because this selection strategy selects points with higher node degrees as the training set, the proportion of unreachable samples is lower than that of Random selection strategy. Therefore, the proportion of unreachable samples that are easier to predict is reduced, so the accuracy is relatively reduced.

Comparing the results of the p2p-Gnutella04 dataset with the ca-GrQc dataset, it can be found that the accuracy on the ca-GrQc dataset is lower than that on the p2p-Gnutella04 dataset. In this paper, the reason for this phenomenon is that the degree of sparsity on different graph structures is different. Because it is easier to classify sparse graph classifiers, it can be preliminarily inferred that no graph algorithm can be applied to all shortest distance problems due to differences in space-time complexity and graph properties. The regression experiment in the next section will further analyze this conclusion. When the shortest path prediction problem is mapped to the classification problem in the neural network model, the structure of the graph needs to be fully considered.

Compared with the traditional algorithm, the learning model framework proposed in this paper only needs to fine-tune the training data set or parameters, without redesigning the algorithm, so it is more universal.

## 5.3  Shortest Path Distance Prediction Based on Regression

In this experiment, polynomial regression is used, and the parameter degree is set to 1.

Different from the classification model, because the prediction results of the regression model are generally in the form of approximate values, this paper uses *Relative Error* $RE = \frac{|l(u,v) - \text{predict}(u,v)|}{l(u,v)}$ to measure the performance of the regression model.

However, because the relative error has the characteristics of normalization, it is not conducive to represent the larger distance error. Thus, this paper also uses the absolute error *Absolute Error* $AE = |l(u,v) - \text{predict}(u,v)|$ measurement method.

For the regression model, the selection of landmarks when constructing the training set defaults to a high priority approach. Since the unweighted graph can be regarded as a weighted graph with a weight of 1, the same data set as the experiment based on the classification model is used here. Table 4 shows the experimental results. In general, the effect of using Node2 vec graph embedding technology is better than that of using LINE. The reason is that in the data set selected in this paper, the shortest path distance is logarithmically distributed [23], and Node2 vec inherits the method of Random walks, which reflects the local structure information of the graph more effectively in the embedding process. In the LINE algorithm, the results of the first-order adjacent objective function and the second-order adjacent objective function are directly spliced [23], so some local graph structure information may be lost. For the acquisition of feature vectors, whether it is a classification model or a regression model, the experimental data does not show a consistent rule with advantages in some way. In the future work, the influence of different feature vector acquisition methods on the prediction results will be further explored.

**Table 4.** MAE and MRE of different modeling methods.

| Graph embedding | Feature vector | Dataset | | | | | |
|---|---|---|---|---|---|---|---|
| | | ca-GrQc | | p2p-Gnutella04 | | wiki | |
| | | MAR | MRE | MAR | MRE | MAR | MRE |
| LINE | Hadamard | 0.5908 | 0.1665 | 0.5695 | 0.1534 | **0.6445** | **0.1828** |
| | Average | 0.6114 | 0.1700 | 0.5755 | 0.1549 | 0.6769 | 0.2082 |
| | Subtraction | 0.6686 | 0.1866 | 0.5733 | 0.1561 | 0.7132 | 0.2090 |
| Node2vec | Hadamar | **0.5406** | **0.1459** | **0.5481** | **0.1449** | 0.6763 | 0.1957 |
| | Average | 0.6256 | 0.1861 | 0.5758 | 0.1576 | 0.6856 | 0.2043 |
| | Subtraction | 0.6404 | 0.1806 | 0.5855 | 0.1607 | 0.7228 | 0.2131 |

This paper selects the representative shortest path prediction algorithm Pow-Cov and ChromLand [2] for comparison. Youtube dataset is widely used in shortest path prediction. These two algorithms also use youtube dataset in experimental evaluation. Accordingly, youtube dataset is selected as the evaluation dataset, feature vector selection Average, graph embedding technology selection Node2 vec, and the evaluation results are shown in Table 5:

**Table 5.** Comparison between our algorithm and PowCov and ChromLand algorithms.

|      | This algorithm | PowCov | ChromLand |
|------|----------------|--------|-----------|
| MAE  | 0.450          | 0.46   | 0.86      |
| MRE  | 0.107          | 0.28   | 0.49      |

It can be seen that the algorithm is superior to the PowCov and ChromLand algorithms when the same number of landmarks is selected, which can reflect the advantage of the strong fitting ability of the learning model. In this paper, the graph embedding and learning model is used for traditional graph theory problems, and only a preliminary exploratory attempt is made in the experiment. This paper believes that with the further discussion of different modeling methods, such as other graph embedding methods, as well as different parameters of feedforward neural networks, and even different neural network models, the prediction results will be further improved.

### 5.4   Shortest Path Distance Prediction on Dynamic Graph

In order to have the value of horizontal comparison on graphs of different sizes, in this experiment, the dynamic change method of each data set is : randomly select two vertices on the graph, and randomly obtain the weights of all edges connected by these vertices in 1 to 5. If the vertice is not connected, the point is retaken. When the neural network model is constructed, the feature vector selects Average, the graph embedding technology selects Node2 vec, and the landmark selection method is high priority. The evaluation results on different datasets are shown in Fig. 2.

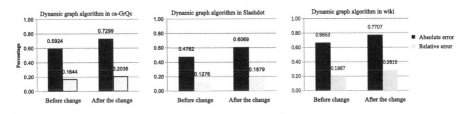

**Fig. 2.** Comparison of errors before and after graph changes on data sets

For the results shown in Fig. 2, when the figure changes dynamically, both the relative error and the absolute error increase, but the magnitude is that a relatively small value under this dimension is tolerable. The threshold used in this experiment is n times the absolute error of the original model, where the value of n is artificially set. Therefore, the accuracy of the algorithm mainly depends on the accuracy of the original model and the setting of n. In order to improve the accuracy of the algorithm on the dynamic graph, the most important thing is to improve the accuracy of the original model.

To further reduce the error, there are two methods:

(1) Improve the complexity of the learning model and improve the prediction accuracy on the original graph.
(2) Improve the threshold, increase the number of re-use of the traditional graph algorithm to calculate the distance.

The target ideas of these two methods are completely different and cannot be used at the same time. In this paper, when querying the distance between a pair of points, scan each changed edge, and calculate the distance between the two points through this changed edge, and the difference between the shortest distance between the two points. It is recalculated only when the minimum value of the difference of each edge scanned is less than a certain threshold.

In addition, the calculation method of the previous paragraph should also be mentioned : n times the absolute error. With these two premises, it is easy to understand the method in (1) : For the polynomial regression model used in this paper, the ability of polynomial fitting complex curves and the prediction accuracy of the model can be improved by increasing the polynomial degree when training the model. The effect of improving the accuracy of prediction on the algorithm on the dynamic graph is : after the graph changes dynamically, in order to judge whether the edge to be queried changes, if the absolute error is small, the path distance between the two points is calculated, and the path passes through the changed edge, which is more accurate, and then more accurately determines whether there is a changed edge on the original shortest path of the two points to be queried. For Method (2), the idea is to reduce the application of the previous learning model for calculation, and more re-use the traditional graph algorithm for calculation. After increasing the threshold, the algorithm is easier to determine the path between two points as a changed edge. This method relatively distrusts the learning model.

These two ideas are applicable in different situations : In Method (1), if a more accurate absolute error is obtained, it can be accurately judged whether a shortest path has changed after the appropriate debugging parameter n, which can reduce unnecessary calculations as much as possible and reduce the computational overhead of the query process. But at the same time, the training process will consume more resources. In method (2), the algorithm prefers to recalculate. This method may cause some unnecessary calculations, which greatly increases the computational overhead of the query process, but relatively speaking, the overhead of the training process is relatively reduced and can ensure accuracy.

In summary, these two methods have their own advantages and disadvantages, which need to be selected in combination with specific application scenarios. According to the above content, it can be explained that the two methods cannot be used together to improve the accuracy of the algorithm.

In the above data set, Table 6 uses the traditional Dijkstra algorithm and the time consumed by using this algorithm. The data in the table are obtained by taking the average value after calculating 6 times on the three data sets, each time.

**Table 6.** Computing time comparison.

|  | Recalculation | Use dynamic graph algorithm |
|---|---|---|
| Ca-GrQc | 0.365 s | 0.454 s |
| wiki | 0.132 s | 0.408 s |
| Slashdot | 18.195 s | 0.542 s |

From the intuitive view of the data in Table 6, for the Ca-GrQc and wiki datasets, the recalculation takes less time than this algorithm, but the results do not indicate that the nature of this algorithm is inferior to recalculation. For Slashdot dataset, the time consumption of using this algorithm is much lower than that of recalculation.

According to the properties of the wiki dataset and the ca-GrQc dataset, it can be considered that the graph connectivity on these two datasets is approximately equal. In the case where the graph properties are basically similar, and the setting goal of the threshold parameter n in this experiment is to minimize the number of recalculations, the reduction of the correct rate can be moderately tolerated. By observing the results in Table 6, it can be seen that for the data set ca-GrQc and wiki, although the former is twice as large as the latter, the time consumption has not increased by an order of magnitude, which is in line with the expectation of time complexity analysis. Since a large number of branch structures are used in the program, it is necessary to repeatedly determine the options of the parameters at runtime, and only direct calculation is required when using the graph algorithm, so the recalculation time is lower than the time of this algorithm. In addition, the time complexity $O(n^2)$ of Dijkstra algorithm is combined to recalculate the scalability problem on large-scale graphs. Considering the size of the graph and the recalculation time of the two in Table 6, it can be seen that the time complexity analysis is expected. It can be expected that the calculation time of the traditional graph algorithm will increase explosively. Therefore, it can be considered that the spatio-temporal performance of the proposed algorithm will be much better than the traditional graph algorithm after the graph size increases.

The performance of Slashdot dataset in Table 6 can verify the analysis of this paper. Calculating the same number of point-to-point distances, the time using this algorithm changes less. According to the size of the data set in Table 2, the data set contains 77360 vertices. Combined with the complexity analysis of the Dijkstra algorithm in the previous paragraph, the actual consumption time is 18.195 s, which is consistent with the expectation, so the correctness of the above analysis can be proved. Based on the above analysis, the experimental results show that the algorithm has better properties.

## 5.5   Time Complexity Analysis

In this paper, the embedding vector of vertices is obtained by graph embedding, and the time complexity is $O(n)$. Then, the landmark method is used to construct the training set at the minimum cost. Assuming that $l$ vertices are selected as landmarks, the training set composed of $l(n - l)$ pairs of shortest paths of vertices can be obtained. For unweighted graphs, if BFS is used to construct the shortest path, the time complexity is $O(l(n + m))$. For weighted graphs, if Dijkstra algorithm is used to construct the shortest path, the time complexity is $O(l(n \log n + m))$. Due to the use of graph embedding, only the feature vectors of two vertices need to be calculated when performing distance prediction, and the time complexity is $O(1)$. Therefore, the time complexity of this algorithm to calculate the shortest distance from a vertice to any other vertice is $O(n + m)$ or $O(n \log n + m)$.

For dynamic graphs, it is assumed that the modification scale is $c$, that is, $c$ edges have changed. According to the description of the algorithm, the time complexity of the algorithm in the worst case is $O(c+n^2)$, where n is the number of vertices, that is, when scanning to the last changed edge, the edge is on the original shortest path between the two points to be queried, so the Dijkstra algorithm needs to be used again. In the best case, the time complexity is $O(c)$, that is, scanning all the changed edges, and no edge is on the original shortest path between the two points to be queried. The time complexity of the above analysis is obtained when the threshold sp_threshold value is reasonable. If the threshold value is too large, too many points will be recalculated, making the time complexity close to the worst case $O(c+n^2)$ ; if the threshold value is too small, although the number of calculations is reduced, the time complexity is close to the best case $O(c)$, but the prediction accuracy is greatly lost. Accordingly, the time complexity and accuracy of the algorithm are determined by the value of the threshold, and the threshold needs to be set according to the needs of the query : if a faster query speed is required, a lower threshold is required, and the accuracy of the query will be reduced. If you need higher query accuracy, you need to increase the threshold, increase the number of recalculations, and the time complexity will increase.

## 6   Conclusion

This paper studies the shortest path distance prediction with the help of machine learning technologies. Graph embeddings and landmarks are used to preprocess graph data, and supervised machine learning models are successfully adapted to solve the shortest path distance prediction problem in the scenarios of graphs with small average shortest path and integer weight, graphs with non-integral weights, and dynamic graphs. They exhibited constant time complexity and good prediction accuracy in the shortest path distance prediction experiments.

# References

1. Thorup, M., Zwick, U.: Approximate distance oracles. J. ACM **52**(1), 1–24 (2005)
2. Bonchi, F., Gionis, A., Gullo, F., Ukkonen, A.: Distance oracles in edge-labeled graphs. In: EDBT, pp. 547–558 (2014)
3. Guyon, I., et al.: Advances in Neural Information Processing Systems 30: Annual Conference on Neural Information Processing Systems 2017 (2017)
4. Zhao, X., Zheng, H.: Orion: Shortest path estimation for large social graphs. In: WOSN, pp. 1–9 (2010)
5. Zhao, X., Sala, A., Zheng, H., Zhao, B.Y.: Efficient shortest paths on massive social graphs. In: CollaborateCom, pp. 77–86. ICST / IEEE (2011)
6. Grover, A., Leskovec, J.: node2vec: Scalable feature learning for networks. In: KDD, pp. 855–864 (2016)
7. Tang, J., Qu, M., Wang, M., Zhang, M., Yan, J., Mei, Q.: LINE: large-scale information network embedding. In: WWW, pp. 1067–1077 (2015)
8. Svozil, D., Kvasnicka, V., Pospichal, J.: Introduction to multi-layer feed-forward neural networks. Chemom. Intell. Lab. Syst. **39**(1), 43–62 (1997)
9. Cormen, T.H., Leiserson, C.E., Rivest, R.L., Stein, C.: Introduction to Algorithms, Second Edition. The MIT Press and McGraw-Hill Book Company (2001)
10. Dijkstra, E.W.: A note on two problems in connexion with graphs. Numer. Math. **1**, 269–271 (1959)
11. Fredman, M.L., Tarjan, R.E.: Fibonacci heaps and their uses in improved network optimization algorithms. J. ACM **34**(3), 596–615 (1987)
12. Fredman, M.L., Willard, D.E.: Trans-dichotomous algorithms for minimum spanning trees and shortest paths. In: FOCS, pp. 719–725 (1990)
13. Chan, E.P.F., Yang, Y.: Shortest path tree computation in dynamic graphs. IEEE Trans. Comput. **58**(4), 541–557 (2009)
14. D'Emidio, M., Forlizzi, L., Frigioni, D., Leucci, S., Proietti, G.: Hardness, approximability, and fixed-parameter tractability of the clustered shortest-path tree problem. J. Comb. Optim. **38**(1), 165–184 (2019)
15. Pallottino, S., Scutellà, M.G.: Dual algorithms for the shortest path tree problem. Networks **29**(2), 125–133 (1997)
16. Rizi, F.S., Schlötterer, J., Granitzer, M.: Shortest path distance approximation using deep learning techniques. In: ASONAM, pp. 1007–1014 (2018)
17. Newman, M.E.J.: The structure and function of complex networks. SIAM Rev. **45**(2), 167–256 (2003)
18. Potamias, M., Bonchi, F., Castillo, C., Gionis, A.: Fast shortest path distance estimation in large networks. In: CIKM, pp. 867–876 (2009)
19. Leskovec, J., Kleinberg, J.M., Faloutsos, C.: Graph evolution: densification and shrinking diameters. ACM Trans. Knowl. Discov. Data **1**(1), 1–44 (2007)
20. Tu, C., Zhang, W., Liu, Z., Sun, M.: Max-margin DeepWalk: discriminative learning of network representation. In: IJCAI, pp. 3889–3895 (2016)
21. Leskovec, J., Lang, K.J., Dasgupta, A., Mahoney, M.W.: Community structure in large networks: natural cluster sizes and the absence of large well-defined clusters. Internet Math. **6**(1), 29–123 (2009)

22. Koch, G., Zemel, R., Salakhutdinov, R.: Siamese neural networks for one-shot image recognition
23. Chopra, S., Hadsell, R., LeCun, Y.: Learning a similarity metric discriminatively, with application to face verification. In: CVPR (1), pp. 539–546 (2005)

# Efficient Regular Path Query Evaluation with Structural Path Constraints

Tao Qiu[1(✉)], Yuhan Wang[1], Meng-xiang Wang[2], Chuanyu Zong[1], Rui Zhu[1], and Xiufeng Xia[1]

[1] School of Computer Science, Shenyang Aerospace University, Shenyang, China
{qiutao,wangyuhan,zongcy,zhurui,xiaxf}@sau.edu.cn
[2] China National Institute of Standardization, Beijing, China
wangmx@cnis.ac.cn

**Abstract.** Regular path query is a technique of using a regular expression (regex) on graph data. Classical methods adopt the finite state automaton to match the regular path query on the graph. Their matching results are the sequences of vertex pairs (i.e., a set of paths), and constraints between paths cannot be satisfied. To solve this problem, we propose a structural regular path query method that satisfies not only regex constraints but also *structural path constraints* a specific constraint needed to be satisfied by the target paths. We first define a structural regular path query and then design the structural automaton to represent this query. Also, we devise an automaton-based matching method using the deep-first traversal on the graph and design optimizations to improve the matching efficiency. Experiments are conducted in real datasets by comparing the proposed method to the traditional methods.

**Keywords:** Regular path query · Graph data · Finite state automaton · Structural path constraints

## 1 Introduction

The graph database uses graphs to store data and can be queried based on semantics. It uses vertices, edges, and attributes to represent data, where vertices represent entities in the real world, and edges between vertices express the relationships between entities. A regex is a pattern with certain grammar rules, typically used to match a target substring in a text. In recent years, a query technique based on regexes for graphs called *regular path query*, has attracted a lot of attention and research and has become an important tool for detecting and analyzing data from graphs. It is widely used in information retrieval [20], social recommendation system [10], and information extraction, etc.

At present, there are two mainstream techniques querying paths from a graph. One is the problem of reachability query with label constraints, and this kind of problem often applies some optimization techniques in the reachability algorithm. The other is the problem of query matching according to the complete regex, which is mainly based on the *non-deterministic finite automata* (NFA). The regex is parsed into NFA, which can further guide the search on the graph and obtain the set of paths satisfying the regex in the graph.

X. Yang et al. (Eds.): ADMA 2023, LNAI 14178, pp. 308–322, 2023.
https://doi.org/10.1007/978-3-031-46671-7_21

Although the problem of regular path query has been actively studied in the area of databases, there are still some limitations. Traditional methods utilize regexes to depict the constraints for the labels of the target path. However, these methods cannot match the paths that have structural path constraints. For example, for the three entities in a social network that have three relationships ("father-child", "mother-child" and "husband-wife"), each relationship is represented by a path and can be depicted by a regular path query. However, the regular path queries cannot guarantee that the matched entities satisfy the three relationships simultaneously since they are matched separately. To match such entities, the paths should satisfy certain constraints, e.g., starting and ending with the same nodes. We call such constraints as structural path constraints.

In this paper, we extend the classical regular path query by introducing a new operator (AND) in the regular repression to support the structural path constraints and call it *structural regular path query*. Then, we design the automaton-based matching method for the structural regular path query and utilize the inverted list to boost the matching efficiency.

The contributions of this paper can be summarized as follows:

- We define a structural regular path query language based on regexes, which describe structural path constraints between paths.
- We extend the classical nondeterministic finite automaton to represent the structural regular path query, named structural nondeterministic finite automaton.
- We propose an automaton-based matching method utilizing the deep first traversal on the graph to match the structural regular path query and optimize the automaton-based matching by selecting good starting matching nodes on the graph.

The rest of the paper is organized as follows. Sect. 2 gives the survey of related work. Sect. 3 introduces necessary backgrounds and gives the problem definition. We present the structural nondeterministic finite automaton in Sect. 4 and present the automaton-based matching method in Sect. 5. Experimental results are presented and analyzed in Sect. 6. Finally, we conclude this paper in Sect. 7.

## 2   Related Work

*(1) Query Processing on Graph*
The regular path query is a typical query for the graph, that finds a set of paths that satisfy regex constraints in a directed graph. Generally, they regard the path of the graph as a sequence of labels, then utilize the automaton to represent the query and use it to verify the label sequence [9,16,18]. Zhang et al. utilize a method of two-phase query processing [25,26]. They first process the longest sequence of fixed predicates in the query, then process the subexpression of the query containing closures. Koschmieder et al. use the rare labels to divide the complete regular path query into several smaller-scale sub-queries by using the dive-and-conquer strategy and also propose the method of bi-directional matching on the self-built graph index [7,8]. Besides, SPARQL is another query widely used for graph data models, e.g., the attribute graph model and the resource description framework (RDF) [2]. SPARQL employs a similar language to the regular path query [1] and it is widely used for some real applications, e.g., gStore [27].

Wilkinson et al. describe a Jena system based on RDF and related semantic web technologies that support large data sets [23]. Bornea et al. describe a new RDF storage and query mechanism that decomposes RDF into relationships based on existing relational representations [3]. Neumann et al. introduce the RDF-3X engine based on the SPARQL implementation, which achieved excellent performance [13]. Weiss et al. propose an RDF storage scheme that enhances the efficiency and scalability of RDF data management by extending triples to six bits in the RDF scheme [21]. Peng et al. propose a technique for processing SPARQL queries in a distributed environment through large RDF graphs by introducing local partial matching as a partial answer in each fragment of the RDF graph [14].

*(2) Subgraph Matching*

Subgraph matching is another data retrieval technique on the graph, while the results are the subgraphs containing the structural constraints. Mhedhbi et al. study the problem of optimizing subgraph queries by changing the order of worst-case best sub plans during query execution [12]. Sun et al. propose a depth-first plus backtracking method to achieve subgraph matching [19], and Wi et al. propose a breadth-first multi-way join method [22]. Also, the dictionary is introduced to optimize storage for the graph. Carlett et al. improve search efficiency by deductively eliminating successor vertices in tree search [4]. Han et al. propose a set intersection algorithm using SIMD instructions to improve the efficiency of subgraph matching [5]. Zeng et al. design an efficient subgraph isomorphism algorithm, which broke through the bottleneck of subgraph isomorphism performance in practical applications by utilizing the characteristics of GPU architecture [24].

## 3    Preliminaries

In this paper, we utilize the typical definition of the regular path query [8], as follows.

**Definition 1.** *(Regular Path Query, abbr. RPQ) An RPQ $Q$ is used to match paths from a directed graph $G$, and $Q$ is a regular expression describing the constraints for the labels in the path of $G$.*

The regular path query needs to be represented using a finite state automaton so that the label sequence of the path in the graph can be checked by this automaton. Upon the final state of the automaton being activated, then a target path is found in the graph. There are two types of finite automata, i.e., non-deterministic finite automata (NFA) [6] and deterministic finite automata (DFA). In this paper, we adopt the NFA to represent the regular path query.

The traditional regular path query uses the regular expression (abbr. regex) to represent the constraints for the labels of the target paths. In order to describe the structural constraints between the target paths, we extend the regular expression to support structural constraints. For a set of labels $x = x_1 \cdots x_k$ $(k \geq 1)$ from the paths of $G$, the extended query is defined recursively as follows:

- **Closure:** The closure $(r_1^*)$ is a regex that $r_1$ matches each label $x_i$ $(0 \leq i \leq k)$ 0 or more times. Since the closure can be represented as $r_1^* = (r_1^+|\epsilon)$, we will use $r_1^+$ for the examples in this paper.

- **Join**: Let $(r_1 \cdot r_2)$ be a regex that can be written as $(x_1 \cdot x_2)$, where $r_1$ and $r_2$ match $x_1$ and $x_2$ respectively, and $\cdot$ denotes label concatenation (i.e., $x_1$ and $x_2$ are the labels of adjacent edges of $G$).
- **Select**: Let $(r_1 \mid r_2)$ be a regex such that $x$ matches $r_1$ or $r_2$.
- **AND**: Let $(r_1 \& r_2)$ be a regex that can be written as $(x_1 \& x_2)$, where $r_1$ and $r_2$ match $x_1$ and $x_2$ respectively, and & denotes that the subpaths containing the labels $x_1$ and $x_2$ have the same starting and ending nodes.

We call the query using the above extended regular expression as *structural regular path query* (SRPQ). There are two types of constraints in the SRPQ. i) The constraints provided by the classical regex operators including **Join**, **Select**, and **Closure**, which have the same syntax as the operators in the classical regular path query. ii) The structural path constraints provided by the newly defined operator **AND**, which require the subpaths with the same starting and ending nodes to be matched by the query.

**Example 1.** *Consider an SRPQ $Q = a \cdot ((b \cdot a) \& (c \mid a)) \cdot b \cdot c^+$ and the graph $G$ in Fig. 1, there are two matching results for $Q$. $(b \cdot a)$ matches the subpath $v_6$-$v_{10}$-$v_7$ and $(c \mid a)$ matches the subpath $v_6$-$v_7$, since these subpaths have the same starting and ending nodes, $((b \cdot a) \& (c \mid a))$ is matched. Since $c^+$ can occur more than one time, the subpaths ending at $v_{12}$ and $v_8$ are the results of $Q$.*

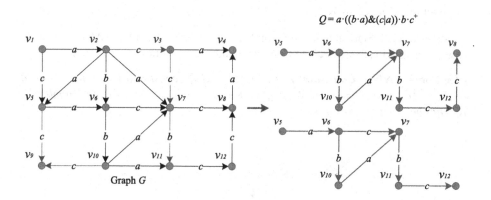

**Fig. 1.** An example of matching an SRPQ on a graph.

In fact, for the SRPQ $Q$, if $Q$ contains the **AND** operator, the matching results of $Q$ are the subgraphs of $G$ consisting of subpaths with structural constraints. Given an SRPQ $Q$ and a directed graph $G$, the problem of structural regular path query evaluation is to find all matching results of $Q$ from $G$.

## 4   Structural Nondeterministic Finite Automaton

In this section, we design a structure to represent the newly proposed structural regular path query, as follows.

**Definition 2.** (*Structural Nondeterministic Finite Automaton, abbr. SNFA*) *A structural nondeterministic finite automaton, denoted as* $\mathcal{A} = (S^S, \Sigma, \delta^S, s_0, F, S^\&)$, *represents a set of subgraph, where* $S^S$ *is a nonempty set of states. For any* $q \in S^S$, $q$ *is a state in* $S^S$. $\Sigma$ *is the alphabet over the regex* $R^S$; $\delta^S$ *is the state transition function of the automaton, which records the transitions between adjacent states.* $s_0$ *is the initial state of the automaton and* $s_0 \in S^S$; $F$ *is the set of final states,* $F \subset S^S$, *and for any* $q \in F$, *call* $q$ *the end state of SNFA;* $S^\&$ *is the* **AND** *constraint function.*

SNFA is defined based on the traditional NFA by involving two new finite states. The finite states are classified into the following types:

- Initial state: the initial state in which the automaton executes.
- Direct state: the new state reached after the transition function $\delta^S$. It is worth noting that since the structural regular path query language may exist empty attribute edge $\varepsilon$ (there is an empty expression in $Q$).
- Indirect state: after reaching the direct state, all empty attribute edges are skipped until $\varepsilon$ does not appear anymore. A direct state may correspond to multiple indirect states, such as branch structures.
- Final state: there are no other states of the automaton after this state.
- AND-Start state: the AND-Start states that multiple subsequent direct states can be reached at the same time, denoted by $\&_{Start}$.
- AND-End state: the states that are the direct successor state of multiple states at the same time, denoted by $\&_{End}$.

For each AND-Start state in the set of AND states, there is a unique AND-End state, We call this relationship as **AND** Constraint and use a function to describe it.

**Definition 3.** (**AND** *Constraint Function*) *An* **AND** *constraint function is denoted as* $S^\& = (q_i, q_j, d)$, *where* $q_i \in \&_{start}$ *and* $q_j \in \&_{End}$, $d$ *denotes the number of branches between the pair of states.*

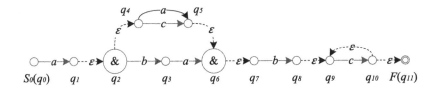

**Fig. 2.** An example of SNFA.

**Example 2.** *Consider the same SRPQ* $Q = a \cdot ((b \cdot a)\&(c \mid a)) \cdot b \cdot c^+$, *the SNFA obtained from* $Q$ *is shown in Fig. 2, where the initial state is* $s_0 = q_0$, *the end state is* $q_{11} \in F$. *In the query, the structural constraint is added by* $((b \cdot a)\&(c \mid a))$, *then the* **AND** *constraint function is* $S^\& = \{q_2, q_6, 2\}$, *where* $q_2$ *is the AND-Start state and* $q_6$ *is the AND-End state, the number of branches is 2 since the least number of subpaths from* $q_2$ *to* $q_6$ *in the matching result is 2.*

The SNFA is used to match the results of $Q$ from $G$. Similar to the classical NFA, an **active state set** (denoted by $S_A$) is utilized to record the activated states during the matching of SNFA. Firstly, the set is initialized by the initial state. If the subsequent state can be activated by the state transition function $\delta^S$ (i.e., the label from the graph $G$), it will be added to the set $S_A$, and the previous activated state will be removed. When the final state is activated, a matched result is found from $G$.

For the example in Fig. 2, the set $S_A = \{q_0\}$ contains the starting states of SNFA. Next, if the accepted label is $a$, the active state set is updated as $S_A = \{q_2\}$, where $q_2$ is the indirect state of $q_0$. For the AND-End state $q_6$, it is activated when $q_5$ and $q_3$ are activated and a label $a$ is used for the state transition of $q_3$.

# 5    Matching Structural Regular Path Query Using SNFA

In this section, we first introduce the matching method using SNFA in Sect. 5.1, then optimize the SNFA-based matching by selecting good starting matching nodes from the graph in Sect. 5.2.

## 5.1    Matching SNFA by Depth-First Traversal

*(1) Determining Starting Nodes*
The idea of automaton-based matching is to treat the automaton as a subgraph and match it from the graph. When the SNFA $\mathcal{A}$ is constructed, $\mathcal{A}$ will be matched from some starting nodes of the graph. To determine the starting nodes, an inverted list for the graph can be utilized, in which an inverted list for a label $l$ records the edges (a pair of nodes) where $l$ appears. Given an SNFA $\mathcal{A}$, we can locate the starting nodes of $\mathcal{A}$'s first label from the graph by the inverted list. For example, the inverted list for the example graph is shown in Fig. 3. For the query in Fig. 2, the list of starting nodes is $l_a$ since $a$ is the first label of $\mathcal{A}$ and $l_a$ is located from the inverted list.

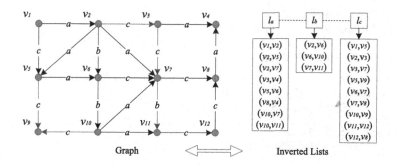

**Fig. 3.** An example of the inverted list for the graph.

*(2) Depth-first Traversal*
As the starting nodes are determined, the SNFA is matched in a depth-first traversal starting from these nodes. In our algorithm, we utilize a similar idea to the method [11]

of matching NFA on a graph which is also implemented in a depth-first traversal, but we need to process the AND state of SNFA.

The skeleton of the algorithm is shown in Alg. 1, the equivalent SNFA $\mathcal{A}$ of $Q$ and the obtained edges (starting nodes) $S_{start}$ are the inputs. A stack $\mathcal{S}$ is utilized to realize the depth-first traversal on the graph. Initially, the edges in $S_{start}$ will be pushed into $\mathcal{S}$ so that only the traversal only starts from the edge whose label equals the first label of $\mathcal{A}$. Here, $\mathcal{S}$ records an edge $e$ and the corresponding status (i.e., the active states) of $\mathcal{A}$ used to match the label of $e$. For example, for the starting nodes, the corresponding statuses are the initial active states of $\mathcal{A}$, i.e., $q_0$ for the query in Fig. 2.

Then, the algorithm iteratively processes the items in $\mathcal{S}$. The obtained active set overrides the active set of $\mathcal{A}$ since $\mathcal{A}$ should back to the status for processing the label of $e$ (lines 6–7). Next, $\mathcal{A}$ updates the states by the label and check if the final state has been activated, if so, a matching result is found and the recorded path is added to the result set $\mathcal{R}$. The algorithm backtracks in three cases. i) The matching of $\mathcal{A}$ becomes to be failed for the current paths, i.e., $\mathcal{A}.S_A = \emptyset$; ii) The traversed path terminates, i.e., $e.nexts() = \emptyset$; iii) A AND-End state of $\mathcal{A}$ is partially activated. Otherwise, the algorithm further processes the path by adding the successor edges of $e$ to the stack (lines 11–13).

---

**Algorithm 1: SNFAMatch**

**Input** : An SNFA $\mathcal{A}$, a set of edge $S_{start}$.
**Output** : The matching results $\mathcal{R}$.

1   initialize a stack $\mathcal{S} \leftarrow \emptyset$;
2   **for** *each edge $e$ in $S_{start}$* **do**
3     $\lfloor$   $\mathcal{S}.push(\langle e, \mathcal{A}.S_A \rangle)$;

    /* depth-first traversal                                  */
4   **while** $\mathcal{S} \neq \emptyset$ **do**
5     $\langle e, S_A \rangle \leftarrow \mathcal{S}.pop()$;
6     $\mathcal{A}.S_A \leftarrow S_A$;
7     $\mathcal{A}.\mathsf{UpdateStates}(e.label)$;
8     **if** $\mathcal{A}.\mathsf{IsFinalActivated}()$ **then**
9       $\lfloor$   $\mathcal{R}.add(\mathcal{A}.match)$;// find a matching result
10      $\lfloor$   $\mathcal{A}.\mathsf{ResetANDEndState}()$;
11     **if** *not* $(\mathcal{A}.S_A = \emptyset$ *or* $e.nexts() = \emptyset$ *or* $\mathcal{A}.\mathsf{IsANDEndPartialActivated}())$ **then**
12       **for** *each edge $e'$ in $e.nexts()$* **do**
13        $\lfloor$   $\mathcal{S}.push(\langle e', \mathcal{A}.S_A \rangle)$;
14     **else if** $\mathcal{A}.S_A = \emptyset$ **then**
15      $\lfloor$   $\mathcal{A}.\mathsf{ResetANDEndState}()$;
16   **return** $\mathcal{R}$;

---

*(3) Processing AND States*

Different from the traditional NFA matching, we need to process the AND states of SNFA in Algorithm 1. According to the definition of SNFA, the subpaths between

AND-Start and AND-End states should be matched simultaneously, e.g., $q_2$-$q_3$-$q_6$ and $q_2$-$q_4$-$q_5$-$q_6$ are required to be matched for the SNFA in Fig. 2. That is, only all subpaths between between AND-Start and AND-End states are matched, we further process the states after the AND-End state.

To do that, we redefine the states of the AND-End, the possible states include *inactive*, *active*, and *partial-active* (traditional automaton only has inactive and active states). For each AND-End state $q_e$ in $\mathcal{A}$, we set a global counter for $q_e$. In the function UpdateStates (line 7), when $q_e$ is activated, we add the counter of $q_e$. Only $q_e$'s counter equals the parameter $d$ of $q_e$ (i.e., the number of branches defined in Definition 3), the state of $q_e$ will be changed to *active*, otherwise, it is assigned a *partial-active* state. As aforementioned, only $q_e$'s state is *active*, we will process the successor states of $q_e$, so the algorithm backtracks when an AND-End state is partially activated (line 11).

Besides, since the counter for an AND-End state $q_e$ is global, we need to reset this counter when matching different paths. For the example in Fig. 4, the algorithm traverses the graph starting from $v_1$, $v_2$, $v_3$, $v_5$, $v_8$, and $v_{10}$, the counter should be reset to 0 for each traversal so that correct results are matched. Therefore, the counter is reset under two cases: a successful matching (line 10) or a failed matching (lines 14–15).

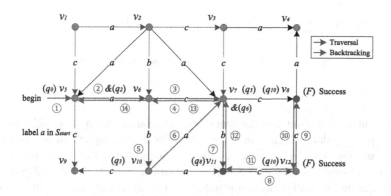

**Fig. 4.** An example of SNFA matching on the graph.

**Example 3.** *Figure 4 shows the example of SNFA matching, the SNFA matching starts from the edge labeled with a. Consider the starting node $v_5$, the set of active states related to each node changes as the depth-first traversal is performed. The structure consisting of nodes $v_6$, $v_7$, and $v_{10}$ is matched by the states between $q_2$ and $q_6$. The circled number indicates the order of accessed labels in the depth-first traversal. Eventually, the final states are activated at nodes $v_8$ and $v_{12}$, corresponding to the two matching results.*

## 5.2   Selecting Good Starting Nodes

Since the SNFA matching starts from the initial node, the algorithm SNFAMatch always matches the subpaths in the graph starting from the edges whose labels equal the first label of SNFA. When the first label of SNFA occurs frequently in the graph, SNFAMatch conducts lots of depth-first traversals which leads to inefficient matching. To solve this issue, we design a method of selecting "good" starting nodes in the graph so that the depth-first traversals are conducted as few as possible. Figure 5 shows the example when the SNFA matching starts from labels a and b in a graph, where the label a results in 8 starting nodes against 3 starting nodes produced by the label b. Obviously, it is more efficient if the SNFA matching starts from the label b.

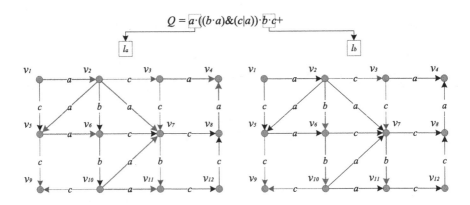

**Fig. 5.** The SNFA matching starts from different nodes.

*(1) Selecting A Starting Unit*

Intuitively, in order to minimize the depth-first traversals as much as possible, the label from the query $Q$ that has the minimal occurrences in the graph $G$ should be used to locate the starting nodes in $G$. Here, we define the *unit* of the SRPQ $Q$ as a label(a *single unit e*) or a sub-regex(a *regex unit L* supporting **Join**, **Select**, and **AND** operators), which is matched by each matching result of $Q$. For the closure unit $e^+$ in $Q$, since $e$ occurs at least one time in the result, we consider $e$ as a single unit. For the running example, we get 3 single units $a$, $b$, and $c$, as well as a regex unit $((b \cdot a)\&(c \mid a))$, as shown in Fig. 6.

Next, we compare the number of occurrences of the units for the query $Q$ and select the unit with the least occurrences as the starting unit for SNFA matching. For the single unit $e$, the occurrences can be obtained directly by the inverted index since the inverted list $l_e$ records the edges (node pairs) where $e$ appears in $G$. As shown in Fig. 6, the inverted lists are obtained from the inverted index for the single units $a$, $b$, and $c$.

For the regex unit, the occurrences in $G$ cannot be obtained from the inverted index since it is a simple sub-regex, rather than a label. To solve this problem, we design a list merging-based method to compute the occurrences for the regex unit.

We use a similar idea of building a plan tree to compute the occurrences for a regex unit [15]. The plan tree is a binary tree structure where the internal node indicates the operator and the leaf node represents the occurrence list of the label in the regex unit. The list merging method is performed based on the plan tree in a bottom-up fashion.

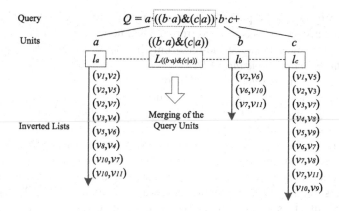

**Fig. 6.** Different units for the SRPQ $Q$.

For example, Fig. 7 shows the plan tree for the running example, the plan tree represents the regex unit $((b \cdot a)\&(c \mid a))$.

For each internal node $v$, the different list merging methods will be performed for the different operators of $v$. Let $l_L$ and $l_R$ be the occurrence list for the left and right children of $v$, the occurrence list of $v$ is computed by merging $l_L$ and $l_R$ in the following way. i) $v$ is the **Join** operator, for the node pairs $(v_m, v_n) \in l_L$ and $(v_i, v_j) \in l_R$, if and only if $v_n = v_i$, then $(v_m, v_j)$ is the node pair in the list of $v$; ii) $v$ is the **Select** operator, for the node pairs $(v_m, v_n) \in l_L$ and $(v_i, v_j) \in l_R$, both of $(v_m, v_n)$ and $(v_i, v_j)$ are the node pairs in the list of $v$; iii) $v$ is the **AND** operator, for the node pairs $(v_m, v_n) \in l_L$ and $(v_i, v_j) \in l_R$, if and only if $v_m = v_i$ and $v_n = v_j$, then $(v_m, v_n)$ is the node pair in the list of $v$. In this way, the list computed from the root of the plan tree is the occurrence list of the regex unit.

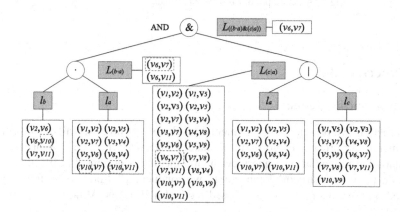

**Fig. 7.** Computing the occurrences of the regex unit.

As shown in Fig. 7, the plan tree generated by the sub-regex $((b \cdot a)\&(c \mid a))$ is given. The occurrence list is computed from the leaf nodes to the root node. The result list $l = \{(v_6, v_7)\}$ means that the sub-regex only has one occurrence $(v_6, v_7)$ in $G$.

*(2) Bi-directional matching using SNFA*

When the starting unit is selected and the corresponding occurrences in the graph are obtained, we need to conduct the SNFA matching starting from these occurrences. However, the starting unit may exist in the middle of the query, which results in the algorithm SNFAMatch cannot be directly applied. Next, we show this issue can be easily solved by conducting a bi-directional matching using SNFA, called SNFABiMatch.

For the case of starting unit existing in the middle of $Q$, $Q$ is divided into left and right sub-queries, denoted by $Q_L$ and $Q_R$ (the starting unit is not included in $Q_L$ and $Q_R$). For the sub-query $Q_R$, we use the same way to build the SNFA $A_R$ and perform SNFAMatch with $A_R$ starting from the second node of occurrence of starting unit. For example, Fig. 8 shows the built SNFAs for $Q_L$ and $Q_R$, $A_R$ starts from $v_7$ to perform the SNFA matching. For the sub-query $Q_L$, we need to build a reversed SNFA $A_L$ for it, since the first node of the occurrence of starting unit indicates the end node of $Q_L$. Also, SNFAMatch runs with $A_L$ in a reversed direction on the graph starting from the first node of the starting unit occurrence, e.g., $A_L$ is the reversed SNFA for the subquery $a$ in Fig. 8, $A_R$ starts from $v_6$ to perform a reversed SNFA matching. Finally, when $A_L$ and $A_R$ find the matching results respectively, a full matching result for $Q$ can be composed by the occurrences of $A_L$, $A_R$, and the starting unit.

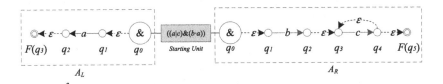

**Fig. 8.** An example of bi-directional matching using SNFA.

# 6    Experiments

## 6.1    Experimental Setting

In this section, we present the experimental results of our proposed algorithms. In the experiments, our proposed algorithms are named SNFAMatch and SNFABiMatch. An automaton-based regular path query evaluation method is presented in [9]. In order to match the paths with structural constraints, we conduct a post-processing for the paths matched by the existing automaton-based method and use NFAMatch+ to represent this method.

We use a real social network data set for experiments. The data set is extracted from the trusted network of Advogato [17] which is an online community platform for developers of free software collated by Stanford University's Large Network Data Set website. It contains 5,200 nodes (users), and 47,100 directed edges representing trust relationships. A user without any trusted certificate is called an observer. It is possible to trust oneself on Advogato, and therefore the network contains loops. A trusted link is called a certification on Advogato, and there are three different levels of certifications, corresponding to three different edge labels: certifications as *apprentice* (weight 0.6, referred to by label $a$), *journeyer* (weight 0.8, referred to by label $b$) and *master* (weight 1.0, referred to by label $c$).

We manually construct structural regular path queries that conform to the labels contained in the data set. According to features of structural path constraints, the queries are divided into four groups, each group contains 20 queries. As shown in Table 1, $Q_1$ to $Q_4$ are four groups of queries with different features.

**Table 1.** Settings for queries.

| Group | Samples | Features of structural path constraints | Quantity |
|-------|---------|------------------------------------------|----------|
| $Q_1$ | $abc(c\|a)$ | None | 20 |
| $Q_2$ | $a(b\&c)c(c\&a)a$ | Label-based | 20 |
| $Q_3$ | $a((ba)\&(c\|a))c(c\&a)a$ | Regex-based | 20 |
| $Q_4$ | $a((b\&a)\&(c\&a))c(c\&a)a$ | Nested forms | 20 |

## 6.2 Comparison of Matching Efficiency

The first experiment tests the matching efficiency of the comparative methods. For each query, the matching time includes the query parsing time and evaluation time. For the group of queries, we average the matching time for queries in a group. The experimental results are plotted in Fig. 9. We can see that SNFABiMatch achieves the best matching performance and SNFAMatch is the runner-up. For example, consider $Q4$ in Fig. 9(a), SNFABiMatch spends 665 ms against the time 793 ms, 866 ms used by SNFAMatch and NFAMatch+, respectively. Also, the comparative methods cost the least matching time for $Q_1$ and cost the most time for $Q_4$, the reason is that the queries in $Q_4$ are more complex than other queries, which results in more matching time.

## 6.3 Scalability

The second experiment tests the scalabilities of the comparative methods by varying the size of the data set. We randomly extract 10%, 30%, 50%, 80%, and 100% edges from the data set for testing the scalabilities. The results are shown in Fig. 10. We can see that the matching time of the methods scales linearly to the size of the data set. For example, for the queries in $Q_3$, SNFAMatch costs the matching time 675 ms, 1264 ms, 3117 ms, 6146 ms, and 7952 ms for 10%, 30%, 50%, 80%, and 100% data set, respectively.

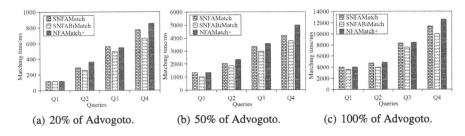

**Fig. 9.** Comparison of matching time on different methods.

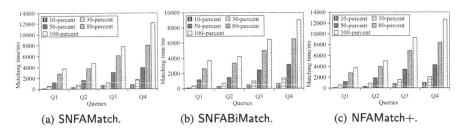

**Fig. 10.** Comparison of matching time on graphs with different sizes.

### 6.4  Cost of Parsing Queries

In this experiment, we test the cost of parsing queries for our proposed methods. For the method SNFAMatch, the query parsing time is the time of building SNFA, while SNFABiMatch includes not only the SNFA building but also the starting unit computation in the phrase of query parsing.

The experimental results are plotted in Fig. 11. No matter for which group of queries, SNFABiMatch always needs more parsing time than SNFAMatch since it needs to compute the starting unit for the queries and perform the bi-directional SNFA matching. Considering the different groups of queries, both SNFAMatch and SNFABiMatch need more parsing time for $Q_4$ than $Q_1$, since the more complex regular path query, more computations are required to build the SNFA.

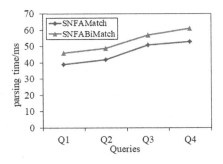

**Fig. 11.** Query parsing time for different queries.

# 7   Conclusion

In this paper, we study the problem of regular path query evaluation. Traditional evaluation methods for regular path queries cannot match the paths with structural constraints. We solve this issue by extending the classical regular path query with a new operator (AND) for the paths so that the matched paths meet the structural constraints defined by the AND operator. A new NFA-based structure is proposed to represent the extended regular path query, named SNFA. Also, the evaluation method is designed using SNFA by conducting depth-first traversals on the graph and an optimization technique of selecting good starting nodes is proposed to further improve the matching efficiency. Experimental results show that our method achieves high performance for evaluating regular path queries.

**Acknowledgements.** This work is partially supported by the National Natural Science Foundation of China (Nos. 62002245, 61802268), the Natural Science Foundation of Liaoning Province (Nos. 2022-BS-218, 2022-MS-303, 2022-MS-302).

# References

1. Arenas, M., Conca, S., Pérez, J.: Counting beyond a yottabyte, or how SPARQL 1.1 property paths will prevent adoption of the standard. In: World Wide Web Conference, pp. 629–638. ACM (2012)
2. Bizer, C., Vidal, M., Weiss, M.: Resource description framework. In: Liu, L., Özsu, M.T. (eds.) Encyclopedia of Database Systems. Springer (2018). https://doi.org/10.1007/978-1-4614-8265-9_905
3. Bornea, M.A., et al.: Building an efficient RDF store over a relational database. In: ACM SIGMOD Conference, pp. 121–132. ACM (2013)
4. Carletti, V., Foggia, P., Saggese, A., Vento, M.: Introducing VF3: a new algorithm for subgraph isomorphism. In: Foggia, P., Liu, C.-L., Vento, M. (eds.) GbRPR 2017. LNCS, vol. 10310, pp. 128–139. Springer, Cham (2017). https://doi.org/10.1007/978-3-319-58961-9_12
5. Han, S., Zou, L., Yu, J.X.: Speeding up set intersections in graph algorithms using SIMD instructions. In: ACM SIGMOD Conference, pp. 1587–1602. ACM (2018)
6. Holzer, M., Kutrib, M.: Nondeterministic finite automata - recent results on the descriptional and computational complexity. Int. J. Found. Comput. Sci. **20**(4), 563–580 (2009)
7. Koschmieder, A.: Cost-based optimization of regular path queries on large graphs. In: Grundlagen von Datenbanken. CEUR Workshop Proceedings, vol. 581. CEUR-WS.org (2010)
8. Koschmieder, A., Leser, U.: Regular path queries on large graphs. Sci. Stat. Database Manag. **7338**, 177–194 (2012)
9. Libkin, L., Vrgoc, D.: Regular path queries on graphs with data. In: International Conference on Database Theory, pp. 74–85. ACM (2012)
10. Ma, H., Zhou, D., Liu, C., Lyu, M.R., King, I.: Recommender systems with social regularization. In: Web Search and Web Data Mining Conference, pp. 287–296. ACM (2011)
11. Mendelzon, A.O., Wood, P.T.: Finding regular simple paths in graph databases. In: International Conference on Very Large Data Bases, pp. 185–193. Morgan Kaufmann (1989)
12. Mhedhbi, A., Salihoglu, S.: Optimizing subgraph queries by combining binary and worst-case optimal joins. VLDB J. **12**(11), 1692–1704 (2019)
13. Neumann, T., Weikum, G.: RDF-3X: a RISC-style engine for RDF. VLDB J. **1**(1), 647–659 (2008)

14. Peng, P., Zou, L., Özsu, M.T., Chen, L., Zhao, D.: Processing SPARQL queries over distributed RDF graphs. VLDB J. **25**(2), 243–268 (2016)
15. Qiu, T., Yang, X., Wang, B., Wang, W.: Efficient regular expression matching based on positional inverted index. IEEE Trans. Knowl. Data Eng. **34**(3), 1133–1148 (2020)
16. Reutter, J.L., Romero, M., Vardi, M.Y.: Regular queries on graph databases. Theory Comput. Syst. **61**(1), 31–83 (2017)
17. Rossi, R.A., Ahmed, N.K.: The network data repository with interactive graph analytics and visualization. In: Proceedings of the Twenty-Ninth AAAI Conference on Artificial Intelligence (2015)
18. Santos, F.C., Costa, U.S., Musicante, M.A.: A Bottom-Up Algorithm for Answering Context-Free Path Queries in Graph Databases. In: Mikkonen, T., Klamma, R., Hernández, J. (eds.) ICWE 2018. LNCS, vol. 10845, pp. 225–233. Springer, Cham (2018). https://doi.org/10.1007/978-3-319-91662-0_17
19. Sun, S., Luo, Q.: In-memory subgraph matching: an in-depth study. In: ACM SIGMOD Conference, pp. 1083–1098. ACM (2020)
20. Wang, M., et al.: PDD graph: bridging electronic medical records and biomedical knowledge graphs via entity linking. In: d'Amato, C., et al. (eds.) ISWC 2017. LNCS, vol. 10588, pp. 219–227. Springer, Cham (2017). https://doi.org/10.1007/978-3-319-68204-4_23
21. Weiss, C., Karras, P., Bernstein, A.: Hexastore: sextuple indexing for semantic web data management. VLDB J. **1**(1), 1008–1019 (2008)
22. Wi, S., Han, W., Chang, C., Kim, K.: Towards multi-way join aware optimizer in SAP HANA. VLDB J. **13**(12), 3019–3031 (2020)
23. Wilkinson, K., Sayers, C., Kuno, H.A., Reynolds, D.: Efficient RDF storage and retrieval in jena2. In: Semantic Web and Databases, pp. 131–150 (2003)
24. Zeng, L., Zou, L., Özsu, M.T., Hu, L., Zhang, F.: GSI: GPU-friendly subgraph isomorphism. In: International Conference on Data Engineering. pp. 1249–1260. IEEE (2020)
25. Zhang, X., den Bussche, J.V.: On the power of SPARQL in expressing navigational queries. Comput. J. **58**(11), 2841–2851 (2015)
26. Zhang, X., Feng, Z., Wang, X., Rao, G., Wu, W.: Context-Free Path Queries on RDF Graphs. In: Groth, P., et al. (eds.) ISWC 2016. LNCS, vol. 9981, pp. 632–648. Springer, Cham (2016). https://doi.org/10.1007/978-3-319-46523-4_38
27. Zou, L., Mo, J., Chen, L., Özsu, M.T., Zhao, D.: Gstore: answering SPARQL queries via subgraph matching. VLDB J. **4**(8), 482–493 (2011)

# EnSpeciVAT: Enhanced SpecieVAT for Cluster Tendency Identification in Graphs

Siqi Xia[1]([✉]) ⓘ, Sutharshan Rajasegarar[1] ⓘ, Christopher Leckie[2] ⓘ, Sarah M. Erfani[2] ⓘ, Jeffrey Chan[3] ⓘ, and Lei Pan[1] ⓘ

[1] Deakin University, Geelong, Australia
{xiasiq,sutharshan.rajasegarar,l.pan}@deakin.edu.au
[2] The University of Melbourne, Melbourne, Australia
{caleckie,sarah.erfani}@unimelb.edu.au
[3] RMIT University, Melbourne, Australia
jeffrey.chan@rmit.edu.au

**Abstract.** Clustering is a process of finding groups of similar objects in a given dataset. Finding clusters in graphs, especially crisp clusters, which have minimal or no overlapping clusters, is challenging. Further, clustering is an ill-defined problem, resulting in multiple possible solutions for the same dataset. Hence, a challenge here is that the possible number of crisp clusters that can be found for a given graph will not be unique. Finding different crisp clusters is useful for modelling patterns or cluster-based prediction tasks. The visual assessment of the clustering tendency (VAT) algorithm, in particular the SpecieVAT algorithm, has been used in the past to identify the number of clusters, i.e., the cluster tendencies, that exist in a graph. These algorithms generate an image of the reordered dissimilarity matrix, and the dark diagonal blocks in the main diagonal reveal the number of clusters that exist in the data. However, this method often fails to show the possible crisp clusters in the graph. We propose a novel algorithm, called EnSpeciVAT, which significantly enhances the SpecVAT algorithm in the context of crisp cluster generation for graph data. It incorporates a fuzzy c-means mechanism to guide the process of extracting crisp clusters in the graph. Our evaluation of six different graph data demonstrates that the EnSpeciVAT can find clear and crisp clusters in graph data.

**Keywords:** Clustering · Graph Clustering Visualisation · Visual Assessment of Tendency

## 1 Introduction

Cluster tendency is the problem of assessing if there exist any clusters in a given dataset. It can help assess the existence of clusters in the data set. The two main methods used to make these assessments are statistical and visual. Statistical methods include the Hopkins test [4], while visual methods include Visual

© The Author(s), under exclusive license to Springer Nature Switzerland AG 2023
X. Yang et al. (Eds.): ADMA 2023, LNAI 14178, pp. 323–337, 2023.
https://doi.org/10.1007/978-3-031-46671-7_22

Assessment of Tendency (VAT) and Improved Visual Assessment of Tendency (iVAT) [20] methods. The VAT algorithm generates possible clusters by reordering the pairwise dissimilarity matrix for a reordered dissimilarity image (RDI). The RDI shows dark blocks along the main diagonal. These can be regarded as the possible clusters in the original data [7,14] (see Fig. 1).

Clustering for graph data has attracted increasing interest in recent years [13]. Graph clustering includes graph cutting and partitioning methods. Spectral methods are a widely used way to facilitate solving graph problems by applying the eigen-decomposition to the node relation matrix of the graph, for example, the Laplacian matrix. Spectral methods can be effective due to their efficiency and ability to provide lower and upper bounds in graph cutting [13]. Spectral methods can also be used with VAT and iVAT to deal with graph data. Combining spectral methods and VAT/iVAT is challenging because calculating distances between vertices to form the dissimilarity matrix in graph data may yield different results. The SpecVAT algorithm provides an effective way of clustering graph data [19]. SpecVAT takes a graph as input, uses the spectral decomposition of the graph adjacency matrix to form a dissimilarity matrix, and then reorders the dissimilarity matrix to show potential clusters for the input graph.

Crisp cluster blocks are defined as clusters with clear boundaries on the original dataset. Crisp cluster blocks can also be visualized in iVAT greyscale images when the boundaries' clarity indicates the cluster crispness level. For example, Fig. 1(a) shows a sample graph, and Fig. 1(b) has clear cluster boundaries in the iVAT image, showing that this clustering method can generate clear clusters. But Fig. 1(c) shows an example without crisp cluster blocks. The crisp cluster blocks are useful for several reasons. Firstly, it is easy to generate separated clusters from the crisp cluster blocks without ambiguity. Furthermore, crisp clusters can facilitate the identification of clusters in datasets with overlapping elements. As in Fig. 2(a), there are overlapping nodes in the graph that determining clusters for the graph can be challenging. With the help of crisp clusters in Fig. 2(b), identifying the anticipated clusters becomes effortless. Thirdly, the crisp cluster blocks can help to find clear nodes distributions in clusters on the original data. This will help to design machine learning-based modelling for each cluster of data separately, which can capture the patterns of behaviour in each cluster effectively. For example, in [1], different deep learning prediction models have been fitted based on pre-determined clusters separately to improve the overall prediction of future energy consumption. Finding crisp clusters in this situation will lead to accurate modelling of the unique patterns of behaviour that exist in each well-separated crisp cluster in the data. Thus, clustering methods that can determine crisp cluster blocks are likely to be more desirable for cluster generation.

As an unsupervised method, the optimal number of clusters for different datasets may not be unique. Different numbers of clusters will generate different cluster patterns from the graph. In many applications, the number of clusters ($k$) is not firmly determined, for example in the Zachary's Karate Club Graph [23].

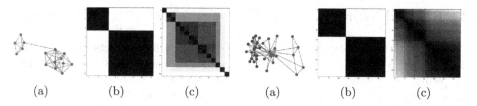

**Fig. 1.** Example graph with (a) An example graph (b) Crisp cluster blocks (c) Non-crisp cluster blocks

**Fig. 2.** Example graph contains overlapping with (a) An example graph (b) Crisp cluster blocks (c) Non-crisp cluster blocks

Different cluster numbers will be suitable for different usage. Previous studies prove that changes in the number of clusters affects the generated cluster patterns. For example, according to [3], the generated cluster patterns with different cluster numbers $k$ are not always able to provide crisp cluster blocks. To generate crisp clusters with respect to different $k$, we propose a novel algorithm named EnSpeciVAT.

We propose a novel clustering algorithm by enhancing the SpecieVAT, named EnSpeciVAT, to generate crisp clusters with changing clustering numbers. EnSpeciVAT has a novel way to generate the weight matrix of graph data based on neighbourhood strength and include fuzzy-c-means (FCM) as guidance in generating cluster blocks. These new designs contribute to generating more easily identified crisp blocks. The different numbers of clusters vary with respect to different implementation requirements. EnSpeciVAT can generate clusters with clear boundaries to facilitate the quick determination of node clusters for various applications. This paper makes the following contributions:

- We design and implement a novel method to generate the weight matrix for graph-structured data in EnSpeciVAT. Neighbourhood relations for nodes and between nodes are used to propose a novel distance measure to estimate the node distances for graph data.
- We adopt fuzzy clustering methods like fuzzy-c-means (FCM) in EnSpeciVAT during the crisp cluster generation process. FCM helps to build the dissimilarity matrix between nodes.
- We thoroughly evaluate EnSpeciVAT on various datasets to evaluate the effectiveness of crisp clustering generation. The distribution of nodes according to the generated clusters is further analysed.

The rest of this paper is organized as follows: Sect. 2 presents the related work. Section 3 elaborates on the three main components of EnSpeciVAT. Section 4 provides the details of our empirical study of EnSpeciVAT to evaluate its effectiveness for crisp cluster generation, and the paper is concluded in Sect. 5.

## 2  Related Work

Clustering is the process of assigning labels to unlabelled objects in $X$ to specific groups [3]. A fuzzy-c-means (FCM) approach uses a fuzzy technique for clustering. FCM generates c-partitions of the data, and each object is assigned a membership degree to each class based on the c-partition. The cluster with the largest membership value can be considered the most plausible cluster, which can be further used in other scenarios. We can extract specific clusters from the results of the FCM method.

Visual Assessment of Tendency (VAT) and Improved Visual Assessment of Tendency (iVAT) are visual methods to find the clustering tendency in data. VAT and iVAT determine the clusters based on the dark blocks along the diagonal in a reordered dissimilarity image (RDI) or cluster heat map after reordering the pairwise dissimilarity matrix [7]. VAT and iVAT do not require prior information about the cluster number. iVAT is an improved version of VAT. It takes the outcome re-ordered weight matrix from VAT and made further re-ordering combining VAT with a path-based distance transform. VAT has seen applications in many recently explored research fields like health [11,17], and social media [12,16].

Many datasets, such as social networks and network flow data, can be represented in a graph structure. Clustering on a graph can be challenging as the normal clustering methods based on distance measures can be difficult with respect to the graph structure. Spectral graph clustering is a traditional way to deal with the graph structure. In [24], spectral clustering has been used on graph convolutional neural networks [24]. Multiview clustering is an important method in partitioning data into different groups based on their heterogeneous features [8]. This method can be applied to contrastive graph clustering [15] and attributed graph [10]. There are additional enhancements based on the multiview method, like filter-based clustering [9]. There are also deep learning approaches for graph clustering, especially applied to deep graph neural networks. In [18], a goal-directed deep learning approach has been proposed to cluster attributed graphs.

Applying the spectral methods on graph clustering based on VAT leading to the algorithm SpecVAT as in Algorithm 1 [13]. SpecVAT and SpecieVAT have been used to study the fuzziness of graph data [3,6]. In this paper, we propose a novel spectral-based algorithm that will help to accurately find the possible crisp clusters that exist in graph data.

## 3  Methodology

A novel spectral-based method, called EnSpeciVAT, is proposed to find crisp clusters in graphs. It improves the SpecieVAT algorithm presented in Algorithm 1 and incorporates the FCM mechanism to arrive at a general method to produce crisp clusters for various graph data sets.

---

**Algorithm 1.** The Original SpecieVAT

---

**Require:** Graph structured data denoted by $G$

   **procedure** GENERATE THE DISSIMILARITY MATRIX

      Calculate the weight matrix $W$

      Calculate the eigenvalues $\{\lambda_1, \lambda_2, \ldots, \lambda_n\}$ and eigenvectors $\{\mathbf{v}_1, \mathbf{v}_2, \ldots, \mathbf{v}_n\}$ for the weight matrix $W$

      Choose the first $k$ eigenvectors $\{\mathbf{v}_1, \mathbf{v}_2, \ldots, \mathbf{v}_k\}$ ($k < n$) based on the decreasing eigenvalues

      Construct a new eigenvector matrix $\tilde{V} \in R^{n \times k}$ by stacking the eigenvectors as $\{\tilde{\mathbf{v}}_1, \tilde{\mathbf{v}}_2, \ldots, \tilde{\mathbf{v}}_k\}$, where $\tilde{\mathbf{v}}_i \in R^k$

      Normalise the row instances from $\tilde{V}$ as $\tilde{V}^*$, where $\tilde{\mathbf{v}}_i^* = \tilde{\mathbf{v}}_i/|\tilde{\mathbf{v}}_i|, \tilde{\mathbf{v}}_i \in \tilde{V}$

      Calculate the dissimilarity between nodes $u$ and $v$ as $d_{uv} = s(u,v) = |\tilde{\mathbf{v}}_u - \tilde{\mathbf{v}}_v|$. The dissimilarity is the element in the dissimilarity matrix $D$

      Output the dissimilarity matrix $D$

   **end procedure**

   **procedure** GENERATE iVAT

      Apply the iVAT algorithm to the dissimilarity matrix to obtain the new reordered matrix

      Plot the greyscale image accordingly

   **end procedure**

---

## 3.1 New Definition for Weight Matrix

The proposed EnSpeciVAT uses the newly defined weight matrix $\mathbf{W}$. Each element of the weight matrix $\mathbf{W}$ is defined as the distance between the nodes calculating from the node strength of the neighbourhood. Let $u, v$ denote two nodes in the graph, the element of weight matrix $\mathbf{W}$ can be denoted as $W(u, v)$. In defining the weight matrix element, we calculate the node strength first.

We use $C(u, v)$ to denote the set of common vertices of $u$ and $v$. $M(u)$ represents the set of vertices of $u$ that are not vertices of $v$ (except $v$), $M(v)$ is the set of vertices of $v$ that are not vertices of $u$ (except $u$). Following [22], we can define the node strength between vertices $u$ and $v$ ($St(u, v)$) in Eq. 1:

$$\begin{aligned}
St(u,v) &= s(M(u), C(u,v)) + s(M(v), C(u,v)) \\
&+ s(C(u,v)) + s(M(u), M(v)) \\
&+ |C(u,v)|/(|M(u)| + |C(u,v)| + |M(v)|)
\end{aligned} \tag{1}$$

In this equation, $s(A, B) = r(A, B)/|A||B|$, where $r(A, B)$ is the number of edges connecting nodes in set A to nodes in set B. In addition, $s(A) = 2r(A, A)/(|A|(|A|) - 1)$ denotes the proportion of edges that connect as a set to themselves. $|A|$ denotes the number of nodes in the set. The aforementioned capability can be employed to establish the distance between nodes, as denoted by $D(u, v) = 1/St(u, v)$, which signifies the connection between these two nodes and the weight matrix element is also defined from this distance $W(u, v) = D(u, v)$.

This distance metric is employed to construct the weight matrix in the EnSpeciVAT algorithm, where each element of the weight matrix represents the strength of the relationship between nodes $u$ and $v$. The initial weight matrix is redefined based on the links and relationships among nodes, which serve as the foundation for determining the final clustering of nodes in the graph, which is influenced by their neighbors and the edges connecting them.

## 3.2   FCM Guide

The Fuzzy C-Means (FCM) method has been introduced in our algorithm to provide guidance to achieve clearer and more informative cluster patterns to generate crisp cluster boundaries. We apply FCM to pre-calculate a fuzzy cluster for each node based on its spectral vectors. The clusters, together with the centres, will be used to discern the dissimilarity between the nodes. The newly defined dissimilarity matrix can sharpen the clustering boundaries produced by the SpecieVAT. Hence this is used with the SpecieVAT for re-ordering and cluster generation. The FCM algorithm has been utilized due to its ability to mitigate the overlaps in the graph and produce more crisp clusters. Unlike other clustering methods such as hard k-means, FCM can effectively identify clusters for fuzzy nodes and minimize the impact of overlaps on clustering outcomes. In comparison to other fuzzy clustering methods, FCM is simple to implement and provides better guidance for reducing fuzziness and generating crisp clusters.

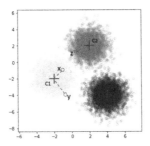

**Fig. 3.** FCM Guidance Value Calculation.

Figure 3 shows an example for the calculation of dissimilarities between different instances. $c_1$, $c_2$, and $c_3$ are clusters calculated from fuzzy c means and $x$, $y$, and $z$ are three instances. If we calculate the FCM value between instances $x$ and $y$ from the same FCM cluster, the FCM value will be $|(x^2 + y^2) - 2c_1^2|$. In contrast, if we calculate the FCM value between $y$ and $z$, the FCM value will be the Euclidean distance between $y$ and $z$, denoted as $\|y, z\|$, if they belong to different FCM clusters.

In summary, the dissimilarity between two different instances, $d(x, y)$ for different scenarios can be written as:

$$d(x, y) = \begin{cases} |(x^2 + y^2) - 2c^2|, & x, y \text{ from the same cluster,} \\ \|x, y\|, & x, y \text{ from different clusters.} \end{cases} \tag{2}$$

Our experiments show that calculating dissimilarity for the same nodes from the same and different clusters will yield different clustering results. We can provide much sharper clusters in SpecieVAT greyscale images if we have a small value for the same cluster scenario and a large value for different cluster scenarios. We aim to define the dissimilarity to provide interpretable clustering patterns with an upper bound for the same clustering scenario and a lower bound for different clustering scenarios.

Hence, we can define the difference between these two scenarios as $\mathbf{dist}(x, y)$. According to our trial experiments, this difference can be given as:

$$\mathbf{dist}(x, y) = |xy - 2c^2| \tag{3}$$

using the definition of dissimilarity in Eq. 2. The difference is small in all the trial experiments, which allows us to balance dissimilarity and clustering.

### 3.3 EnSpeciVAT Algorithm

The algorithm for EnSpeciVAT is listed in Algorithm 2. The novel algorithm embeds the neighbourhood strength and FCM in the SpecieVAT image to generate crisp clusters. After we determined dissimilarity matrix, we further re-order it according to the iVAT algorithm. In [2], a dissimilarity measure based on paths was proposed that if two objects are situated at a considerable distance from each other (reflected by a large value of the distance metric), but a path exists that connects them via other objects, such that the distances between each successive object in the path are small, then the value of the distance should be reduced to a smaller value to account for this connection. Following this reordering procedure, iVAT can offer an improved approach to visualizing the clustering tendency and generate more accurate clusters. This algorithm can help reduce the influence on crispness from the number of clusters.

## 4 Experiments

### 4.1 Dataset

Three types of data have been used in the experiments to obtain the results to investigate the algorithm's performance.

**Simple Graphs.** First of all, two simple synthetic graphs are included to evaluate how the algorithm reacts to simple data (like Fig. 4(a) and Fig. 4(b)). The first graph is a simple ten-node graph divided into two parts. Within each part, the nodes are strongly connected; between these two parts, only a single edge weakly connects them. The second graph is slightly more complex in structure. For some nodes, the connections are complex, which can be considered more compact, while some nodes have weak connections with other parts.

---

**Algorithm 2.** EnSpeciVAT:

---

**Require:** Graph structured data $G$

  **procedure** GENERATE DISSIMILARITY MATRIX

    Calculate the neighbourhood strength between nodes as the weight matrix $W$ as in Eq. 1

    Calculate the eigenvalues $\{\lambda_1, \lambda_2, \ldots, \lambda_n\}$ and eigenvectors $\{\mathbf{v}_1, \mathbf{v}_2, \ldots, \mathbf{v}_n\}$ for the weight matrix $W$

    Choose the first $k$ eigenvectors $\{\mathbf{v}_1, \mathbf{v}_2, \ldots, \mathbf{v}_k\}$ $(k < n)$ based on the decreasing eigenvalues

    Build up as new eigenvector matrix $\tilde{V} \in R^{n \times k}$ by stacking the eigenvectors as $\{\tilde{\mathbf{v}}_1, \tilde{\mathbf{v}}_2, \ldots, \tilde{\mathbf{v}}_k\}$, where $\tilde{\mathbf{v}}_i \in R^k$

    Determine the $k$ fuzzy clusters from the eigenvector matrix $\tilde{V}$ using FCM

    Determine the centres as $\mathbf{c} = c_1, c_2, \ldots, c_k$ from the FCM final updated centres after iterations

    Each instance from $\tilde{V}$ will be assigned to one cluster based on the maximum probability of FCM updates

    **for** <nodes $u$ and $v$ from graph> **do**

      Calculate the dissimilarity between nodes $u$ and $v$ from the $u$-th and $v$-th rows of the eigenvector matrix $V$ and centers $\mathbf{c}$ based on $\mathbf{dist}(u, v)$ as the entries for dissimilarity matrix $D$

    **end for**

    Normalise the dissimilarity matrix $D$

    Output the dissimilarity matrix $D$

  **end procedure**

  **procedure** GENERATE GREYSCALE IMAGE

    Reorder the $D$ according to iVAT

    Draw the greyscale image

  **end procedure**

---

**Community Graphs.** To investigate the performance of the algorithm on community graphs, the community graphs (Fig. 4(c) and Fig. 4(d)) with clear community patterns are included. The two synthetic community graphs are the windmill and caveman [21] graphs. Both graphs have six distinct patterns of connections between the nodes. In the case of the windmill graph, the patterns are connected by a node, and in the case of the caveman graph, the patterns are connected by edges.

**Real-World Graphs.** Finally, to test the algorithm's performance in real-world and more complex situations, two additional real-world graphs were used, the Zachary's Karate Club [23] graph (Fig. 4(e)), and the Les Miserables character [5] graph (Fig. 4(f)). Zachary's Karate Club graph is a widely used graph data set to study fuzzy communities. There is still disagreement about the exact number of clusters for this dataset. The Les Miserables character graph is generated from the relationships between the characters in the novel *Les Miserables*.

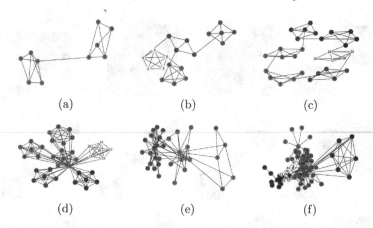

**Fig. 4.** Graph Datasets Used in Experiments with (a) Simple Graph 1 (b) Simple Graph 2 (c) Community Graph 1 (d) Community Graph 2 (e) Zachary's Club Graph (f) Les Miserables Graph.

## 4.2   Experiment Settings

The experiments are conducted on six datasets, with the number of clusters varying from 2 to 5. Two different results are generated and examined. Firstly, the clustering greyscale images from the EnSpeciVAT are displayed to show the final clustering results. These images demonstrate how well the crisp clusters are generated using the proposed method. These results are compared with the previous work in [3]. Secondly, we use the spectral method from the clustering results to determine the actual clusters for the graph nodes that reflect the cluster distributions on the original datasets. It allows us to see how these sharp clusters are distributed on the dataset, i.e., it helps us to interpret the clustering better.

## 4.3   Experiment Results

**Generated Clusters Based on SpecieVAT Images** According to the images shown in Fig. 5, EnSpeciVAT can generate crisp clusters for different datasets generally. For the synthetic sample data sets, Simple Graph 1 shows some fuzzy parts for $k$ greater than two. This is because Simple Graph 1 has a simple and clear cluster of the original data set. The fuzzy and ambiguous clustering has resulted from forcing the simple dataset to be split into a large number of clusters. Although the improved method can generate some clear clusters, some fuzzy parts remain. The improved method can generate crisp clusters according to the required number of clusters as for Simple Graph 2. The method works well with data that has clear patterns. It can divide the data into crisp clusters.

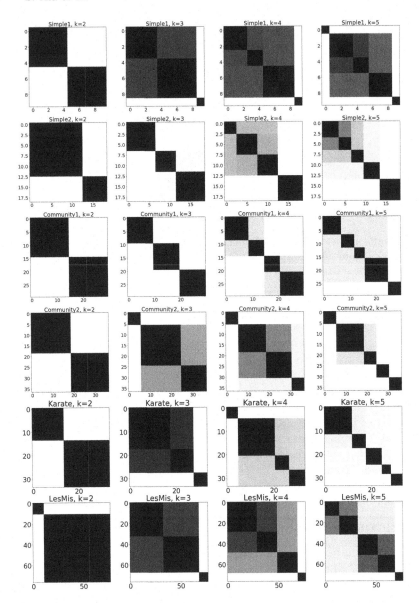

**Fig. 5.** Greyscale Images with Generated Clusters for Different Datasets with Cluster Numbers 2–5.

The results are similar for the two data sets of the community graphs. Both community graphs have clear initial clustering patterns, and the generated greyscale images can provide clear clusters according to the pre-defined clustering numbers. EnSpeciVAT also produces clear clusters for the graph data with different cluster numbers for the real-world datasets—Zachary's Karate

Club data, and the Les Miserables Character data. The cluster for $k = 3$ may have some fuzzy parts for both data sets. However, it can still produce some clear clusters. For $k = 2, 4, 5$, the clusters generated from the greyscale image are crisp. Hence, the corresponding clusters are easily determined for further use from these crisp blocks.

These greyscale images demonstrate that EnSpeciVAT found clear clusters with different numbers of clusters and with less ambiguity in the clustering than the exiting methods. EnSpeciVAT works for synthetic graph data with clear patterns and for real-world data sets.

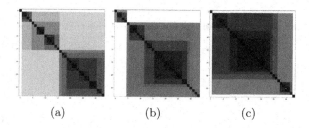

(a)                    (b)                    (c)

**Fig. 6.** SpecieVAT images with different cluster numbers (a) k = 2 (b) k = 3 (c) k = 4, in study [3] for Zachary's Karate Club graph.

**Cluster Distributions on the Original Graph Data.** According to [3], the experiments are conducted on Zachary's Karate Club graph. In [3], the Laplacian matrix has been used as the weight matrix on the original SpecieVAT method. In addition, different numbers of clusters have been applied and discussed how they could influence the clustering results. Figure 6, indicates that crisp clusters cannot be easily formed using the existing methods when the cluster numbers vary. That is, we have to pre-determine the optimised cluster numbers for clear clusters. Hence, the cluster number needs to be fixed.

Figure 7 illustrates the distribution of clustered nodes in the graph datasets. In general, the proposed EnSpeciVAT method provides very clear and comprehensive clustering results for generated graph data with clear patterns and clusters. Taking Simple Graph 2, Community 1 and 2 as examples, when the number of clusters required increases, the method provides a clear number of clusters with clear boundaries, as seen from the output images. The clusters are consistent with the original cluster patterns when looking at the corresponding clusters in the original graph. For smaller clustering numbers, nodes and communities with more links are likely to be clustered together, and as the number changes, the larger clusters split into smaller ones but do not violate the original patterns. EnSpeciVAT can provide clean clusters, and its results are close to the original graph cluster patterns for these generated data with known patterns.

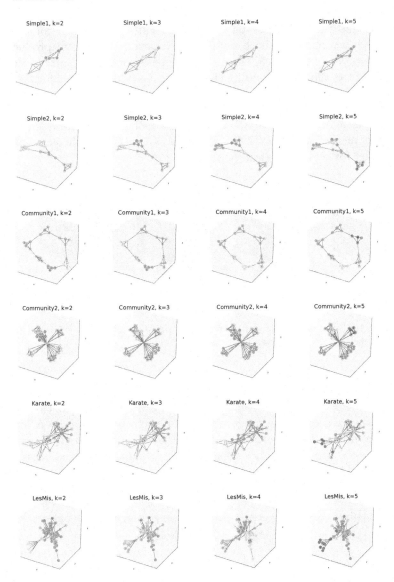

**Fig. 7.** Vertices cluster distributions on original datasets with Cluster Number 2–5

Simple Graph 1 seems to yield an overly complex situation when the cluster numbers go to a relatively high level as a generated graph with clear patterns. When the cluster counts are low, the method can show clear clusters; while for high cluster counts, the clusters generated are limited to one or two nodes. This is because the number of clusters will be too large for the number of nodes in the original graph. Although greyscale can still provide clean clusters, the results may not be as usable.

When we look at real-world data, such as Zachary's Karate Club and Les Miserable Character data, the distribution analysis for various cluster numbers can become more complex. We can generate the clusters we need from the results of the greyscale images because there are clear clusters with different cluster numbers. The proposed method EnSpeciVAT still provides clean clusters based on the required cluster numbers when faced with messy, difficult, and uninterpretable data. Looking at the distributions on the graph, for Zachary's Karate Club data, as the number of clusters increases, the clusters generated may be less intuitive compared to when $C = 2$ or $C = 3$. For Zachary's Karate Club data, many researchers originally believed that this graph should be decomposed into two clusters, but many studies disagree [3]. In our results, the crisp clusters are still available even with a higher number of clusters. However, the algorithm tends to cluster nodes with the same edge connections instead of creating clusters for nodes that are located together. This is because we use the neighbourhood strength to build the weight matrix, implying that the nodes with similar connections are more likely to be clustered together.

We investigate the Les Miserables data. This data has more complex edges between nodes and less fixed community definitions than Zachary's Karate Club data. We can observe that the clusters generated are crisp and easy to determine from the greyscale image. We can generate clusters with a defined number of clusters directly from the image. Examining the graph distributions, there are different cluster patterns for different cluster numbers, and the patterns match the original character relationships to some extent. As depicted in the colour patterns of the distributed graph, the majority of characters that are generally believed to belong to the same community can be clustered together. While a definite ground truth to define appropriate clusters may not exist, the presented outcomes suggest possible clustering criteria. Moreover, by cross-checking with the original characters for the nodes, these clusters align with the original novel details. This example demonstrates the applicability of our method to non-fixed cluster criteria, and the resulting clusters with varying sizes can be beneficial for diverse analytical purposes in further investigations, given that the original node communities are not fixed.

**Modularity Index.** Modularity index is introduced in [3] as a validity index for partitions of graph. For a graph $G = (V, E, W)$, $V$ denotes a set of $n$ vertices, and $W$ denotes the weight matrix. The most popular form of modularity assumes that $W$ is organized as an $(n \times n)$ positive, symmetric edge weight matrix of $G$. Let $V$ be partitioned into $c$ crisp subsets of vertices (indices), say $V_1, V_2, \ldots, V_c$, the modularity index $Q_h$ can be defined as:

$$Q_h = \frac{1}{\|W\|} \sum_{k=1}^{c} \left( \sum_{i,j \in V_k} \left[ w_{ij} - \frac{m_i m_j}{\|W\|} \right] \right) \tag{4}$$

where $\|W\| = \sum_{j=1}^{n} \sum_{i=1}^{n} w_{ij}$, $m_i = \sum_{n}^{k=1} w_{ik}$. We use this to compare the quality of clusters between SpecieVAT and EnSpeciVAT for the selected real-world dataset in Table 1. Our results indicate that EnSpeciVAT for generating crisp clusters can yield high modularity index values for both datasets, when using different numbers of clusters. Furthermore, our approach can produce crisp clusters irrespective of any changes in the number of clusters considering large indices for various cluster numbers.

**Table 1.** Modularity index based on SpecieVAT and EnSpeciVAT for Karate and LesMis Dataset

| Dataset | SpecieVAT | | | | EnSpeciVAT | | | |
|---|---|---|---|---|---|---|---|---|
| | 2 | 3 | 4 | 5 | 2 | 3 | 4 | 5 |
| Karate | 1.01 | 0.96 | 0.85 | 0.77 | 0.46 | 0.82 | 0.56 | 0.37 |
| LesMis | 0.62 | 0.95 | 0.84 | 1.32 | 0.09 | 0.18 | 0.42 | 1.08 |

## 5  Conclusion

Generating crisp clusters across different graph datasets with varying cluster numbers is a challenging research problem. We propose a novel algorithm named EnSpeciVAT to enhance the SpecVAT algorithm. Neighbourhood strength and FCM are embedded in the dissimilarity matrix construction in EnSpeciVAT. It is the first time that the VAT and FCM methods have been combined for clustering. For various synthetic and real-world graph data, EnSpeciVAT generates crisp clusters with different cluster numbers. In the future, how the different crisp clusters in the graph data can be used for machine learning-based modelling and prediction will be investigated.

**Acknowledgements.** This research was funded by the Australian Research Council. ARC - Discovery Project, DP200101960.

## References

1. Fahiman, F., Erfani, S.M., Rajasegarar, S., Palaniswami, M., Leckie, C.: Improving load forecasting based on deep learning and K-shape clustering. In: Proceedings of the 2017 International Joint Conference on Neural Networks, pp. 4134–4141. IEEE (2017)
2. Fischer, B., Zöller, T., Buhmann, J.M.: Path based pairwise data clustering with application to texture segmentation. In: Figueiredo, M., Zerubia, J., Jain, A.K. (eds.) EMMCVPR 2001. LNCS, vol. 2134, pp. 235–250. Springer, Heidelberg (2001). https://doi.org/10.1007/3-540-44745-8_16
3. Havens, T.C., et al.: Clustering and visualization of fuzzy communities in social networks. In: Proceedings of the 2013 IEEE International Conference on Fuzzy Systems (FUZZ-IEEE), pp. 1–7. IEEE (2013)
4. Janofsky, J.S., McCarthy, R.J., Foistein, M.F.: The Hopkins competency assessment test: a brief method for evaluating patients' capacity to give informed consent. Psychiatr. Serv. **43**(2), 132–136 (1992)

5. Knuth, D.E.: The Stanford GraphBase: a platform for combinatorial computing. ACM (1993)
6. Komarasamy, G., Wahi, A.: Finding the number of clusters using visual validation VAT algorithm. Int. J. Eng. Technol. **5**(5), 3951–3957 (2013)
7. Kumar, D., Bezdek, J.C.: Visual approaches for exploratory data analysis: a survey of the visual assessment of clustering tendency (VAT) family of algorithms. IEEE Syst. Man Cybern. Mag. **6**(2), 10–48 (2020)
8. Li, X., Zhang, H., Wang, R., Nie, F.: Multiview clustering: a scalable and parameter-free bipartite graph fusion method. IEEE Trans. Pattern Anal. Mach. Intell. **44**(1), 330–344 (2020)
9. Lin, Z., Kang, Z.: Graph filter-based multi-view attributed graph clustering. In: Proceedings of the 2021 IJCAI, pp. 2723–2729 (2021)
10. Lin, Z., Kang, Z., Zhang, L., Tian, L.: Multi-view attributed graph clustering. IEEE Trans. Knowl. Data Eng. (2021)
11. Liu, Q., et al.: Assessing the global tendency of Covid-19 outbreak. MedRxiv, p. 2020-03 (2020)
12. Meng, L., Tan, A.-H., Wunsch II, D.C.: Adaptive Resonance Theory in Social Media Data Clustering. AIKP, Springer, Cham (2019). https://doi.org/10.1007/978-3-030-02985-2
13. Nascimento, M.C., De Carvalho, A.C.: Spectral methods for graph clustering-a survey. Eur. J. Oper. Res. **211**(2), 221–231 (2011)
14. Palaniswami, M., Rao, A.S., Kumar, D., Rathore, P., Rajasegarar, S.: The role of visual assessment of clusters for big data analysis: from real-world Internet of Things. IEEE Syst. Man Cybern. Mag. **6**(4), 45–53 (2020)
15. Pan, E., Kang, Z.: Multi-view contrastive graph clustering. Adv. Neural. Inf. Process. Syst. **34**, 2148–2159 (2021)
16. Rachunok, B.A., Bennett, J.B., Nateghi, R.: Twitter and disasters: a social resilience fingerprint. IEEE Access **7**, 58495–58506 (2019)
17. Rajendra Prasad, K., Mohammed, M., Noorullah, R.: Visual topic models for healthcare data clustering. Evol. Intel. **14**(2), 545–562 (2021)
18. Wang, C., Pan, S., Hu, R., Long, G., Jiang, J., Zhang, C.: Attributed graph clustering: a deep attentional embedding approach. arXiv:1906.06532 (2019)
19. Wang, L., Geng, X., Bezdek, J., Leckie, C., Kotagiri, R.: SpecVAT: enhanced visual cluster analysis. In: Proceedings of the 2008 8th IEEE International Conference on Data Mining, pp. 638–647. IEEE (2008)
20. Wang, L., Nguyen, U.T.V., Bezdek, J.C., Leckie, C.A., Ramamohanarao, K.: iVAT and aVAT: enhanced visual analysis for cluster tendency assessment. In: Zaki, M.J., Yu, J.X., Ravindran, B., Pudi, V. (eds.) PAKDD 2010. LNCS (LNAI), vol. 6118, pp. 16–27. Springer, Heidelberg (2010). https://doi.org/10.1007/978-3-642-13657-3_5
21. Watts, D.J.: Networks, dynamics, and the small-world phenomenon. Am. J. Sociol. **105**(2), 493–527 (1999)
22. Yang, S., Luo, S., Li, J.: A novel visual clustering algorithm for finding community in complex network. In: Li, X., Zaïane, O.R., Li, Z. (eds.) ADMA 2006. LNCS (LNAI), vol. 4093, pp. 396–403. Springer, Heidelberg (2006). https://doi.org/10.1007/11811305_44
23. Zachary, W.W.: An information flow model for conflict and fission in small groups. J. Anthropol. Res. **33**(4), 452–473 (1977)
24. Zhu, H., Koniusz, P.: Simple spectral graph convolution. In: Proceedings of the 2021 International Conference on Learning Representations (2021)

# Pessimistic Adversarially Regularized Learning for Graph Embedding

Mengyao Li, Yinghao Song, Long Yan, Hanbin Feng, Yulun Song,
Yang Li, and Gongju Wang

Data Intelligence Division, Unicom Digital Technology Co., Ltd., Beijing 100032,
China
wanggj129@chinaunicom.cn

**Abstract.** Autoencoder frameworks have been effectively employed
for graph embedding, resulting in successful analysis of graph in low-
dimensional space. Recently, generative models (GANs), which learn
data distribution of the adversarial method have been increasingly
applied to graph autoencoders (GAEs). Despite the effectiveness of current research, many GAEs lack the ability to provide instantaneous
feedback and ensure stable updates within the GAN component. In
particular, the MiniMax Multi-Agent Deep Deterministic Policy Gradient (M3DDPG) has demonstrated that using a 1-step gradient descent
can enhance the performance, which can also be leveraged to train the
encoder to further improve the adaptability of graph embedding. Motivated by this, we propose the Pessimistic Graph Autoencoder (PGAE),
and its variational version Pessimistic Variational Graph Autoencoder
(PVGAE). These methods reduce the output probability of the discriminator module through pessimistic parameters which make the feature
distribution generated by encoder restore maximally the actual distribution of the original graph. Furthermore, we employ graph embedding
to reconstruct the original graph information and constrain the generation of embedding vectors to preserve topological structure and node
content of the original graph. Our approaches yield competitive results
in node clustering and node classification tasks, outperforming numerous
state-of-the-art graph autoencoders across three benchmark datasets.

**Keywords:** Graph Autoencoder · Graph Embedding · MiniMax
Multi-Agent Deep Deterministic Policy Gradient

## 1 Introduction

A graph (a.k.a. network) is a paramount language for describing and modeling
complex relational systems [1]. Graph analyses and applications have become
prevalent in various fields. Effective graph analytics can entirely acquire and
quantitatively interpret the information hidden in the graph structure and node

---

M. Li and Y. Song—Equal contribution.

© The Author(s), under exclusive license to Springer Nature Switzerland AG 2023
X. Yang et al. (Eds.): ADMA 2023, LNAI 14178, pp. 338–351, 2023.
https://doi.org/10.1007/978-3-031-46671-7_23

content, which can be utilized in downstream prediction and decision tasks. However, scaling the graph up inevitably increase the number of nodes, and its complexity grows accordingly. As a result, graph analysis tasks face the challenges of high computational complexity and heavy storage burdens. To address issues, graph embedding (also known as node embedding) has been proposed. It efficiently converts the node content, topology, and other information of the graph into a low-dimensional continuous vector space [2]. Compared to direct analysis in high-dimensional graphs, this learning pattern is not only faster and more accurate, but also facilitates existing machine learning methods for analysis (such as SVM, k-nearest neighbors, and k-means).

Graph embedding algorithms can be broadly classified into three categories, i.e., matrix factorization-based graph embeddings, random walk-based graph embeddings, and graph neural network-based (GNN-based) graph embeddings [3]. While matrix factorization-based methods learn node embeddings from nodes by constructing higher-order adjacency matrices [4], they only capture the linear relation and lack extensive exploration of deep information. On the other hand, random walk-based methods optimize node embeddings by extracting pairs of nodes that occur simultaneously in a short-range random walk [5] with a focus on the representation of neighbors. Studies [6] have shown that these methods are similar to factorization-based techniques. However, GNN-based methods often outperform both of these approaches [7]. According to model architecture, they can be roughly divided into recurrent graph neural networks (RecGNNs), convolutional graph neural networks (ConvGNNs), graph autoencoders (GAEs), and spatial-temporal graph neural networks (STGNNs) [8].

Despite their effectiveness, most GNN-based methods rely heavily on the quality of the adjacency matrix. When the adjacency matrix is incomplete or under attacked, the resulting embedding vectors tend to perform poorly. To tackle this issue, there has been growing interest in employing the autoencoder framework to obtain vector space representations of node content and graph structure, especially GAEs. GAE and VGAE [9] first introduce the autoencoder framework into the graph embedding. MGAE [10] proposes a marginalized single-layer autoencoder in the graph clustering task. GALA [11] treats the encoding and decoding steps as Laplacian smoothing and Laplacian sharpening, respectively. Although GAEs preserve graph structure information and reconstruction loss, they often ignore the latent representations, making the resulting embedding vectors vulnerable to injection attacks and unreliable.

It should be noted that while GANs [12] have been extended to GAEs to solve the problem of vulnerable embedding vectors, their potential in improving graph embedding tasks is still being explored. GANs are mostly composed of a discriminative module and a generative module, where the discriminative module distinguishes the "fake" data generated by the generation module and real data. The generation module aims to simulate the real data as closely as possible. Recently, adversarial learning mechanism has been applied to graph embedding with ARGA and ARVGA being among the first to introduce adversarial mechanisms into the graph autoencoder framework. Other works, such as GANE [13],

WARGA [14] and DBGAN [13] have also demonstrated the effectiveness and reliability of adversarial networks in graph embedding tasks. However, GANs have the potential to improve graph embedding tasks up to now.

Furthermore, in order to further improve the performance of GANs in graph embedding representations, inspired by the MiniMax Multi-agent Deep Deterministic Policy Gradient (M3DDPG) algorithm [15], we implement a perturbation within the adversarial autoencoder. Subsequently, we propose an adversarial autoencoder based on pessimistic estimation.

The results output by the discriminant module is intervened to reasonably reduce the output probability to guide the generation of latent representations, that is, to interfere with the generated "fake data" to improve the effectiveness and robustness of graph embedding. Compared to ARGA and ARVGA, the algorithm proposed in this paper has better results on node clustering and connection prediction tasks. Our contributions are two-fold:

- We propose the Pessimistic Graph Autoencoder (PGAE), its variational version Pessimistic Variational Graph Autoencoder (PVGAE). The introduction of pessimistic estimation reasonably interferes with the results of the discriminative module, improves the learning ability of graph autoencoder, and narrow the gap between the generated distribution and the real data.
- Experiments on benchmark datasets show that the proposed methods outperform most existing methods in node classification, node clustering and visualization tasks.

## 2   Related Work

### 2.1   Graph Embedding

Graph embedding algorithms can be broadly classified into three categories: matrix factorization-based graph embeddings, random walk-based graph embeddings, and GNN-based graph embeddings. Matrix factorization-based methods derive the embedding representation by analyzing the relation matrix of the node. Graph factorization (GF) [16] employs the node vector inner product to capture edge relations between nodes. HOPE [4] leverages asymmetric transitivity between directed edges to capture graph content and calculate the graph embedding. Random walk-based methods enable the application of high-dimensional data to machine learning algorithms. Deepwalk [17] is a combination of random walk [18] and word2vec [19]. Node2vec enhances deepwalk by introducing a bias weight for the walk, resulting in sophisticated graph embeddings. GNN-based methods formulate certain strategies on nodes and edges to transform graph-structured data into standard representations, which are then fed it into various neural networks to be trained for diverse tasks. GNN-based methods are typically classified into four categories. RecGNNs learn the node embedding by iteratively propagating neighbor content, often consuming more computational resources, such as SSE [20]. ConvGNNs utilize the node's adjacency matrix, degree matrix, and feature vector to calculate the embedding. The advantage of these models is that structural content can be shared among layers, like DCNNs [21]. GAEs are unsupervised learning frameworks that encode graph structure and nodes into latent vector space and then reconstruct graph data, such as DRNE [22]. STGNNs consider both temporal and spatial information and seek hidden patterns, which can be applied in practical scenarios [23].

## 2.2 Graph Autoencoders

As a subset of GNN-based graph embedding methods, the basic belief of GAEs is to encode and decode the graph content based on autoencoder and then complete the nonlinear embedding representation of the input data. GraphEncoder [24] encodes and reconstructs the graph similar matrix, where the resulting embeddings preserve the sparseness of the original data. SDNE [25] utilizes the principle of autoencoder to reconstruct the first-order proximity of nodes, by forming node embedding. VGAE [26] adopts an autoencoder with GCN as the encoder to compute the embedding representation of the undirected graph. More recently, GANs have been applied to GAEs. ARGA [27] combines topology structure and node content to form graph embedding and uses an adversarial training approach to match the prior distribution. GANE [13] also adopts GAN, with the generative module attempting to generate node pairs and the discriminative module distinguishing vertex pairs from real links. WARGA [14] directly regularizes the latent distribution of node embedding to target distribution by applying the Wasserstein metric. DBGAN [28] uses GCN to encode and performs bidirectional mapping between the latent representation and the original data to ensure that the distribution of node embeddings remains consistent with the original distribution.

# 3 Adversarially Pessimistic Graph Learning

## 3.1 Problem Definition

Given a graph $G = \{V, E, X\}$, where $V = \{v_1, v_2, \ldots, v_n\}$ is the set of nodes in the graph. The relationships between nodes can be described by the adjacency matrix $A = [a_{i,j}] \in R^{n \times n}$, where $a_{i,j}$ represents the relationship between nodes $v_i$ and $v_j$. The node features are represented by the feature matrix $X = [x_1, x_2, .., x_n]^T \in R^{n \times b}$, where $x_i$ represents the feature content of node $v_i$ and $b$ is the dimension of the node features after preprocessing.

The goal of graph embedding is to learn a low-dimensional representation $Z \in R^{n \times d}$ of the node features. This is achieved by training a function $f(A, X) = Z$ that captures both the topology information of the graph and the feature content of the nodes. The obtained representation $Z$ can be easily adapted for various downstream applications.

For the classification task, the feature set $C$ is defined as the set of all node features. The final node feature set $C^{fin}$ is obtained by intersecting the feature sets with low missing rates for each label, and has $b$ elements. For the clustering task, the final node feature set $C^{fin}$ is equal to the original feature set $C$ with $b^c$ elements. In the classification task, the node embedding $Z$ is divided into training set $Z^{train}$ and testing set $Z^{test}$.

## 3.2 Preprocessing Modules

The PGAE/PVGAE data processing has three modules: general data preprocessing, graph embedding computing, and node clustering/node classification. The general data preprocessing module prepares the initial feature matrix $X$ for the graph embedding computing module, which inputs the preprocessed $X$ and adjacency matrix $A$ into the PGAE/PVGAE to calculate the graph embedding representation $Z$. This $Z$ is then used for node clustering and node classification tasks (Fig. 1).

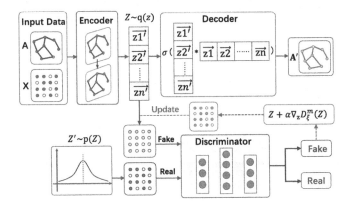

**Fig. 1.** The architecture of the pessimistic graph autoencoder consists of a graph convolution encoder on the top left, which takes the adjacency matrix $A$ and feature matrix $X$ as input and outputs the node embedding $Z$. On the top right is an inner product decoder $D^e$ that takes the embedding vector $Z$ as input and produces the reconstructed adjacency matrix $A$. The lower right corner has an $A$ discriminator $D^m$, which takes both the prior distribution's output $Z'$ and the node embedding $Z$ as input and outputs the parameters for training the graph convolutional encoder with the pessimistic gradient $\nabla_Z D^m(Z)$.

## 3.3 Graph Convolutional Encoder Model $G(X, A)$

The graph convolution network is used to calculate the graph embedding vector. It accepts graph information including adjacency matrix $A$ and feature matrix $X$. The graph convolution utilizes the function $f\left(Z^{(l)}, A \mid W^{(l)}\right)$, $W^{(l)}$ means the embedding representation of input data during the $l$ graph convolution, and $l$ embedding vector $Z^{(l+1)}$ can be expressed as (1), where the embedding vector $Z^{(0)}$ of the first convolution is feature matrix $X$ of the node, $Z^{(0)} = X$.

$$Z^{(l+1)} = f\left(Z^{(l)}, A \mid W^{(l)}\right) \tag{1}$$

The calculation of each layer of the graph convolutional network can be expressed as (2), where $\tilde{A} = A + I$, $\tilde{D}_{ii} = \sum_j \tilde{A}_{ij}$, $I$ is the identity matrix of $A$, and $\phi$ is the activation function of convolutional layer, like $Relu(x) = max(0, x))$ or $sigmoid(x) = 1/(1 + e^x)$.

$$f\left(Z^{(l)}, A \mid W^{(l)}\right) = \phi\left(\tilde{D}^{-\frac{1}{2}} \tilde{A} \tilde{D}^{-\frac{1}{2}} Z^{(l)} W^{(l)}\right) \tag{2}$$

Variational graph encoder can be expressed as (3) and (4), where $\mu$ is the mean of vector $z_i$. The calculation can be done by $\log \sigma = f_{linear}(X, A \mid W'^{(1)})$, and needs to use $W^{(0)}$ to calculate the corresponding parameters.

$$q(Z \mid X, A) = \prod_{i=1}^{n} q\left(z_i \mid X, A\right) \tag{3}$$

$$q\left(z_i \mid X, A\right) = N\left(z_i \mid \mu_i, \mathrm{diag}\left(\sigma^2\right)\right) \tag{4}$$

### 3.4 Inner Product Decoder

The task of the decoder is to reconstruct the graph information. The reconstruction process is shown in (5). Our algorithms will operate even if the embedding vector does not have the node feature matrix $X$.

$$\hat{A} = sigmoid(ZZ^T) \tag{5}$$

The reconstruction loss of PGAE and PVGAE is defined as (6), where $a_{i,j}$ represents the relation between nodes $v_i$ and $v_j$ in adjacency matrix after reconstruction. In addition, PVGAE has a Kullback-Leibler distance (KL divergence) [29] loss, expressed in $KL[q(\cdot)\|p(\cdot)]$, which is derived from the VAE, so we also specify a Gaussian Prior distribution $p(Z) = \prod_i p(z_i) = \prod_i N(z_i|0, I)$.

$$\begin{aligned} L_{rec} &= E_{q(Z|(X,A))}(\log p(\hat{A} \mid Z)) \\ &= \sum_{i=1}^{n} \sum_{j=1}^{n} (a_{i,j} * \log\left(\hat{a}_{i,j}\right) + (1 - a_{i,j}) * \log\left(1 - \hat{a}_{i,j}\right)) \end{aligned} \tag{6}$$

$$L_{kl} = E_{q(Z|(X,A))}(KL[q(Z|X,A)|p(Z)]) \tag{7}$$

### 3.5 Training

The algorithm trains an discriminator to make the generated embedding vector $Z$ match the prior distribution. The discriminator, $D_\xi^m$, is a 3-layer fully connected structure with a sigmoid output activation that judges the difference between the graph embedding vector $z \in Z$ from the encoder and the vector $z^{'} \in Z^{'}$ from the prior distribution. $z^{'}$ is labeled as 1 and $z$ is labeled as 0. The discriminator is trained by minimizing the binary cross-entropy loss as given in Eq. (8) using $z^{'}$, a vector sampled from the prior distribution.

$$L_{D_\xi^m} = E_{z' \sim p(z')} \log D_\xi^m(z_i^{'}) + E_{x,a \sim p(x,a)} \log(1 - D_\xi^m(G_\theta(x,a))) \quad (8)$$

The loss function for training the $G$ parameters based on the GAN idea can be written as shown in Eq. (9), which means to make the generated graph embedding vector close to the samples of the prior distribution.

$$L_{G_\theta} = E_{x,a \sim p(x,a)} log(D_\xi^m(G_\theta(x,a))) \quad (9)$$

For simplicity, we label $Z = G(X, A)$. Now, we re-write Eq. (9) and change the input of $D^m$ from $Z$ to $Z_{pess}$, which is disturbed. The form of $Z_{pess}$ is shown in Eq. (10), where the value of the disturbance is defined as $\nabla_Z D_\xi^m(Z)$, $\alpha$ is the degree of pessimism. Maximizing the output of $D^m$ by $Z_{pess}$ increases the output of the loss function, which is why it is referred to as a pessimistic input. The revised formula is shown in Eq. (11). It tells the graph embedding generator $G$ that the generated graph embedding $Z$ still has a gap with the prior distribution and needs further adjustment of parameters. Therefore, $L_{G_\theta}$ and $L_{pess}$ satisfy the relationship shown in inequality (12).

$$Z_{pess} = Z + \alpha \nabla_Z D_\xi^m(Z) \quad (10)$$

$$L_{pess} = E_{x,a \sim p(x,a)} log(D_\xi^m(z_{pess})) \quad (11)$$

$$L_{G_\theta} \leq L_{pess} \quad (12)$$

### 3.6    Algorithm Explanation

Algorithm 1 outlines the steps of our method. The first step is to initialize the parameters $\xi$ and $\theta$ of the discriminator $D^m$ and the encoder (generator) $G$, respectively (Lines 1–2). The second step is to generate the graph embedding calculation result set $Z$ using the encoder (Line 3). After that, $m$ samples are sampled from $Z^{'}$, which is obtained from the result set $Z$ and the prior distribution $p_z$, to prepare for training (Lines 7–8). The parameters of $\xi$ are then updated (Line 10). In the end, PGAE updates the $\theta$ parameter using the pessimistic loss of the discriminator (Line 14) and the reconstruction loss (Line 16), while PVGAE updates the $\theta$ parameter using the pessimistic loss (Line 14), reconstruction loss (Line 16), and KL divergence loss (Line 22).

---

**Algorithm 1.** Pessimistic Adversarially Regularized Graph Autoencoder

---

1: Initialize discriminator $D^m$'s parameters with $\xi$
2: Initialize encoder $G$'s parameters with $\theta$
3: Get input graph data $H = \{V, X\}$
4: **for** $epoch = 1, 2, 3...E$ **do**
5:     Generate latent variables matrix Z refer to $G_\theta(X, A) = Z$
6:     **for** $iterator = 1, 2, ..., K$ **do**
7:         Sample a random minibatch of m samples $\{z_1, z_2, ..., z_m\}$ from latent matrix $Z$
8:         Sample a random minibatch of m samples $\{z_1', z_2', ..., z_m'\}$ from prior distribution $p_z$
9:         Update the $\xi$ by minimizing the loss:
10:            $L_{D_\xi^m} = E_{z' \sim p(z')} \log D_\xi^m(z_i') + E_{x,a \sim p(x,a)} \log(1 - D_\xi^m(G_\theta(x, a)))$
11:        Get pessimistic input:
12:            $Z_{pess} = Z + \alpha \nabla_Z D_\xi^m(Z)$
13:        Get pessimistic loss:
14:            $L_{pess} = E_{x,a \sim p(x,a)} log(D_\xi^m(z_{pess}))$
15:        Get reconstruction loss:
16:            $L_{rec} = \sum_{i=1}^n \sum_{j=1}^n (a_{i,j} * \log(\hat{a}_{i,j}) + (1 - a_{i,j}) * \log(1 - \hat{a}_{i,j}))$
17:        For PGAE:
18:        Update the $\theta$ by minimizing the loss:
19:            $L_G = L_{rec} - L_{pess}$
20:        For PVGAE:
21:        Update the $\theta$ by minimizing the loss:
22:            $L_G = (L_{rec} - L_{kl}) - L_{pess}$
23:     **end for**
24: **end for**

---

## 4  Experiment

### 4.1  Datasets and Evaluation Metrics

To evaluate the effectiveness of PGAE and PVGAE, we conduct experiments on three standard benchmark citation datasets as shown in Table 1.

In order to evaluate the effect of the algorithm more comprehensively, this paper adopts Accuracy (Acc), Normalized Mutual Information (NMI) [30], F1-score, Precision and Adjusted Rand Index (ARI) to evaluate the node clustering task. Accuracy is used as the evaluation metric for classification tasks.

### 4.2  Experimental Settings

We apply all competitors to 3 datasets to compute graph embeddings, and use the results for node clustering task and node classification to verify quality. This section discusses PGAE, PVGAE and the pessimistic parameter $\alpha$ involved. The parameters of baseline methods compared were set as what was used by their authors. The evaluation index is recorded in 10 iterations, and the maximum value is reserved as the final performance of the algorithm.

In our experiments, we apply the following variants of PGAE and PVGAE. First, methods that solely employ features for graph embedding including K-means and Spectral. Secondly, Node2vec, LINE, Deepwalk, and DNGR singular use topological information for graph embedding. Finally, the GAEs that use both node features and topology information for graph embedding learning include GAE, VGAE, ARGA, and ARVGA.

**Table 1.** Summary of benchmark datasets

| Dataset | Classes | Nodes | Edges | Features |
|---------|---------|-------|-------|----------|
| Cora | 7 | 2708 | 5429 | 1433 |
| Citeseer | 6 | 3312 | 4732 | 3703 |
| Pubmed | 3 | 19717 | 44338 | 500 |

## 4.3 Experiment Results

Our objective is to allocate distinct clusters to individual nodes, and the outcomes are presented in Table 2. It can be inferred that methods that employ both features and topology show better performance than others. Compared with all baselines, PGAE and PVGAE achieve remarkable performance on all datasets, especially on Cora. PVGAE outperforms most other competitors on Citeseer. In summary, PGAE exhibits better performance on clustering tasks than PVGAE.

**Table 2.** Comparison of node clustering task on Acc, NMI, F1, Precious and ARI (Bold is the best result, underlined is the competitor)

| Algorithm | Cora (%) | | | | | Citeseer (%) | | | | | Pubmed (%) | | | | |
|-----------|------|------|------|------|------|------|------|------|------|------|------|------|------|------|------|
| | Acc | NMI | F1 | Pre | ARI | Acc | NMI | F1 | Pre | ARI | Acc | NMI | F1 | Pre | ARI |
| K-means | 49.2 | 32.1 | 36.8 | 36.9 | 23.0 | 54.0 | 30.5 | 40.9 | 40.5 | 27.9 | 39.8 | 0.1 | 19.5 | 57.9 | 0.2 |
| Spectral | 36.7 | 12.7 | 31.8 | 19.3 | 3.1 | 19.3 | **56.0** | 29.9 | 17.9 | 1.0 | 40.3 | 4.2 | 27.1 | 49.8 | 0.2 |
| Node2vec | 56.3 | 42.0 | – | – | – | 40.8 | 13.0 | – | – | – | 65.6 | 25.0 | – | – | – |
| LINE | 30.7 | 10.1 | – | – | – | 25.0 | 5.6 | – | – | – | 43.1 | 7.2 | – | – | – |
| Deepwalk | 48.4 | 32.7 | 39.2 | 36.1 | 24.3 | 33.7 | 8.8 | 27.0 | 24.8 | 9.2 | 68.4 | 27.9 | 67.0 | <u>68.6</u> | 0.3 |
| DNGR | 41.9 | 31.8 | 34.0 | 26.6 | 14.2 | 32.6 | 18.0 | 30.0 | 20.0 | 4.4 | 45.8 | 15.5 | 46.7 | 62.9 | 0.5 |
| GAE | 59.6 | 42.9 | 59.5 | 59.6 | 34.7 | 40.8 | 17.6 | 37.2 | 41.8 | 12.4 | <u>67.2</u> | 27.7 | 66.0 | 68.4 | 27.9 |
| VGAE | 60.9 | 43.6 | 60.9 | 60.9 | 34.6 | 34.4 | 15.6 | 30.8 | 34.9 | 9.3 | 63.0 | 22.9 | 63.4 | 63.0 | 21.3 |
| ARGA | 64.0 | 44.9 | 61.9 | 64.6 | 35.2 | 57.3 | 35.0 | 54.6 | 57.3 | **34.1** | 66.8 | **30.5** | 65.6 | **69.9** | 29.5 |
| ARVGA | 63.8 | 45.0 | 62.7 | 62.4 | 37.4 | 54.4 | 26.1 | 52.9 | 54.9 | 25.4 | 58.8 | 18.4 | <u>67.8</u> | 69.4 | <u>30.6</u> |
| PGAE | **73.8** | **51.4** | **73.8** | **73.8** | **49.8** | <u>57.6</u> | 32.2 | <u>58.7</u> | <u>59.9</u> | 30.8 | **69.9** | <u>30.1</u> | **69.9** | **69.9** | **32.2** |
| PVGAE | <u>0.673</u> | <u>49.5</u> | <u>67.4</u> | <u>67.3</u> | <u>44.2</u> | **61.3** | <u>35.2</u> | **61.3** | **61.3** | <u>33.4</u> | 64.0 | 21.5 | 64.0 | 64.0 | 22.1 |

Our goal is to classify each node into one of the multiple labels for node classification task, with the outcome displayed in Table 3. Our methods yield slightly improvements over the baseline models ARGA and ARVGA, which PVGAE demonstrating superior performance on the Cora dataset, surpassing the base model by 2.3%. One the other hand, PGAE exhibits the most effective performance on Pubmed, exceeding the baseline model by 2.6%.

**Table 3.** Node classification task Accuracy (Bold is the best result, underlined is the competitor)

| Algorithm | Cora (%) | Citeseer (%) | Pubmed (%) |
|---|---|---|---|
| Raw | 47.87 | 49.33 | 69.11 |
| Node2vec | 64.20 | 46.00 | 67.00 |
| LINE | 57.20 | 34.40 | 59.10 |
| Deepwalk | 70.66 | 44.30 | 64.00 |
| GAE | 71.53 | <u>65.77</u> | 72.14 |
| VGAE | <u>75.24</u> | **69.05** | <u>75.29</u> |
| ARGA | 73.30 | 52.2 | 74.12 |
| ARVGA | 74.38 | 63.80 | 74.69 |
| PGAE | 73.00 | 59.20 | **76.30** |
| PVGAE | **76.10** | 62.10 | 71.10 |

Figure 2 records the Acc of the PGAE/PVGAE and competitors on classification task by using XGBoost classifier. The solid line part is the average performance of the algorithm, and the shaded part represents the interval of the 10 runs results, including the maximum and minimum. Figure 2(a) shows the results on Cora. Before Epoch is 30, Acc of PVGAE is slightly higher than that of ARVGA on Citeseer (Fig. 2(b)). Figure 2(c) shows the results on Pubmed. The average performance of PGAE's Acc is the highest. In summary, PVGAE is slightly better than PGAE in classification tasks.

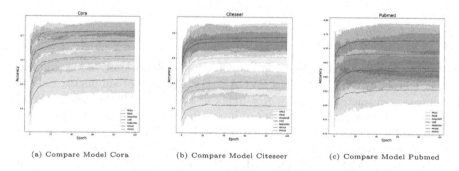

(a) Compare Model Cora        (b) Compare Model Citeseer        (c) Compare Model Pubmed

**Fig. 2.** Accuracy on different epoch for classification

A signification application of graph embedding is to generate visualizations of networks in two-dimensional spaces. Therefore, we demonstrate the graph embeddings of baseline models on 3 datasets using t-SNE [31]. For nodes corresponding to distinct categories of papers, we adopt different colors. Therefore, the more clustered the nodes of the same color, the superior the visualization result. As shown in Fig. 3, we can see that PGAE and PVGAE achieve superior node clustering and separation than the baseline model, and boundaries of each cluster become clearer.

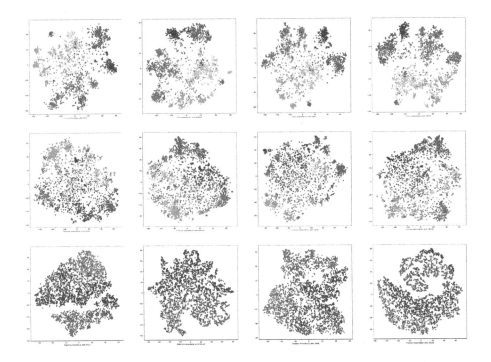

**Fig. 3.** Graph Embedding Visualization

### 4.4  Parameter Analysis

In this section, we will take node classification task to analyze the influence of parameters of the PGAE and PVGAE. For the sake of simplicity, only the pessimistic parameter $\alpha$ on the Acc index is analyzed. All the results are the average of 10 runs, and the number of iterations is specified as 100 Epoch.

Figure 4(a) shows the performance of PGAE. On Cora, set $\alpha$ to 1.0 performs the best, and Acc is 64.4%. On Citeseer, set $\alpha$ to 5 is the best performance, and Acc value is 44.88%. On Pubmed, setting $\alpha$ to 0.1 performed the best, with value of 72.2%. Overall, PGAE has a stable performance on Cora and Pubmed, and the most fluctuating performance on Citeseer.

Figure 4(b) shows the performance of PVGAE. On Cora, set $\alpha$ to 1 performs the best, and Acc value is 64.4%. On Citeseer, set $\alpha$ to 0.5 is the best performance, and Acc value is 54.08%. On Pubmed, set $\alpha$ to 0.5 is the best performance, with value of 68.36%. Overall, PVGAE's performance on Cora and Pubmed is relatively stable, while its performance fluctuates on Citeseer.

To sum up, PGAE and PVGAE need to adjust the corresponding $\alpha$ parameters for different datasets.

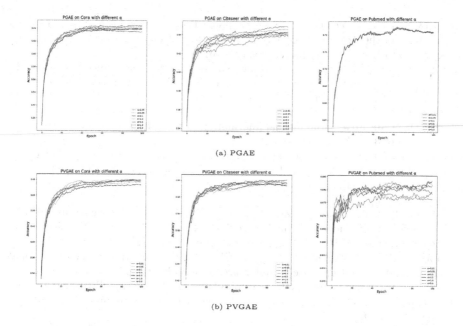

(a) PGAE

(b) PVGAE

**Fig. 4.** $\alpha$ Impact of parameters on the clustering model

## 5    Conclusion

In this paper, we propose an original graph embedding method based on graph autoencoders, Pessimistic Graph Autoencoder (PGAE), and its variant Pessimistic Variational Graph Autoencoder (PVGAE). We introduce pessimism into the adversarial training procedure to reasonably disturb the "fake" data generated by the generator. Specifically, we reduce the discriminative module's strategy to distinguish samples as real, resulting in more robust representations. Experimental conducted on datasets proved that our method outperforms baseline algorithms in node clustering, node classification, and graph visualization tasks.

## References

1. Xu, M.: Understanding graph embedding methods and their applications. SIAM Rev. **63**(4), 825–853 (2021)
2. Cai, H., Zheng, V.W., Chang, K.C.C.: A comprehensive survey of graph embedding: problems, techniques, and applications. IEEE Trans. Knowl. Data Eng. **30**(9), 1616–1637 (2018)
3. Cui, P., Wang, X., Pei, J., Zhu, W.: A survey on network embedding. IEEE Trans. Knowl. Data Eng. **31**(5), 833–852 (2018)
4. Ou, M., Cui, P., Pei, J., Zhang, Z., Zhu, W.: Asymmetric transitivity preserving graph embedding. In: Proceedings of the 22nd ACM SIGKDD International Conference on Knowledge Discovery and Data Mining, pp. 1105–1114 (2016)

5. Grover, A., Leskovec, J.: node2vec: scalable feature learning for networks. In: Proceedings of the 22nd ACM SIGKDD International Conference on Knowledge Discovery and Data Mining, pp. 855–864 (2016)

6. Qiu, J., Dong, Y., Ma, H., Li, J., Wang, K., Tang, J.: Network embedding as matrix factorization: unifying DeepWalk, LINE, PTE, and node2vec. In: Proceedings of the Eleventh ACM International Conference on Web Search and Data Mining, pp. 459–467 (2018)

7. He, D., et al.: Adversarial mutual information learning for network embedding. In: IJCAI, pp. 3321–3327 (2020)

8. Scarselli, F., Gori, M., Tsoi, A.C., Hagenbuchner, M., Monfardini, G.: The graph neural network model. IEEE Trans. Neural Networks **20**(1), 61–80 (2008)

9. Kipf, T.N., Welling, M.: Variational graph auto-encoders. arXiv preprint arXiv:1611.07308 (2016)

10. Wang, C., Pan, S., Long, G., Zhu, X., Jiang, J.: MGAE: marginalized graph autoencoder for graph clustering. In: Proceedings of the 2017 ACM on Conference on Information and Knowledge Management, pp. 889–898 (2017)

11. Park, J., Lee, M., Chang, H.J., Lee, K., Choi, J.Y.: Symmetric graph convolutional autoencoder for unsupervised graph representation learning. In: Proceedings of the IEEE/CVF International Conference on Computer Vision, pp. 6519–6528 (2019)

12. Goodfellow, I., et al.: Generative adversarial networks. Commun. ACM **63**(11), 139–144 (2020)

13. Zheng, S., Zhu, Z., Zhang, X., Liu, Z., Cheng, J., Zhao, Y.: Distribution-induced bidirectional generative adversarial network for graph representation learning. In: Proceedings of the IEEE/CVF Conference on Computer Vision and Pattern Recognition, pp. 7224–7233 (2020)

14. Liang, H., Gao, J.: Wasserstein adversarially regularized graph autoencoder. arXiv preprint arXiv:2111.04981 (2021)

15. Li, S., Wu, Y., Cui, X., Dong, H., Fang, F., Russell, S.: Robust multi-agent reinforcement learning via minimax deep deterministic policy gradient. In: Proceedings of the AAAI Conference on Artificial Intelligence, vol. 33, pp. 4213–4220 (2019)

16. Ahmed, A., Shervashidze, N., Narayanamurthy, S., Josifovski, V., Smola, A.J.: Distributed large-scale natural graph factorization. In: Proceedings of the 22nd International Conference on World Wide Web, pp. 37–48 (2013)

17. Wu, Z., Pan, S., Chen, F., Long, G., Zhang, C., Philip, S.Y.: A comprehensive survey on graph neural networks. IEEE Trans. Neural Netw. Learn. Syst. **32**(1), 4–24 (2020)

18. Veličković, P., Cucurull, G., Casanova, A., Romero, A., Lio, P., Bengio, Y.: Graph attention networks. arXiv preprint arXiv:1710.10903 (2017)

19. Hamilton, W., Ying, Z., Leskovec, J.: Inductive representation learning on large graphs. In: Advances in Neural Information Processing Systems, vol. 30 (2017)

20. Dai, H., Kozareva, Z., Dai, B., Smola, A., Song, L.: Learning steady-states of iterative algorithms over graphs. In: International Conference on Machine Learning, pp. 1106–1114. PMLR (2018)

21. Atwood, J., Towsley, D.: Diffusion-convolutional neural networks. In: Advances in Neural Information Processing Systems, vol. 29 (2016)

22. Tu, K., Cui, P., Wang, X., Yu, P.S., Zhu, W.: Deep recursive network embedding with regular equivalence. In: Proceedings of the 24th ACM SIGKDD International Conference on Knowledge Discovery & Data Mining, pp. 2357–2366 (2018)

23. Li, Y., Yu, R., Shahabi, C., Liu, Y.: Diffusion convolutional recurrent neural network: data-driven traffic forecasting. arXiv preprint arXiv:1707.01926 (2017)

24. Tian, F., Gao, B., Cui, Q., Chen, E., Liu, T.Y.: Learning deep representations for graph clustering. In: Proceedings of the AAAI Conference on Artificial Intelligence, vol. 28 (2014)
25. Wang, D., Cui, P., Zhu, W.: Structural deep network embedding. In: Proceedings of the 22nd ACM SIGKDD International Conference on Knowledge Discovery and Data Mining, pp. 1225–1234 (2016)
26. Kingma, D.P., Welling, M.: Auto-encoding variational Bayes. arXiv preprint arXiv:1312.6114 (2013)
27. Pan, S., Hu, R., Long, G., Jiang, J., Yao, L., Zhang, C.: Adversarially regularized graph autoencoder for graph embedding. arXiv preprint arXiv:1802.04407 (2018)
28. Liu, X., Du, H., Xu, J., Qiu, B.: DBGAN: a dual-branch generative adversarial network for undersampled MRI reconstruction. Magn. Reson. Imaging **89**, 77–91 (2022)
29. Goldberger, J., Gordon, S., Greenspan, H., et al.: An efficient image similarity measure based on approximations of KL-divergence between two Gaussian mixtures. In: ICCV, vol. 3, pp. 487–493 (2003)
30. Guo, L., Dai, Q.: Graph clustering via variational graph embedding. Pattern Recogn. **122**, 108334 (2022)
31. Wattenberg, M., Viégas, F., Johnson, I.: How to use t-SNE effectively. Distill **1**(10), e2 (2016)

# M²HGCL: Multi-scale Meta-path Integrated Heterogeneous Graph Contrastive Learning

Yuanyuan Guo[1,2], Yu Xia[1,2], Rui Wang[1,2], Rongcheng Duan[3], Lu Li[3], and Jiangmeng Li[1,2(✉)]

[1] Institute of Software Chinese Academy of Sciences, Beijing, China
{guoyuanyuan2022,jiangmeng2019}@iscas.ac.cn
[2] University of Chinese Academy of Sciences, Beijing, China
[3] Defence Industry Secrecy Examination and Certification Center, Beijing, China

**Abstract.** Inspired by the successful application of contrastive learning on graphs, researchers attempt to impose graph contrastive learning approaches on heterogeneous information networks. Orthogonal to homogeneous graphs, the types of nodes and edges in heterogeneous graphs are diverse so that specialized graph contrastive learning methods are required. Most existing methods for heterogeneous graph contrastive learning are implemented by transforming heterogeneous graphs into homogeneous graphs, which may lead to ramifications that the valuable information carried by non-target nodes is undermined thereby exacerbating the performance of contrastive learning models. Additionally, current heterogeneous graph contrastive learning methods are mainly based on initial meta-paths given by the dataset, yet according to our deep-going exploration, we derive empirical conclusions: only initial meta-paths cannot contain sufficiently discriminative information; and various types of meta-paths can effectively promote the performance of heterogeneous graph contrastive learning methods. To this end, we propose a new multi-scale meta-path integrated heterogeneous graph contrastive learning (M²HGCL) model, which discards the conventional heterogeneity-homogeneity transformation and performs the graph contrastive learning in a joint manner. Specifically, we expand the meta-paths and jointly aggregate the direct neighbor information, the initial meta-path neighbor information and the expanded meta-path neighbor information to sufficiently capture discriminative information. A specific positive sampling strategy is further imposed to remedy the intrinsic deficiency of contrastive learning, i.e., the hard negative sample sampling issue. Through extensive experiments on three real-world datasets, we demonstrate that M²HGCL outperforms the current state-of-the-art baseline models.

**Keywords:** Graph contrastive learning · Heterogeneous information network · Graph neural network · Meta-path

## 1 Introduction

HINs are complex networks containing multiple types of nodes and edges, which possess different attributes and features. Compared with traditional homogeneous

X. Yang et al. (Eds.): ADMA 2023, LNAI 14178, pp. 352–367, 2023.
https://doi.org/10.1007/978-3-031-46671-7_24

information networks, HINs contain richer semantic and structural information, which further empowers HINs to reflect more sophisticated relationships in the real world, thereby having wider applications, such as recommendation systems, social network analysis, and bioinformatics. In recent years, graph neural networks (GNNs) have achieved impressive success. The variants of GNNs, heterogeneous graph neural networks (HGNNs) [3,8,17,20,24], have also attracted widespread attention. However, existing HGNNs require a mass of annotation information for downstream tasks, which is expensive to obtain in practice. To address this issue, many researchers attempt to leverage self-supervised learning to capture discriminative information from HINs [9,14].

Self-supervised learning is a training approach that allows models to learn from raw input data by utilizing the inherent structure or patterns in the data without explicit supervision. Contrastive methods, a type of self-supervised learning, aim to create effective representations by bringing semantically similar pairs together and pushing dissimilar pairs apart. Inspired by the success of contrastive learning in computer vision [1,6], many researchers have applied contrastive learning to homogeneous networks, called graph contrastive learning. [9,10,15] extend such a learning paradigm to HINs and propose heterogeneous graph contrastive learning (HGCL). However, in practice, HGCL suffers from two critical issues: 1) existing methods are restricted to the initial meta-paths given by the dataset; 2) when implementing graph contrastive learning by pre-transforming heterogeneous graphs into homogeneous graphs, conventional methods inevitably discard valuable information from non-target nodes.

Regarding the first issue, we reckon that if HGCL is limited to short-chain information learned from the initial meta-paths, there is a problem of lacking global structural and contextual information. It may also fail to fully describe the semantic relationships between nodes, thus potentially unable to capture more complex semantic information. For example, in a movie recommendation system, if models only consider movies with the same director based on the initial 2-hop meta-paths, the other movies of the same genre may be unexpectedly ignored. Solely leveraging the long-chain information learned from the expanded meta-paths based on the initial meta-paths can improve the model to obtain global structural and contextual information, but such an approach may introduce noise and redundant information in the learned representations. For example, in a social network consisting of three types of nodes (people, companies, and schools) and two types of relationships (people-company/school), if an expanded 4-hop meta-path, e.g., people-company-people-company-people, is solely leveraged to learn the features of people, the learned features may contain certain redundant information because the third person node in this meta-path may have the same features as the second person node. Given the aforementioned issues, we propose a hypothesis that jointly aggregating the initial meta-path neighbors and the expanded meta-path neighbors may remedy the limitations of solely using one type of meta-paths, which further promotes the model's capacity to better capture the diverse semantic information of nodes. To verify our hypothesis, we design and conduct exploratory experiments on HeCo [21]. The

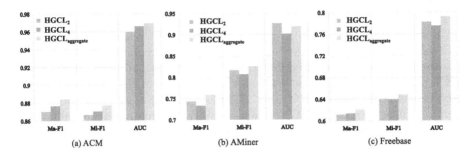

**Fig. 1.** Experiments to analyze the effects of meta-paths with different lengths on a specific HGCL model, i.e., HeCo [21]. $HGCL_2$ represents the model using initial 2-hop meta-paths. $HGCL_4$ represents the model using expanded 4-hop meta-paths instead of the initial meta-paths. $HGCL_{aggregate}$ denotes the model jointly using the initial and expanded meta-paths.

experimental results, as shown in Fig. 1, demonstrate that the performance of the model aggregating the initial and expanded meta-paths exceed the performance of models solely using one type of meta-paths.

Considering the second issue, we provide the corresponding analyses that the conventional approach of directly transforming heterogeneous graphs into homogeneous graphs often involves a certain discriminative information loss. Different types of nodes and edges in HINs contain distinct semantic information, but after the transformation, only the information of relationships between nodes and edges of the same type is reserved, thereby resulting in the loss of specific discriminative information. For example, HGCML [22] transforms HINs into multiple subgraphs based on multiple meta-paths, ignoring the rich heterogeneous information contained by the non-target nodes. Despite utilizing heterogeneous information in HINs for contrastive learning, HeCo [21] only aggregates the neighbor information of the target type in downstream tasks, disregarding the neighbor information of non-target types, thereby leading to information loss.

In light of the analyses of existing issues in the field of HGCL, we propose Multi-scale Meta-path integrated Heterogeneous Graph Contrastive Learning, dubbed $M^2HGCL$. Concretely, the contributions of this paper are three-fold:

- We introduce a novel multi-scale meta-path integrated heterogeneous graph contrastive learning approach, i.e., $M^2HGCL$, which conducts graph contrastive learning in a joint manner, without relying on the conventional heterogeneity-homogeneity transformation.
- To sufficiently exploit the diverse semantic relationships among different types of nodes and edges in HINs for heterogeneous graph contrastive learning, we propose a positive sampling strategy specific to HINs.
- We perform comprehensive experiments on three widely-used HIN datasets to evaluate the effectiveness of our proposed model. The experimental results demonstrate that $M^2HGCL$ outperforms the current state-of-the-art base-

lines, indicating the superiority of our approach for heterogeneous graph contrastive learning tasks.

## 2   Related Work

### 2.1   Heterogeneous Graph Neural Networks

Heterogeneous graph neural network (HGNN) has recently received much attention and there have been some models proposed. For example, HAN [20] argues that different types of edges should be assigned different weights and that different neighbor nodes should have distinct weights within the same type of edge. To achieve this, HAN employs node-level attention and semantic-level attention. Building on HAN, MAGNN [3] generates node embeddings by applying node content transformation, meta-path aggregation, and inter-meta-path aggregation. To model large-scale heterogeneous graphs found on the web, scholars have proposed HGT [8]. HetGNN [24], operating in an unsupervised setting, leverages BiLSTM to capture both the heterogeneity of structure and content and is applicable to both transductive and inductive tasks. While the aforementioned methods have been successful, the majority of them are semi-supervised or supervised learning and depend heavily on the labeled data.

### 2.2   Graph Contrastive Learning

Graph contrastive learning (GCL) has emerged as one of the most critical techniques for unsupervised graph learning. Typically, a graph contrastive learning framework consists of a graph view generation component that generates positive and negative views, and a contrastive target that distinguishes positive pairs from negative pairs. DGI [19] achieves contrastive learning by maximizing the mutual information between the local representations and corresponding global representations of the graph. Besides, GraphCL [23] employs augmentation techniques to create two views and then performs local-to-local contrast, while proposing four distinct augmentation methods. Moreover, to mitigate GraphCL's reliance on graph augmentation and negative samples, BGRL [18] draws inspiration from BYOL [4] and employs only basic graph augmentation techniques, thus eliminating the need for negative samples.

### 2.3   Heterogeneous Graph Contrastive Learning

Some researchers have extended graph contrastive learning to heterogeneous graphs and proposed the concept of heterogeneous graph contrastive learning. For instance, HDGI [15] applies DGI on heterogeneous graphs by leveraging meta-paths to model the semantic structure and by utilizing graph convolutional modules and semantic-level attention mechanisms to capture node representations. DMGI [14] introduces a contrastive learning approach based on single-view and multi-meta-path to enhance learning efficacy. It guides the fusion of different meta-paths through consensus regularization to capture complex relationships in graph data. HGCML [22] generates multiple subgraphs as multiple views using meta-paths and proposes a contrastive objective induced by meta-paths between views. HeCo [22] proposes to learn node embeddings by utilizing the network pattern and meta-path view of HIN to capture both local and high-order structures. However, while HGCML employs contrastive learning on heterogeneous graphs, it actually relies on contrastive learning on homogeneous graphs and does not fully utilize the rich heterogeneous information present in heterogeneous graphs. HeCo, on the other hand, employs heterogeneous information in contrastive learning training but does not do so in the final node embeddings.

## 3   Preliminary

In this section, we introduce the related concepts of heterogeneous graphs.

**Definition 1 Heterogeneous Information Network(HIN).** *HIN is defined as a graph network $G = (V, E, A, R)$, where $V$ denotes the set of nodes, $E$ denotes the set of edges, and $A$ and $R$ are the type sets of nodes and edges respectively, noting that $|A| + |R| > 2$. Moreover, the set of nodes and edges $V, E$ are associated with a node-type mapping function $\phi : V \rightarrow A$ and an edge-type mapping function $\varphi : E \rightarrow R$.*

**Definition 2 Meta-path.** *Meta-path $P_m$ is a path composed of multiple relations between nodes in a HIN, defined as $a_1 \rightarrow a_2 \rightarrow ... \rightarrow a_{n+1}$ (abbreviated as $a_1 a_2 a_{n+1}$), describing the compound relation between node $a_1$ and $a_{n+1}$. Note that $P_m \in P_M$, where $P_M$ is the set of meta-paths. For example, Fig. 2 (c) shows three meta-paths existing in the dataset Freebase: MDM represents two movies with the same director, MAM represents two movies with the same actor, and MWM represents two movies with the same writer.*

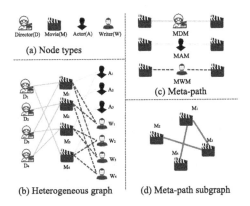

**Fig. 2.** An example of heterogeneous information networks (Freebase).

**Definition 3 Meta-path Neighbors.** *Given the meta-path $P_m$, the meta-path neighbors $N_{P_m}^{MP}$ is a set of nodes connected to the target node through the meta-path $P_m$. For example, in Fig. 2 (b), the meta-path neighbors of movie $M_3$ based on the meta-path MDM are movie $M_1$ and $M_2$.*

**Definition 4 Meta-path Subgraph.** *Given the meta-path $P_m$, the subgraph $G^{P_m}$ represents the homogeneous graph including only the target node. Edges between any two nodes in $G^{P_m}$ depend on whether there exists meta-path $P_m$ between the corresponding nodes in the heterogeneous graph. For example, in Fig. 2, (d) represents the subgraph $G^{MDM}$ based on the meta-path MDM.*

## 4  Methodology

In this section, we propose M²HGCL, a novel heterogeneous graph contrastive learning method that integrates multi-scale meta-paths. The overall architecture of our model is shown in Fig. 3. We first generate the subgraph based on every initial meta-paths and aggregate the direct neighbors, the initial meta-path neighbors, and the expanded meta-path neighbors of the target node as the node representation of this subgraph. Moreover, considering the specificity of HINs, we further design a positive sampling strategy for HINs.

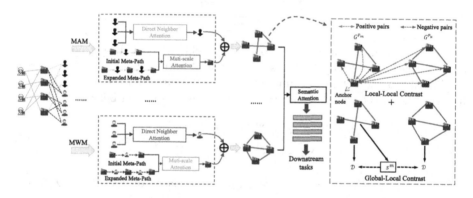

**Fig. 3.** The overall architecture of M²HGCL.

### 4.1  Node Feature Transformation

Due to the existence of multiple types of nodes in HIN, each of which is in different feature spaces, they cannot be directly used for training. Therefore, we refer to HeCo [22] and firstly transform the features of all types of nodes into the same latent vector space. Specifically, for a node $i$ with type $\phi_i$, we use a matrix $\mathbf{W}_{\phi_i}$ specific to the node type to perform the following transformation on the initial feature $x_i$ of this node:

$$h_i = \sigma(\mathbf{W}_{\phi_i} \cdot x_i + \mathbf{b}_{\phi_i}),\tag{1}$$

where $h_i$ is the initial node embedding after transformation, $\sigma$ is an activation function, and $\mathbf{b}_{\phi_i}$ denotes as vector bias.

## 4.2   Muti-scale Meta-path Neighbors Aggregation

In this subsection, we discard the conventional heterogeneity-homogeneity transformation, and instead propose a joint manner for each subgraph generated based on the initial meta-path that aggregates the direct neighbors, the initial meta-path neighbors, and the expanded meta-path neighbors to fully capture the discriminative information.

Since the meta-path is given in each subgraph, the node type of per target node's direct neighbors based on this meta-path is the same. For example, in Fig. 2, the direct neighbors of the movie $M_3$ based on the meta-path MAM are actors $A_1$ and $A_3$, and they are both of the type 'actor'. HAN [20] believes that same-type neighbor nodes do not have uniform contributions to the target node. In other words, it is important to consider the unique impact of each neighbor node on the target node, rather than treating them as completely identical or equivalent. Therefore, we adopt the attention mechanism to aggregate different neighbor node information to the target node. Specifically, we do the following aggregation:

$$h_{i,P_m}^{OH} = \sigma\left( \sum_{j \in N_{P_m}^{OH}} \zeta_{i,j}^{OH} \cdot h_j \right), \tag{2}$$

where $N_{P_m}^{OH}$ is the set of the direct one-hop neighbors of the target node $i$ based on meta-path $P_m$. $h_j$ is the transformed initial node embedding of the direct neighbor, $\sigma$ is the activation function, and $\zeta_{i,j}^{OH}$ represents the attention value between the direct neighbor and the target node $i$, as shown below:

$$\zeta_{i,j}^{OH} = \frac{e^{\text{LeakyReLU}(\gamma_{OH,P_m}^T \cdot [h_i || h_j])}}{\sum_{t \in N_{P_m}^{OH}} e^{\text{LeakyReLU}(\gamma_{OH,P_m}^T \cdot [h_i || h_t])}}, \tag{3}$$

where $\gamma_{OH,P_m}$ is the learnable parameter vector based on meta-path $P_m$. $||$ is the concatenate operation.

Then, based on the given initial meta-path, we expand it to a 4-hop meta-path. For example, for the meta-path MAM, the expanded 4-hop meta-path is MAMAM. Next, we encode the subgraph based on initial and expanded meta-paths separately using GCN [11] encoder $\varepsilon$, as illustrated below:

$$\mathbf{H}^{I-P_m} = \varepsilon(\mathbf{H}, \mathbf{X}^{I-P_m}), \tag{4}$$

$$\mathbf{H}^{E-P_m} = \varepsilon(\mathbf{H}, \mathbf{X}^{E-P_m}), \tag{5}$$

where $\mathbf{H}$ is the initial node feature matrix, and $\mathbf{X}^{I-P_m}$ and $\mathbf{X}^{E-P_m}$ are the adjacency matrices of subgraphs generated based on the given initial meta-path and its expanded meta-path, respectively.

And we use muti-scale attention to aggregate the representations of nodes based on the initial and expanded meta-path, which can be expressed as follows:

$$h_{i,P_m}^{AGG} = \omega_{I-P_m} \cdot h_i^{I-P_m} + \omega_{E-P_m} \cdot h_i^{E-P_m}, \tag{6}$$

here, $\omega_{I-P_m}$ and $\omega_{E-P_m}$ represent the importance weights of the initial meta-path and expanded meta-path, respectively. And $\omega_{I-P_m} + \omega_{E-P_m} = 1$. Note that $h_i^{I-P_m} \in \mathbf{H}^{I-P_m}$ and $h_i^{E-P_m} \in \mathbf{H}^{E-P_m}$. Taking $\omega_{I-P_m}$ as an example, its computation is shown as follows:

$$\begin{aligned} \omega_{I-P_m} &= \frac{e^{\eta_{I-P_m}}}{e^{\eta_{I-P_m}} + e^{\eta_{E-P_m}}}, \\ \eta_{I-P_m} &= \frac{1}{|V|} \sum_{i \in V} \tanh(\gamma^T \cdot [\mathbf{W}_{MP} h_i^{I-P_m} + \mathbf{b}_{MP}]), \end{aligned} \tag{7}$$

where $\mathbf{W}_{MP}$ and $\mathbf{b}_{MP}$ are the learnable parameters, and $\gamma$ represents the learnable shared attention vector.

Finally, the node representation of each subgraph consists of two parts: the direct neighbors and the aggregated representation of neighbors based on the initial meta-path and expanded meta-path. These two parts are concatenated to form the node representation of each subgraph. The method is as follows:

$$\mathbf{H}^{P_m} = [h_{0,P_m}^{OH} || h_{0,P_m}^{AGG}, h_{1,P_m}^{OH} || h_{1,P_m}^{AGG}, ..., h_{|V_{P_m}|-1,P_m}^{OH} || h_{|V_{P_m}|-1,P_m}^{AGG}], \tag{8}$$

where $|V_{P_m}|$ denotes the number of nodes in meta-path subgraph $G^{P_m}$.

## 4.3   Contrastive Learning with Positive Sampling Strategy

To overcome the issue of hard negative samples in the classic GCL, we propose to select the most similar node as the positive sample from a semantic perspective. We take all subgraphs as positive and negative views in a pairwise manner. Then, we consider nodes that have meta-paths to the anchor node in the positive view and the nodes corresponding to the anchor node in the negative view as positive samples, while the remaining nodes are considered negative samples. Specifically, if there exists a meta-path $P_m$(limited to the initial meta-path) between the anchor node and $k$ other nodes in the positive subgraph $G^{P_m}$, then the number of positive samples for the anchor node is $k + 1$, which is different from the conventional GCL method that only utilizes one positive sample per anchor node. We illustrate the significance of this approach with practical examples. For instance, in the ACM dataset, two articles written by the same author are likely to belong to the same field. Therefore, we consider these two articles as a positive sample pair during the contrastive learning process, aiming to reduce the distance between them and classify them as closely as possible.

Moreover, in order to learn rich semantic information from different meta-path subgraphs, we propose a new contrastive objective, which consists of two parts: one maximizes the local and global mutual information, and the other discriminates between local and local.

Firstly, we aim to maximize the mutual information between node representations (i.e., local representations) and the summary vector that captures the global information content of the entire graph. To achieve this goal, we define the local-global contrast objective as follows:

$$\mathcal{L}_{global}^{m,n}(i) = -log(\mathcal{D}(h_i^m, s^m)) - log(\mathcal{D}(h_i^n, s^m)), \tag{9}$$

where $s^m$ is the summary vector obtained from the subgraph $G^{P_m}$ through the READOUT($\cdot$) function (mean pooling in this paper). $h_i^m$ and $h_i^n$ denote the representation of node $i$ in the meta-path subgraph $G^{P_m}$ and $G^{P_n}$, and $\mathcal{D}(h_i^m, s^m) = \sigma(h_i^m \mathbf{W} s^m)$. $\mathbf{W}$ is a learnable matrix.

Secondly, the local-local contrast objective aims to learn discriminative node representations in order to enhance the usability of node representations in downstream tasks. Specifically, based on the proposed positive sampling strategy, the local-local contrast objective is defined as follows:

$$\mathcal{L}_{local}^{m,n}(i, \mathbb{P}_i) = -log \frac{pos\_sim}{pos\_sim + neg\_sim},$$
$$pos\_sim = \sum_{v \in \mathbb{P}_i} e^{sim(h_i^m, h_v)/\tau}, \tag{10}$$
$$neg\_sim = \sum_{v \notin \mathbb{P}_i, v \in V_{P_m}} e^{sim(h_i^m, h_v)/\tau} + \sum_{v \notin \mathbb{P}_i, v \in V_{P_n}} e^{sim(h_i^n, h_v)/\tau},$$

where $\tau$ denotes the temperature parameter that controls the distribution of the data. $h_i^m$ and $h_i^n$ are the representation of node $i$ in the meta-path subgraph $G^{P_m}$ and $G^{P_n}$. $\mathbb{P}_i$ represents the set of positive samples corresponding to node $i$ based on the proposed positive sampling strategy. $V_{P_m}$ and $V_{P_n}$ are the sets of nodes in the $G^{P_m}$ and $G^{P_n}$.

Finally, the overall objective:

$$\mathcal{J} = \sum_{m \in P_M} \sum_{n \in P_M} \sum_{i \in V} (\alpha \cdot \mathcal{L}_{global}^{m,n}(i) + (1 - \alpha) \cdot \mathcal{L}_{local}^{m,n}(i, \mathbb{P}_i)), \tag{11}$$

where $P_M$ is the set of meta-paths. $\alpha$ is a weight coefficient and $\alpha \in [0, 1]$. $m, n$ denote $G^{P_m}$ and $G^{P_n}$. Ultimately, we use a semantic-level attention mechanism to integrate all subgraphs generated based on the initial meta-paths, producing the final representation of the target node for downstream tasks.

## 5   Experiment

### 5.1   Experimental Setup

**Datasets.** To demonstrate the superiority of our model over state-of-the-art methods, we conduct extensive experiments on three HIN datasets. Dataset statistics and details are summarized in Table 1.

**Baselines.** We compare M²HGCL with ten baselines, including six unsupervised heterogeneous methods, one semi-supervised heterogeneous method, and three unsupervised homogeneous methods. The unsupervised heterogeneous methods are Mp2vec [2], HERec [16], HetGNN [24], DMGI [14], HGCML [22] and HeCo [21]. The semi-supervised method is HAN [20]. The unsupervised homogeneous methods are GraphSAGE [5], GAE [12], and DGI [19].

Table 1. The statistics of the datasets.

| Dataset | Node | Meta-path | Target node | Num of class |
|---------|------|-----------|-------------|--------------|
| AMiner [7] | paper(P):6,564 author(A):13,329 reference(R):35,890 | PAP PRP | paper | 4 |
| ACM [25] | paper(P):4,019 author(A):7,167 subject(S):60 | PAP PSP | paper | 3 |
| Freebase [13] | movie(M):3,492 actor(A):33,401 director(D):2,502 writer(W):4,459 | MAM MDM MWM | movie | 3 |

**Implementation Details.** The expanded meta-path denotes the 4-hop meta-path obtained by expanding the initial 2-hop meta-path. For optimizing our model, we utilize Glorot initialization to initialize our model parameters and employ Adam as the optimization algorithm during training. For the temperature parameter $\tau$ and $\alpha$ in the overall objective, we test ranging from 0.1 to 0.9 with step 0.1 to select the optimal value. For the number of hidden layer nodes, AMiner and Freebase are 64, and ACM is 128. Besides, we determine the learning rate for our model, setting it at $3e-3$ for the AMiner dataset, $5e-4$ for the ACM dataset, and $1e-3$ for the Freebase dataset. In meta-path neighbor encoding, we leverage 1-layer GCN as the encoder. For a fair comparison, we randomly run the experiments 5 times and report the average results with standard deviations. For HGCML, we conduct experiments with 40% and 60% labeled nodes, and report the results by following the official implementation. HeCo and other competitors' results are reported from [21].

## 5.2 Node Classification

We utilize the learned node embeddings to train a linear classifier for evaluating our model. We randomly select 40% and 60% of labeled nodes from each dataset for training, and reserve 1,000 nodes for validation and another 1,000 nodes for testing. Our evaluation metric for node classification on the test set is Macro-F1, Micro-F1, and AUC, which are reported as the average results with

**Table 2.** Quantitative results on node classification with 40% labeled nodes. The best results are marked in bold, and the second-best are underlined.

| Methods(40%) | Data | AMiner | | | ACM | | | Freebase | | |
|---|---|---|---|---|---|---|---|---|---|---|
| | | Ma-F1 | Mi-F1 | AUC | Ma-F1 | Mi-F1 | AUC | Ma-F1 | Mi-F1 | AUC |
| HAN | X,A,Y | 63.85 | 76.89 | 80.72 | 87.47 | 87.21 | 94.84 | 59.63 | 63.74 | 77.74 |
| | | ±1.5 | ±1.6 | ±2.1 | ±1.1 | ±1.2 | ±0.9 | ±2.3 | ±2.7 | ±1.2 |
| GraphSAGE | X,A | 45.77 | 52.10 | 74.44 | 55.96 | 60.98 | 71.06 | 44.88 | 57.08 | 66.42 |
| | | ±1.5 | ±2.2 | ±1.3 | ±6.8 | ±3.5 | ±5.2 | ±4.1 | ±3.2 | ±4.7 |
| GAE | X,A | 65.66 | 71.34 | 88.29 | 61.61 | 66.38 | 79.14 | 52.44 | 56.05 | 74.05 |
| | | ±1.5 | ±1.8 | ±1.0 | ±3.2 | ±1.9 | ±2.5 | ±2.3 | ±2.0 | ±0.9 |
| Mp2vec | X,A | 64.77 | 69.66 | 88.82 | 62.41 | 64.43 | 80.48 | 57.80 | 61.01 | 75.51 |
| | | ±0.5 | ±0.6 | ±0.2 | ±0.6 | ±0.6 | ±0.4 | ±1.1 | ±1.3 | ±0.8 |
| HERec | X,A | 64.50 | 71.57 | 88.70 | 61.21 | 62.62 | 79.84 | 59.28 | 62.71 | 76.08 |
| | | ±0.7 | ±0.7 | ±0.4 | ±0.8 | ±0.9 | ±0.5 | ±0.6 | ±0.7 | ±0.4 |
| HetGNN | X,A | 58.97 | 68.47 | 83.14 | 72.02 | 74.46 | 85.01 | 48.57 | 53.96 | 69.48 |
| | | ±0.9 | ±2.2 | ±1.6 | ±0.4 | ±0.8 | ±0.6 | ±0.5 | ±1.1 | ±0.2 |
| DGI | X,A | 54.72 | 63.87 | 77.86 | 80.23 | 80.41 | 91.52 | 53.40 | 57.82 | 72.97 |
| | | ±2.6 | ±2.9 | ±2.1 | ±3.3 | ±3.0 | ±2.3 | ±1.4 | ±0.8 | ±1.1 |
| DMGI | X,A | 61.92 | 63.60 | 88.02 | 86.23 | 86.02 | 96.35 | 49.88 | 54.28 | 70.77 |
| | | ±2.1 | ±2.5 | ±1.3 | ±0.8 | ±0.9 | ±0.3 | ±1.9 | ±1.6 | ±1.6 |
| HGCML | X,A | 74.13 | 85.11 | 92.87 | 90.36 | **90.32** | 96.29 | 62.57 | 70.49 | 79.12 |
| | | ±0.6 | ±0.2 | ±0.1 | ±0.1 | **±0.3** | ±0.2 | ±0.1 | ±0.2 | ±0.2 |
| HeCo | X,A | 73.75 | 80.53 | 92.11 | 87.61 | 87.45 | 96.40 | 61.19 | 64.03 | 78.44 |
| | | ±0.5 | ±0.7 | ±0.6 | ±0.5 | ±0.5 | ±0.4 | ±0.6 | ±0.7 | ±0.5 |
| M²HGCL | X,A | **76.04** | **86.79** | **94.65** | **91.48** | 90.10 | **97.19** | **64.04** | **72.50** | **81.09** |
| | | **±0.2** | **±0.6** | **±0.5** | **±0.2** | ±0.2 | **±0.3** | **±0.1** | **±0.4** | **±0.3** |

standard deviations. The results are shown in Table 2 and 3. It is evident that M²HGCL generally achieves the best performance among all baseline methods and all splits, even outperforming semi-supervised learning methods, i.e., HAN. This indicates the effectiveness of fully utilizing heterogeneous information in HIN and the importance of expanded meta-paths. However, we consider that the ACM dataset's 'subject' nodes may have had a relatively smaller representation compared to other node types, resulting in only marginal differences in the node embeddings. This could be a contributing factor to our model's underperformance on the ACM dataset when classifying. Notably, the AMiner dataset suffers from label imbalance, with the number of objects in the label with the maximum number of nodes being 7 times greater than that in the label with the minimum number of nodes. This demonstrates the effectiveness of the proposed model in practical scenarios with label imbalance.

## 5.3 Node Clustering

We further perform K-means clustering to verify the quality of learned node embeddings. We use two widely-used clustering evaluation metrics, namely normalized mutual information (NMI) and adjusted rand index (ARI), to compare the performance of our proposed model with other baseline methods. The results are presented in Table 4. As we can see, our proposed model achieves significant improvements over other baseline methods on the AMiner and Freebase datasets. Specifically, on the Freebase dataset, our model outperforms approximately 36% on NMI and 49% on ARI compared to HGCML, demonstrating the superiority and effectiveness of our approach. However, the results on the ACM dataset fall short of other baseline methods. As analyzed in Sect. 5.2, the relatively smaller number of 'subject' nodes in the ACM dataset compared to other node types, which may have resulted in only marginal differences in the final node representations, is the primary reason for our underperformance on the ACM dataset when clustering.

**Table 3.** Quantitative results on node classification with 60% labeled nodes. The best results are marked in bold, and the second-best are underlined.

| Methods(60%) | Data | AMiner | | | ACM | | | Freebase | | |
|---|---|---|---|---|---|---|---|---|---|---|
| | | Ma-F1 | Mi-F1 | AUC | Ma-F1 | Mi-F1 | AUC | Ma-F1 | Mi-F1 | AUC |
| HAN | X,A,Y | 62.02 | 74.73 | 80.39 | 88.41 | 88.10 | 94.68 | 56.77 | 61.06 | 75.69 |
| | | ±1.2 | 1.4 | ±1.5 | ±1.1 | ±1.2 | ±1.4 | ±1.7 | ±2.0 | ±1.5 |
| GraphSAGE | X,A | 44.91 | 51.36 | 74.16 | 56.59 | 60.72 | 70.45 | 45.16 | 55.92 | 66.78 |
| | | ±2.0 | ±2.2 | ±1.3 | ±5.7 | ±4.3 | ±6.2 | ±3.1 | ±3.2 | ±2.0 |
| GAE | X,A | 63.74 | 67.70 | 86.92 | 61.67 | 65.71 | 77.90 | 50.65 | 53.85 | 71.75 |
| | | ±1.6 | ±1.9 | ±0.8 | ±2.9 | ±2.2 | ±2.8 | ±0.4 | ±0.4 | ±0.4 |
| Mp2vec | X,A | 60.65 | 63.92 | 85.57 | 61.13 | 62.72 | 79.33 | 55.94 | 58.74 | 74.78 |
| | | ±0.3 | ±0.5 | ±0.2 | ±0.4 | ±0.3 | ±0.4 | ±0.7 | ±0.8 | ±0.4 |
| HERec | X,A | 65.53 | 69.76 | 87.74 | 64.35 | 65.15 | 81.64 | 56.50 | 58.57 | 74.89 |
| | | ±0.7 | ±0.8 | ±0.5 | ±0.8 | ±0.9 | ±0.7 | ±0.4 | ±0.5 | ±0.4 |
| HetGNN | X,A | 57.34 | 65.61 | 84.77 | 74.33 | 76.08 | 87.64 | 52.37 | 56.84 | 71.01 |
| | | ±1.1 | ±2.2 | ±0.9 | ±0.6 | ±0.7 | ±0.7 | ±0.8 | ±0.7 | ±0.5 |
| DGI | X,A | 55.45 | 63.10 | 77.21 | 80.03 | 80.15 | 91.41 | 53.81 | 57.96 | 73.32 |
| | | ±2.4 | ±3.0 | ±1.4 | ±3.3 | ±3.2 | ±1.9 | ±1.1 | ±0.7 | ±0.9 |
| DMGI | X,A | 61.15 | 62.51 | 86.20 | 87.97 | 87.82 | 96.79 | 52.10 | 56.69 | 73.17 |
| | | ±2.5 | ±2.6 | ±1.7 | ±0.4 | ±0.5 | ±0.2 | ±0.7 | ±1.2 | ±1.4 |
| HGCML | X,A | 76.26 | 86.10 | 93.41 | 91.02 | 91.10 | 97.49 | 65.17 | 71.08 | 79.24 |
| | | ±0.7 | ±0.8 | ±0.6 | ±0.2 | ±0.4 | ±0.4 | ±0.3 | ±0.1 | ±0.1 |
| HeCo | X,A | 75.80 | 82.46 | 92.40 | 89.04 | 88.71 | 96.55 | 60.13 | 63.61 | 78.04 |
| | | ±1.8 | ±1.4 | ±0.7 | ±0.5 | ±0.5 | ±0.3 | ±1.3 | ±1.6 | ±0.4 |
| M²HGCL | X,A | **77.56** | **87.17** | **94.88** | **91.41** | **91.29** | **98.43** | **66.61** | **73.02** | **81.96** |
| | | **±0.7** | **±0.6** | **±0.3** | **±0.1** | **±0.2** | **±0.3** | **±0.4** | **±0.2** | **±0.2** |

## 5.4 Ablation Study

We conduct the ablation study on our model, analyzing the effectiveness of its constituent components. To demonstrate the effectiveness of incorporating expanded meta-path neighbor information, we limit our model to the initial meta-path, referring to this variant as $\text{M}^2\text{HGCL}_{w/o\ expanded}$. To underscore the importance of incorporating the direct neighbor information in final embeddings, we eliminate the direct neighbor information and solely employ meta-path neighbor information for the embedding of the subgraph, denoting this variant as $\text{M}^2\text{HGCL}_{w/o\ direct}$. Additionally, we individually remove the local-global and local-local contrast objective to evince the indispensability of these two components in the contrastive objective and dub these two variants as $\text{M}^2\text{HGCL}_{w/o\ global}$ and $\text{M}^2\text{HGCL}_{w/o\ local}$. Finally, to highlight the efficacy of our positive sampling strategy in enhancing model performance, we substitute our proposed strategy with the traditional GCL positive sample definition method, calling this variant $\text{M}^2\text{HGCL}_{w/o\ p-samp}$. We present the experimental results of all variants in Table 5, and the results evince that $\text{M}^2\text{HGCL}$ outperforms all other variants, thereby substantiating the efficacy and indispensability of each component of our proposed model.

**Table 4.** Quantitative results on node clustering. The best results are marked in bold, and the second-best results are underlined.

| Datasets | AMiner | | ACM | | Freebase | |
|---|---|---|---|---|---|---|
| Metrics | NMI | ARI | NMI | ARI | NMI | ARI |
| GraphSage | 15.74 | 10.10 | 29.20 | 27.72 | 9.05 | 10.49 |
| GAE | 28.58 | 20.90 | 27.42 | 24.49 | 19.03 | 14.10 |
| Mp2vec | 30.80 | 25.26 | 48.43 | 34.65 | 16.47 | 17.32 |
| HERec | 27.82 | 20.16 | 47.54 | 35.67 | 19.76 | 19.36 |
| HetGNN | 21.46 | 26.60 | 41.53 | 34.81 | 12.25 | 15.01 |
| DGI | 22.06 | 15.93 | 51.73 | 41.16 | 18.34 | 11.29 |
| DMGI | 19.06 | 20.09 | 51.66 | 46.64 | 16.98 | 16.91 |
| HGCML | <u>36.10</u> | <u>35.29</u> | **65.13** | **62.77** | 15.46 | 14.29 |
| HeCo | 32.26 | 28.64 | 56.87 | 56.94 | <u>20.38</u> | <u>20.98</u> |
| $\text{M}^2\text{HGCL}$ | **39.14** | **37.00** | <u>59.19</u> | <u>58.11</u> | **20.96** | **21.23** |

**Table 5.** Ablation study of $\text{M}^2\text{HGCL}$ for pretext tasks on node classification.

| Variants | AMiner | | | ACM | | | Freebase | | |
|---|---|---|---|---|---|---|---|---|---|
| $\text{M}^2\text{HGCL}_{w/o\ expanded}$ | 73.12 | 79.21 | 93.17 | 88.74 | 88.54 | 96.92 | 61.33 | 64.21 | 79.34 |
| $\text{M}^2\text{HGCL}_{w/o\ direct}$ | 73.14 | 78.30 | 93.37 | 85.91 | 85.53 | 94.82 | 59.94 | 63.19 | 78.34 |
| $\text{M}^2\text{HGCL}_{w/o\ global}$ | 74.26 | 80.06 | 94.03 | 76.22 | 76.04 | 88.77 | 61.57 | 64.00 | 80.46 |
| $\text{M}^2\text{HGCL}_{w/o\ local}$ | 71.89 | 79.02 | 93.23 | 86.19 | 85.88 | 95.05 | 56.23 | 58.17 | 77.11 |
| $\text{M}^2\text{HGCL}_{w/o\ p-samp}$ | 74.73 | 80.93 | 93.64 | 80.47 | 81.17 | 89.95 | 58.68 | 61.29 | 78.22 |
| $\text{M}^2\text{HGCL}$ | **77.56** | **87.17** | **94.88** | **91.41** | **91.29** | **98.43** | **66.61** | **73.02** | **81.96** |

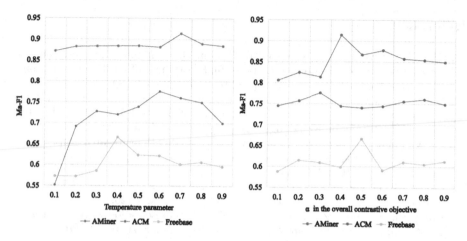

**Fig. 4.** Analysis of the temperature parameter $\tau$ and the weight coefficient $\alpha$ in the overall contrastive objective.

## 5.5 Analysis of Hyperparameter

In this section, we analyze two crucial hyperparameters, the weight coefficient $\alpha$ in the overall contrastive objective and the temperature parameter $\tau$, that play a crucial role in model training. To explore their sensitivity, we conduct a comprehensive evaluation of the model's performance on node classification.

Figure 4 demonstrates the model's performance on three datasets for different values of $\tau$ and $\alpha$. From Fig. 4, we can observe that as $\tau$ and $\alpha$ increase, the performance of the model exhibits an initial rise in volatility, followed by a decrease across all three datasets, among which the ACM dataset exhibits a smaller variation range with increasing $\tau$, but there still exists a peak performance point. Specifically, the optimal values for the AMiner dataset are $\tau = 0.6$ and $\alpha = 0.3$, for the ACM dataset are $\tau = 0.7$ and $\alpha = 0.4$, and for the Freebase dataset are $\tau = 0.4$ and $\alpha = 0.5$. $\tau$ is observed to be less than or equal to 0.5 across all datasets, indicating the dominant impact of global-local loss over local-local loss based on Eq. 11. This insight provides valuable information on the contribution of different loss terms to our model performance.

## 6 Conclusion

We propose a novel model for heterogeneous graph contrastive learning that integrates multi-scale meta-paths to address existing issues in HGCL. Our model represents target nodes using direct neighbor information, the initial neighbor information, and the expanded meta-paths neighbor information. We introduce a positive sampling strategy for HINs and weight the local-local and local-global contrast objectives to enhance model performance. Experiments show that our proposed model outperforms other state-of-the-art methods.

# References

1. Chen, T., Kornblith, S., Norouzi, M., Hinton, G.: A simple framework for contrastive learning of visual representations. In: Proceedings of the 37th International Conference on Machine Learning, pp. 1597–1607. PMLR (2020)
2. Dong, Y., Chawla, N.V., Swami, A.: Metapath2vec: scalable representation learning for heterogeneous networks. In: Proceedings of the 23rd ACM SIGKDD International Conference on Knowledge Discovery and Data Mining, pp. 135–144 (2017)
3. Fu, X., Zhang, J., Meng, Z., King, I.: MAGNN: metapath aggregated graph neural network for heterogeneous graph embedding. In: Proceedings of The Web Conference 2020, pp. 2331–2341 (2020)
4. Grill, J.B., et al.: Bootstrap your own latent-a new approach to self-supervised learning. Adv. Neural. Inf. Process. Syst. **33**, 21271–21284 (2020)
5. Hamilton, W., Ying, Z., Leskovec, J.: Inductive representation learning on large graphs. In: Advances in Neural Information Processing Systems 30 (2017)
6. He, K., Fan, H., Wu, Y., Xie, S., Girshick, R.: Momentum contrast for unsupervised visual representation learning. In: Proceedings of the IEEE/CVF Conference on Computer Vision and Pattern Recognition, pp. 9729–9738 (2020)
7. Hu, B., Fang, Y., Shi, C.: Adversarial learning on heterogeneous information networks. In: Proceedings of the 25th ACM SIGKDD International Conference on Knowledge Discovery & Data Mining, pp. 120–129 (2019)
8. Hu, Z., Dong, Y., Wang, K., Sun, Y.: Heterogeneous graph transformer. In: Proceedings of the Web Conference 2020, pp. 2704–2710 (2020)
9. Hwang, D., Park, J., Kwon, S., Kim, K., Ha, J.W., Kim, H.J.: Self-supervised auxiliary learning with meta-paths for heterogeneous graphs. Adv. Neural. Inf. Process. Syst. **33**, 10294–10305 (2020)
10. Jiang, X., Lu, Y., Fang, Y., Shi, C.: Contrastive pre-training of GNNs on heterogeneous graphs. In: Proceedings of the 30th ACM International Conference on Information & Knowledge Management, pp. 803–812 (2021)
11. Kipf, T.N., Welling, M.: Semi-supervised classification with graph convolutional networks. arXiv preprint arXiv:1609.02907 (2016)
12. Kipf, T.N., Welling, M.: Variational graph auto-encoders. arXiv preprint arXiv:1611.07308 (2016)
13. Li, X., Ding, D., Kao, B., Sun, Y., Mamoulis, N.: Leveraging meta-path contexts for classification in heterogeneous information networks. In: 2021 IEEE 37th International Conference on Data Engineering (ICDE), pp. 912–923. IEEE (2021)
14. Park, C., Kim, D., Han, J., Yu, H.: Unsupervised attributed multiplex network embedding. In: Proceedings of the AAAI Conference on Artificial Intelligence, vol. 34, pp. 5371–5378 (2020)
15. Ren, Y., Liu, B., Huang, C., Dai, P., Bo, L., Zhang, J.: Heterogeneous deep graph infomax. arXiv preprint arXiv:1911.08538 (2019)
16. Shi, C., Hu, B., Zhao, W.X., Philip, S.Y.: Heterogeneous information network embedding for recommendation. IEEE Trans. Knowl. Data Eng. **31**(2), 357–370 (2018)
17. Tang, J., Qu, M., Mei, Q.: PTE: predictive text embedding through large-scale heterogeneous text networks. In: Proceedings of the 21th ACM SIGKDD International Conference on Knowledge Discovery and Data Mining, pp. 1165–1174 (2015)
18. Thakoor, S., et al.: Large-scale representation learning on graphs via bootstrapping. arXiv preprint arXiv:2102.06514 (2021)

19. Velickovic, P., Fedus, W., Hamilton, W.L., Liò, P., Bengio, Y., Hjelm, R.D.: Deep graph infomax. ICLR (Poster) **2**(3), 4 (2019)  ·

20. Wang, X., et al.: Heterogeneous graph attention network. In: Proceedings of the 2019 World Wide Web Conference, pp. 2022–2032 (2019)

21. Wang, X., Liu, N., Han, H., Shi, C.: Self-supervised heterogeneous graph neural network with co-contrastive learning. In: Proceedings of the 27th ACM SIGKDD Conference on Knowledge Discovery & Data Mining, pp. 1726–1736 (2021)

22. Wang, Z., Li, Q., Yu, D., Han, X., Gao, X.Z., Shen, S.: Heterogeneous graph contrastive multi-view learning. arXiv preprint arXiv:2210.00248 (2022)

23. You, Y., Chen, T., Sui, Y., Chen, T., Wang, Z., Shen, Y.: Graph contrastive learning with augmentations. Adv. Neural. Inf. Process. Syst. **33**, 5812–5823 (2020)

24. Zhang, C., Song, D., Huang, C., Swami, A., Chawla, N.V.: Heterogeneous graph neural network. In: Proceedings of the 25th ACM SIGKDD International Conference on Knowledge Discovery & Data Mining, pp. 793–803 (2019)

25. Zhao, J., Wang, X., Shi, C., Liu, Z., Ye, Y.: Network schema preserving heterogeneous information network embedding. In: Proceedings of the 29th International Joint Conference on Artificial Intelligence, pp. 1366–1372 (2020)

# Author Index

X. Yang et al. (Eds.): ADMA 2023, LNAI 14178, pp. 369–370, 2023.
https://doi.org/10.1007/978-3-031-46671-7

Printed in the United States
by Baker & Taylor Publisher Services